2004

RELATIVISM, SUFFERING AND BEYOND

RELATIVISM, SUFFERING AND BEYOND

Essays in Memory of Bimal K. Matilal

edited by
P. BILIMORIA
J. N. MOHANTY

DELHI
OXFORD UNIVERSITY PRESS
CALCUTTA CHENNAI MUMBAI
1997

Oxford University Press, Walton Street, Oxford OX2 6DP

Oxford New York
Athens Auckland Bangkok Calcutta
Cape Town Chennai Dar es Salaam Delhi Florence
Hong Kong Istanbul Karachi Kuala Lumpur Madrid
Melbourne Mexico City Mumbai Nairobi Paris
Singapore Taipei Tokyo Toronto

and associates in

Berlin Ibadan

ISBN 0 19 563858 1

Typeset by Rastrixi, New Delhi 110 070
Printed in India at Pauls Press, New Delhi 110 020
and published by Manzar Khan, Oxford university Press
YMCA Library Building, Jai Singh Road, New Delhi 110 001

Preface

The authors whose essays are collected in this volume share with the editors their admiration, respect, and affection for the late Professor Bimal Krishna Matilal. They have written essays, as tokens of their gratitude, on themes in which he was deeply interested. Also included in this volume are a near complete list of his publications (prepared by Heeraman Tiwari) and a photograph (courtesy Mrs Karabi Matilal).

We hope that this volume will be a fitting memorial to a great scholar, kind friend and genial colleague.

J.N.M.
P.B.

Note on Contributors

SIBAJIBAN BHATTACHARYYA was formerly Acharya B.N. Seal Professor of Philosophy at Calcutta University.

PURUSHOTTAMA BILIMORIA is Professor of Philosophy, Deakin University of Melbourne, and Visiting Professor, University of California (Berkeley and Santa Barbara).

ARINDAM CHAKRABARTI is Professor of Philosophy at Delhi University.

MARGARET CHATTERJEE was Professor of Philosophy at Delhi University and then Director, Indian Institute of Advanced Study, Simla. She currently teaches at Westminster College, Oxford.

D.P. CHATTOPADHYAYA is a National Fellow of the Indian Council of Philosophical Research, New Delhi, and Director of the Project of the History of Indian Science, Philosophy and Culture.

ELIOT DEUTSCH is Professor of Philosophy at the University of Hawaii at Manoa.

BRENDAN S. GILLON is Assistant Professor at the Department of Linguistics, McGill University, Montreal.

ROBERT P. GOLDMAN is Professor of Sanskrit and Chairman of the Department of South and Southeast Asian Studies at the University of California at Berkeley.

RICHARD GOMBRICH is Boden Professor of Sanskrit at Oxford University.

WILHELM HALBFASS is Professor in the Department of South Asia Studies at the University of Pennsylvania, Philadelphia.

MASAAKI HATTORI is Professor Emeritus at Kyoto University, Kyoto.

RICHARD P. HAYES is on the faculty of Religious Studies at McGill University, Montreal.

HANS HERZBERGER is Professor of Philosophy at the University of Toronto, Canada.

RADHIKA HERZBERGER is Director of the Rishi Valley Education Centre, Andhra Pradesh, India.

MICHAEL KRAUSZ is the Milton C. Nahm Professor and Chairman of the Department of Philosophy at Bryn Mawr College, USA.

JULIUS LIPNER is University Lecturer in the Divinity Faculty at Cambridge. He is also a Director of the Dharam Hinduja Institute of India Research in the faculty's Centre for Advanced Religious and Theological Studies.

R.A. MALL is Professor in the Department of Philosophy at the University of Bremen, Germany.

J.N. MOHANTY is Professor of Philosophy at Temple University, Philadelphia, and Vice-President, American Philosophical Association.

MARTHA NUSSBAUM is Ernst Freund Professor of Law and Ethics at the University of Chicago.

C. RAM-PRASAD is Gordon Milburn Junior Research Fellow in Philosophy and Theology at Trinity College, Oxford.

MARK SIDERITS is Professor in the Department of Philosophy at the Illinois State University, Normal, Illinois.

Contents

Introduction 1
J.N. Mohanty

1. Gadādhara's Theory of Meaning of Pronouns 16
Sibajiban Bhattacharyya

2. The Earliest Brahmanical Reference to Buddhism? 32
Richard Gombrich

3. Scepticism Revisited: Nāgārjuna and Nyāya via Matilal 50
D.P. Chattopadhyaya

4. Matilal on Nāgārjuna 69
Mark Siderits

5. Relativism and Beyond: A Tribute to Bimal Matilal 93
Michael Krausz

6. Whose Experience Validates What for Dharmakīrti? 105
Richard P. Hayes

7. Seeing Daffodils, Seeing as Daffodils and Seeing Things Called 'Daffodils' 119
Arindam Chakrabarti

8. Negative Facts and Knowledge of Negative Facts 128
Brendan S. Gillon

9. Happiness: A Nyāya-Vaiśeṣika Perspective 150
Wilhelm Halbfass

10. Causal Connections, Cognition and Regularity: Comparativist Remarks on David Hume and Śrī Harṣa 164
C. Ram-Prasad

11. *Eṣa Dharmaḥ Sanātanaḥ:* Shifting Moral Values and the Indian Epics 187
Robert P. Goldman

12. A Note on Identity and Mutual Absence in Navya-nyāya 224
Kamaleswar Bhattacharya

13. Emotions as Judgements of Value and Importance 231
Martha Nussbaum

14. On Śaṅkara's Attempted Reconciliation of 'You' and 'I': *Yuṣmadasmatsamanvaya* 252
Purushottama Bilimoria

15. Two Truths, or One? 278
Radhika Herzberger and Hans G. Herzberger

16. Śaṅkara on *Satyaṃ Jñānam Anantaṃ Brahma* 301
Julius Lipner

17. Some Indian Strands of Thought Relating to the Problem of Evil 319
Margaret Chatterjee

18. Outline of an Advaita Vedāntic Aesthetics 336
Eliot Deutsch

19. Religiophilosophical Meditations on the Rgvedic Dictum: *Ekam Sad Viprā Bahudhā Vadanti* 348
R.A. Mall

20. The Buddhist Theory Concerning the Truth and Falsity of Cognition 361
Masaaki Hattori

List of Publications by B.K. Matilal 372

Introduction: Bimal Matilal, the Man and the Philosopher

J.N. Mohanty

Bimal Krishna Matilal, Spalding Professor of Eastern Religion and Ethics in the University of Oxford, and Fellow of All Souls, at the time of his death on 8 June 1991, was born in Joynagar, West Bengal, India on 1 June 1935. He moved to Calcutta at the age of 14 for his education, read for his BA at the Islamia College and moved to Sanskrit College and the Sanskrit Department of the University of Calcutta for his graduate study. Besides earning the college degrees of BA and MA and many medals and prizes Matilal continued to study Nyāya under traditional paṇḍits and earned the title of Tarkatīrtha ('Master of Dialectics') in 1962. Already in 1957, he had been appointed a lecturer at the Sanskrit College. In 1962, he went to Harvard to study under Daniel Ingalls, who had then become famous for having published a remarkable book on Navya-nyāya. During his Harvard years, he not only pursued his research under Ingalls on Navya-nyāya, he also learnt mathematical logic with Quine and read voraciously and critically in contemporary Western philosophy. This gave him a perspective from which he was to launch an effort to change the way Indian philosophy was being looked at in the Western world (and also in India). Convinced as he became that ancient and medieval Indian philosophy had much to say and contribute to the on-going philosophical discussion in the West, he devoted his academic life to establishing dialogues between the texts of the Hindu, Buddhist, and Jain schools and the most eminent philosophers of the contemporary West.

He earned his Harvard Ph.D. in 1965, and briefly returned to

India and took up teaching positions at the Sanskrit College and later in Santiniketan where he attracted the attention and earned the respect of the eminent philosopher Kalidas Bhattacharyya. But he soon left India in order to take up a teaching position at the University of Toronto. With a break of one year at the University of Pennsylvania and another year at the School of Oriental and African Studies in London, he spent eleven years at the University of Toronto, and became a Canadian citizen. In 1976 he was appointed to the Spalding Chair at Oxford. He continued editing his *Journal of Indian Philosophy* for another fifteen years, writing numerous articles for scientific journals and books, participating in conferences around the world, organizing conferences at Oxford, meeting and talking to philosophers, and playing host to those he encountered at All Souls. In course of time, he became a major partner in the new style of conversation between the East and the West. It was a tragedy that such a productive and almost inspired existence should be cut short so prematurely.

I met Bimal Matilal for the first time when I joined the faculty of Calcutta Sanskrit College's Postgraduate and Research Department in 1960 (although I had known of him at the university as a brilliant student in the Sanskrit Department). He came to my office one day to ask if I would like to read and comment on an essay he had written on the Nyāya analysis of empty terms (such as 'rabbit's horn'). As I was looking through it, I was struck by the way he had succeeded in showing how the Nyāya paraphrasing of sentences containing empty terms had anticipated, by many centuries, Russell's attempt to do the same. (I am not sure if that essay was ever published, but I now recognize it as the ancestor of Chapter 4 of *Epistemology, Logic and Grammar in Indian Philosophical Analysis.*) That occasion was the beginning of our friendship which later grew into a sort of intellectual partnership. Quine's *Methods of Logic* came to his attention, I believe, in 1961, and we worked through it together. At that time I was studying Gaṅgeśa's *Tattvacintāmaṇi* with Pandit Ananta Kumar Tarkatīrtha. Although his principal mentor was Pandit Madhusūdana Nyāyācārya (who later replaced Pandit Ananta Kumar as Professor of Nyāya upon the latter's moving to the Research Department of the College), Matilal was also studying some texts with Pandit

Ananta Kumar. Amongst others there, Gaurinath Sastri, then Principal of the College (with whom I had earlier studied Sanskrit at the Presidency College) was reading *Kiraṇāvali*, and Kalidas Bhattacharyya was reading *Advaitasiddhi*. We would all stay on to attend each other's classes. What a great time we had, working through these great texts under a superb teacher! Then Matilal left for Harvard. Although I did not see much of him for a number of years — I remember, after his return from the United States, he came to my house, and I took him to Kalidas Bhattacharyya to introduce him to the great philosopher — he was in our minds as one who, already a Tarkatīrtha, was learning modern logic with Quine, and was soon going to be a star among us. And a star he became.

Belonging as he did to the same generation as I, there is, or rather was, a common intellectual situation which all of us, and he, encountered when we started doing Indian philosophy. It would be appropriate to recall some features of that situation. That situation may be described as consisting of a set of problems and concerns as to how to do Indian philosophy. On the one hand, we all realized that it was imperative to do Indian philosophy by returning (from the English expositions and commentaries) to the Sanskrit sources. But how to interpret the texts? The way our past generation of Indian philosophers did that had proven unsatisfactory to us. We were searching for a new way. While some of us started by primarily studying Western philosophy and then went over to the paṇḍits to study the Sanskrit source material, Bimal started out as a Sanskritist, then became a Tarkatīrtha, and gradually acquired, first in Calcutta by his own efforts, and then at Harvard, the tools of Western logic and philosophy. Moving in opposite directions, we met somewhere in the middle.

There were, perhaps there still are, three ways of doing Indian philosophy, three traditions among which we, including Matilal, found ourselves. First, there was the long, continuous tradition of scholarship of the paṇḍits. At the other extreme, there was several hundred years of European Indological scholarship. There was also the way the English-educated university professors in India did Indian philosophy. This last group was the most unsatisfactory for us, with the exception of some very special cases.

The paṇḍit tradition preserved the study of Sanskrit texts in a rigorous manner along the lines of distinct schools, Nyāya, Mīmāṃsā, Vedānta, etc. Here you study the texts under a master, and transmit the learning to the next generation. A certain way of reading, perhaps separately characterizing each lineage (as in the case of the distinct *gharāṇās* of Indian music), was being preserved and transmitted, and in each case, a certain rhetoric was being preserved and used. By our times, creativity had come to an end, again with some rare exceptions. But we were grateful for the masterly way in which a rich and complicated tradition had been preserved and was being transmitted. The paṇḍits are indispensable, and he regretted, as did I, that they were a vanishing species.

There were two subtle changes taking place, which philosophers might miss; two seemingly insignificant changes, but only seemingly so. For one thing, a certain mode of living had been dissociated from the mode of thinking. For another, the Indian philosophers were writing in English. There was a break with tradition on both counts. One could be grateful to the paṇḍits, but one could not go on doing just the way they did what they were doing.

For a Harvard-educated scholar, one option was to become an Indologist. Indology has been a curious discipline. Born in eighteenth-century Germany in the wake of the discovery of the Indo-European language group as such, and fostered under the aegis of the European Romanticism's search for its 'other', the Indologist perpetuated these two in an exemplary manner. On the one hand, there grew a great tradition of philological work, of minute philological-cum-historical-cum-grammatical research. There is no gainsaying that this tradition has established itself as a solid research tradition, and I know how Matilal admired it. But what was clearly deficient in it, from the point of view of a philosopher, was a lack of concern for ideas as such, for the philosophical content of the texts, and in this regard the philological was influenced by the romantic. For the European, the Orient was the 'other'. At worst, India was the dream-land of childlike naivety in relation to the Western world; at its highest, Indian culture was enthused by the idea of a mystic union with the One. But as Hegel bluntly put it, the Indians did not *think*, they did not raise their intuitions to concepts, so how could they have philosophy? Philosophy, as two great European minds since Hegel — Husserl and Heidegger — held, was a Greek invention, a European destiny. The Indologist therefore studied Śaṅkara, not to focus on his theories and argumentations or evidence in support, but to locate Śaṅkara beliefs in the context of Indian culture, especially religious history. The Indian philosopher of today could learn from the methodical philological researches in order to correct some of his ahistorical preconceptions, but he cannot use their results for *doing* Indian philosophy. Matilal parted company.

The Indologist had another deficiency: he regarded the traditional paṇḍit interpretation of the texts as biased and without adequate philological basis. In reading the Vedic texts, for example, he wanted to return to the 'texts themselves' without the *via media* of the commentarial tradition. Here the new hermeneutics of Gadamer et al. goes against him. You cannot return to the 'original intention'. You have to go through the density of interpretive history. We have to read Śaṅkara through *Bhāmatī* or the *Vivaraṇa*.

Now to come to the third strand. Indian philosophers educated in Western thought developed a myopic vision of Indian thought. They characterized it in such global terms as 'spiritual' and 'transcendental'. Our professors in Calcutta — with perhaps the exception of Rash Vihary Das and Kalidas Bhattacharyya — talked about Indian philosophy in edifying language. Not that they did not know the texts. They wanted to instil in us the perception that Indian philosophy was superior to Western. One respect in which this superiority was explicated was by claiming that Indian philosophy was practical (i.e. aiming at the removal of pain and suffering, leading eventually to *mokṣa*) and spiritual (in a rather undefined sense of the term, and we all felt we knew what it was about), culminating in a mystic union with the truth. All this was contrasted with the alleged theoretical, intellectual, and scientific nature of Western thought. It struck me much later as strange that Husserl, in his Vienna lectures, drew a similar contrast, but used the alleged theoretical character of Western philosophy to show its superiority over the practically oriented Eastern thought. This only confirmed my suspicion that such contrasts must be spurious. We did not want edifying discourse. Navya-nyāya confirmed our hunch that Indian thinking was rigorously theoretical and relentlessly intellectual.

It is surprising that these clichés have survived at least a century of intellectual contact, during which it should have become clear to scholars on each side having some acquaintance of the other that they are just not true — that neither is Indian philosophy based on intuition, for example, nor is Western thinking materialistic (to emphasize one way of drawing the contrast, often to be found in India); that Indian logical theories were no less logical, and epistemological theories no less analytical, than their counterparts in the West; and that Indian ontological theories no more or less 'secular' than the Western. (For two other such stereotypes, think of the claim that Indians held a cyclic view of time, as opposed to the linear view of time of the Judaeo-Christian West; or the claim that Indians did not accept the law of non-contradiction.)

It became Matilal's life's mission to demonstrate that such contrasts are mistaken. He spent his life pressing the point and demonstrating that Indian thought, even in its most metaphysical and soteriological concerns, was rigorously analytical, logical and discursive. One should bear in mind that he never claimed that all Indian thinking was logical analysis. He was fully aware and cognizant of the mystical and metaphysical dimensions. But, as he said in his Inaugural Lecture at Oxford, even the mystical illumination had a logico-linguistic aspect and basis. He strongly argued for 'the seriousness and professionalism of Indian mystical philosophers'. Using Nāgārjuna's *Vigrahavyāvartanī* and Śri Harṣa's *Khaṇḍanakhaṇḍakhādya*, he brings out the logic behind the so-called mystical positions. Note that he never held those positions himself. But with characteristic generosity, he writes: 'My personal philosophical view does not, I must admit, coincide with that of either Mahāyāna Buddhism or Advaita Vedānta. But I must emphasize at the same time that these two philosophical systems of the East were not the work of fools.'[1] In emphasizing the analytical content (and form) of the Sanskrit philosophical texts, then, he was not saying that this was all that there was in them, but wanted to bring to the forefront an aspect which had been neglected by most Indian (and Western) writers, and even denied by many.

Once the familiar clichés about East and West are rejected, there is another option open and that is a sort of relativism which would insist that Eastern logic, for example, is not only a different logic, but is, as logic, different. This would amount to saying that the Eastern and Western modes of thinking and standards of rationality are just incommensurable. On this view, Eastern and Western philosophies are simply not the same sorts of enterprise. Each has its own standards of logical and rational assessment. For various reasons, which I cannot give here, I consider such a relativism totally wrong. But one amongst them is particularly important: the entities designated by 'Eastern Philosophy' and 'Western Philosophy' are themselves constructs from other more basic entities which themselves tend to resolve into 'systems of differences'. So even asking what characterizes Eastern Philosophy (or, even, Indian Philosophy), or what characterizes Western Philosophy, is in the long run a question that cannot be straightforwardly answered. The relativistic answer briefly stated above presupposes that these constructs are entities about which one could say things like 'Eastern Philosophy is characterized by the property 'φ', adding that this property 'φ' belongs only to Eastern philosophy. The fact seems to be that there is no such thing as Eastern Philosophy, or even no such thing as Indian Philosophy. Taking Indian Philosophy into consideration, one finds innumerable schools, sub-schools, individual philosophers and their texts, anonymous texts, texts that effectively moved into Tibet and China where they had their efficacy; there are texts in local languages which do not even find a mention in the official histories of Indian Philosophy. It is only at great risk of over-simplification that one sets up a global entity called 'Indian Philosophy', and then proceeds to ascribe to it features, relativistically or otherwise.

But leaving this aside, there is a more sober alternative to the relativistic answer: Indian philosophers raised a large number of fundamental questions which were closely similar to some of the questions which exercised thinkers in the West. They also asked questions which had never been asked in the West. Likewise, they did not ask some of those questions with which Western philosophers have been concerned. We should also note that sometimes when they seem to be asking the same questions as their counterparts in the West, the Indian thinkers give those questions a formulation, a twist, even an interpretation which you never find in Western philosophy. In dealing with those common questions, the Indians often produced theories and arguments which are much like the theories and arguments of their Western counterparts. But they also saw aspects of those problems and came up with analyses and arguments which are not to be found in the Western tradition. Both traditions supply viable alternative answers to certain questions, just as thinkers belonging to one tradition may very well learn from those belonging to the other how not to make certain mistakes and how to avoid certain conceptual muddles and how to ask certain questions more perspicuously. There should be no place for national, geographical, or cultural chauvinism in philosophy.

I think this was Bimal Matilal's overall point of view. This is, I suspect, in his view, the value of comparative philosophy. He thought he was doing a sort of comparative philosophy. We often talked about this, and I expressed my usual misgivings about the *philosophical* value of doing comparative philosophy. He would agree with me that there is a way of doing comparative philosophy which is superficial and of no philosophical value. This is what I have often called 'tagging theories on one side to theories on the other'. What is the philosophical point of doing this except to demonstrate that two traditions produced similar theories? But, as Donald Davidson once said in course of a conversation, why should one read Indian philosophy if the Indian philosophers have given the same sort of answers to precisely the same questions as the Western philosophers do? Matilal's response to this challenge would be, I imagine, twofold: first, comparative philosophy in a certain sense is unavoidable for one who writes about Indian philosophy in English. He writes in his Preface to *Epistemology, Logic and Grammar:*

> I believe that anyone who wants to explain and translate systematically from Indian philosophical writings into a European language will, knowingly or unknowingly, be using the method of 'comparative philosophy'. In other words, he cannot help but compare and contrast Indian philosophical concepts with those of Western philosophy, whether or not he is conscious of so doing. Otherwise, any discourse on Indian philosophy in a Western language would, in my opinion, be impossible. Thus,

'comparative philosophy' in this minimal sense, should no longer be treated as a derogatory phrase.[2]

The same situation, I am sure, would obtain if English or German philosophy texts were rendered into Sanskrit, or if, I would suppose, Kant is translated into English. Comparative philosophy would cut across the East–West dichotomy.

Secondly, and this, I am sure, would be his response to Donald Davidson, Matilal would insist that in spite of the similarities he was so good at bringing out, Indian philosophers did not ask many of the questions which Western philosophers asked, and vice versa. If we keep all such possibilities of questioning in mind, then

the study of Indian philosophy is not simply necessary from a cross-cultural point of view, or from the viewpoint of understanding the 'Indian Mind' (if there is such a thing), but that it is most urgently needed for increasing creativity and comprehensiveness in the philosophic endeavours of modern professional philosophers.[3]

In other words, Indian philosophy could contribute to the formation of a global philosophy, not in the sense of a philosophical theory acceptable to all (for that would not be philosophical), not in the sense of a common project to which all different traditions can contribute, but as a common discourse in which they can participate — in other words, a conversation of [hu]mankind (not a conversation of the West or of the East by itself).

It would not be an exaggeration to say that Matilal devoted his life to making such a dialogue possible.

Being as good a philosopher as he was, Matilal also made substantial contributions to the project of interpreting Indian philosophy, while at the same time advancing and defending a philosophical position of his own. He devoted a lot of attention to the controversy between Nyāya and Buddhism, i.e. to the issues and argumentations that were debated by the two schools. These issues are logical (concerning the theory of inference), semantic (concerning the relation between language and the world), epistemological (concerning primarily the nature and validity of perception), and metaphysical (concerning the controversy between realism and idealism). Matilal thought a great deal about the relation between language and the world, and consequently thought about the positions of Bhartṛhari and Nāgārjuna who occupy two extreme positions in the scale of possible views on the matter. In his writings on these matters, Matilal did not simply play the role of an expositor and interpreter, he also defended a sort of Nyāya realism as opposed to the Buddhist phenomenalism, idealism, and constructivism. Even when he defended Nyāya realism, he — true to the spirit of Indian philosophical writings — always sought to strengthen the defence of the *pūrvapakṣas* — Buddhism and Bhartṛhari — as much as possible and nowhere gives the impression of scoring an easy victory. The realism is defended through a series of choices: preferring the theory of 'mixture of *pramāṇas*' (*pramāṇasamplava*) so that not only touch and vision, but also perception and inference, may cognize the same object; showing that the theory of extrinsic truth of cognitions (*paratahprāmāṇya*) is consistent with all the facts; establishing a causal route from the intentional objects back to material bodies; arguing the thesis that the object of sensory awareness is both an intentional object and a material object (thereby rendering a separate domain of intentional objects superfluous); and, in accordance with Navya-nyāya, distinguishing between entities posited for analysis of cognitions and entities belonging to the 'inner circle' of the ontology. In doing all this, especially the last, Matilal not only defends Nyāya realism, but considerably revises the classical Nyāya, even the Navya-nyāya ontology. Thus he retains only natural kind universals and dismisses artefact universals (such as *potness*) as bogus universals. In doing all this, he proved himself to be a creative interpreter.

In his *magnum opus, Perception: An Essay on Classical Indian Theories of Knowledge*, Matilal transforms the presentation of the Indian theories into a marvellous conversation in which Oxford philosophers Ayer, Strawson, Dummett, and Mackie, and American analytic philosophers Quine, Sellars, and Chisholm, converse with Diṅnāga, Dharmakīrti, Śāntarakṣita, and Jñānaśri amongst the Buddhists, and Uddyotakara, Vācaspati, Udayana, and Gaṅgeśa amongst the Nyāya philosophers. Familiar philosophical questions — what is the distinction between seeing and seeing as? Is there anything purely given? To what extent do language and memory modify perception? How would an uncompromising realism like the Nyāya's account for perceptual illusion? Can perception be construed as an inference as the Buddhist wants to do? Are pleasure and pain themselves experiences or are they, rather, possible objects of experience? — are discussed in considerable detail. We have both comparative philosophy and creative thinking.

In the last years of his life, Matilal was keenly aware of the need to counter the reigning relativisms and also to take into account the deconstructionist modes of thinking. During the spring of 1988, he spent a semester with us at Temple as a Visiting Professor, and stayed in our house. This gave us an opportunity for long conversations on philosophical matters of common interest. We had been working together on a volume of translations of basic source material of Indian philosophy (a project which, alas, went into abeyance with his illness). Besides comparing our notes and checking each other's translations, we talked on a wide range of issues. He wanted to get out of the narrow confines of Oxford analytic philosophy, which led us to discussions of Brentano, Husserl, Heidegger, and the hermeneutic tradition in general. We talked a great deal about relativism (on which he and Michael Krausz had agreed to write a book).[4] Largely owing to the influence of Gayatri Chakraborty Spivak, he wanted to try his hand at deconstructing the traditional way of doing Indian philosophy. There was no limit to his courage in thinking. It is difficult to say where he would have gone in his thinking, had not his life been cut short so prematurely.

There are several questions which may be raised regarding Matilal's way of doing Indian philosophy. Let me formulate three of them in the order of their severity. First, one may ask, is all Indian philosophy analytic? Are not there other facets of Indian thinking? Second, why interpret Indian philosophy 'in the light of' contemporary Western analytic philosophy? Why not introduce, in one's discourse about Indian philosophy, the other contemporary philosophical styles, methodologies, and figures? Third, why interpret Indian philosophy 'in the light of' Western philosophy at all? Why not, for example, do just the reverse, i.e. interpret, translate, and criticize Western thought from the point of view of Indian thought? Let me briefly respond to these questions, as I think Matilal would have done (gathering threads from the innumerable conversations we have had over the years).

I think Matilal knew Indian philosophical literature too well to hold the view that all Indian philosophy is analytical. The emphasis he often placed on this aspect was intended, as he explicitly says in a passage, to correct another one-sided but more misleading emphasis placed on the religious, alogical and non-discursive aspects of that tradition. But at a certain point, philosophical thinking — in India as well as in the West — rises above the cultural milieu from which it has come and from which it derives its nourishment, achieving a certain level of idealization, and it is then that its discourse tends to be universal discourse. A philosophical thesis is then sought to be grounded in arguments, reasonings, and empirical evidence. It can then be called analytical in the broad sense. To say that a large part of Indian philosophy is analytical is not to assign that part to any standard school of analytical philosophy (which by itself, as we know, is enormously variegated). Perhaps one could say that philosophy was regarded in the Indian tradition as a hard-headed, rigorous discipline where definitions, arguments, and disputations prevailed, and which made use of grammar, philology, etymology, analysis of ordinary language (*lokavyavahāra*), and an appeal to ordinary (and extraordinary) experience and textual hermeneutics for vindicating or refuting philosophical claims. In other words, philosophy was a most serious theoretical enterprise. Even the mystic who held that reality was ineffable sought to ground his thesis in logical reasoning — as Matilal argued in his Oxford Inaugural Lecture.

The second question was forcefully pressed, in a recent discussion in Calcutta, by Sibajiban Bhattacharyya. Sibajiban's point was that since 'fashions' in philosophy change, interpretive stances geared to the present style will inevitably make room for the latest to arrive on the scene. Stcherbatsky, for example, interpreted Buddhism using the jargon of the prevailing neo-Kantianism of the late nineteenth century. Analytical philosophy itself has undergone great transformations. What do we do, then, to ensure that interpretations of Indian philosophy, in the light of any current trend, will last? Now to this I will give the following response on behalf of Matilal. For one thing, one can only do something best, and one can only interpret, as Gadamer insisted, from one's present historical situation and not *sub specie aeternitatis*. There is no guarantee that one's interpretation will outlast time and history. For another, as far as Matilal was concerned, although he was primarily thinking in terms of the analytical tradition (Quine, Strawson, Dummett, Davidson et al.), he had an open mind towards Brentano (some of whose ideas he used as early as his Harvard dissertation), Husserl (about whom he learned in course of time) and Heidegger (whom he tried to understand through the 1980s).

I think a more radical questioning remains: why try at all to 'interpret Indian philosophy' — or, for that matter, Chinese philosophy — from the point of view of Western thought? Is not this asymmetry — for Western philosophy is not studied, expounded, and criticized from the point of view of Oriental thought — a sign of the cultural hegemony of the West, of what Husserl called the 'Europeanization of the Earth'? This is indeed a very difficult question to answer, but a question no one who is caught up in this asymmetry should avoid. This is not the occasion to deal with this question in detail, but let me — in retrospect, not being sure how Matilal would have responded — make some preliminary suggestions. First, it is a contingent historical situation which accounts for the fact that even self-characterized purists about Indian philosophy, the 'orthodox' interpreters of Indian thought — the gurus, saints and professors alike — write unhesitatingly on Indian philosophy in the English language (without worrying if such discourse does not entail a surreptitious interpretation, the sort of interpretation that is being questioned). The fact that the professional philosophers in India do not generally write in Sanskrit or in any of the modern Indian languages (although there have been some feeble attempts to do so), but rather write in English, shows that the alleged 'asymmetry' is due not to any self-consciously adopted and defended methodological stance, but due to a historical contingency over which we did not hold any sway. (Exactly out of the same sort of historical contingency, European thinkers do not have to write in Sanskrit or in any of the modern Indian languages.) Secondly, Indian philosophers of past generations whose interpretive stance Matilal opposed, no less thought from a Western perspective; only, they used the language of a Kant, a Hegel, a Bradley, or some philosopher of that breed. Third, there is a growing attempt in India — highly commendable and instructive — to interpret, talk about, and criticize some very fundamental concepts of Western thought in the language of Indian philosophy. A very good example of this sort of work is to be found in the volume *Samvāda.*[5] Finally, the goal should be — as it certainly was Matilal's — to overcome this contingency, this asymmetry, and instead of interpreting one in the light of the other, to evolve a discourse and a conversation in which the partners would be Plato, Bhartṛhari, Aristotle, Gautama, Vātsyāyana, Diṅnāga, Quine, Dharmakīrti, and Carnap — to name only a few. This goal is far off. Even members of the contemporary philosophical community — those at Oxford and at Freiburg, for example — do not have unimpeded communication amongst themselves. How can we expect them to admit such 'alien' figures from ancient and medieval India into a communicative community which knows no national, geographical, linguistic, and cultural bounds? But that is at least what we may aim at, if philosophy is to be a rational enterprise and not hopelessly culture-bound. Nobody contributed more to the advancement of that goal than Matilal.

The point was never how to present Indian writings to the West, to make the illimitable content and form of Sanskrit philosophical texts intelligible to the Western readers. That is a spurious motive. Why should we? The point was to let them converse. Conversation to be genuine, requires from both partners humility, willingness to listen and understand, to step into the other's point of view. The conversation, in this case, was facilitated, because the mediator, Matilal himself, was a genuinely humble person, not arrogant to want to preach, but always willing to learn and to listen. Even he had moments of frustration when Western parochialism stubbornly confronted him, as also when Orientalists here and there, even at Oxford, opposed his 'philosophization' of the texts. The Indologist's prejudice that philosophy is Western in conception, origin, and execution was, even for him, a hard nut to crack. If you are a Sanskritist, you do philology, grammar, religion, study mythology, rituals, witchcraft, magics and the like. Why philosophy? They did not realize that Matilal's mission was precisely to destroy that prejudice by revealing (not constructing) the philosophical content of those texts. In doing this, he was not seeking something Western in the core of Sanskrit texts. But he believed that 'philosophy' had a global *sense*, and that geographical, political, and cultural modifications of that sense should be fundamentally suspect. What he wanted was to actualize this global sense in actual *practice*, not in a favoured theory.

Bimal Matilal was a gentle, courteous, unassuming and discerning person, a friend, colleague, co-researcher, but above all a fine person. David Hume had advised, 'amidst all your philosophy, be a man'. Matilal was both a good philosopher and a good man — no mean, though certainly a rare, combination. For him as a person and as a friend many of us have been mourning within. On this occasion, I wish to remember him as a philosopher. He would have, I am sure, liked this most.

NOTES

1. B.K. Matilal, *The Logical Illumination of Indian Mysticism*, Inaugural Lecture at the University of Oxford, Oxford: Oxford University Press, 1978.
2. B.K. Matilal, *Epistemology, Logic and Grammar in Indian Philosophical Analysis*, The Hague: Mouton, 1971, p. 13.
3. Ibid., p. 12.
4. See Michael Krausz's essay in this volume.
5. *Samvāda: A Dialogue between Two Philosophical Traditions*, ed. by Daya Krishna et al., Delhi: Indian Council of Philosophical Research and Motilal Banarsidass, 1991.

Chapter 1

Gadādhara's Theory of Meaning of Pronouns*

Sibajiban Bhattacharyya

I GADĀDHARA'S THEORY OF WORD-MEANING

In order to explain Gadādhara's theory of meaning of pronouns, it is necessary first to briefly state his theory of word-meaning. In Indian philosophy in general and in Navya-nyāya in particular, theories of meaning of sentences are regarded as a part of theories of origin of true cognition. Hence in discussing theories of meaning, Indian philosophical systems almost exclusively consider how a hearer acquires information second-hand from what a speaker tells him. Indian theories of meaning refer to the speaker's intention only in so far as it is necessary for the hearer to cognize truly what the speaker says.

According to Navya-nyāya, a word means an object only under a mode of presentation. Anyone who knows the meaning of the word knows the object under this mode on hearing, remembering, or otherwise cognizing, the word. The problem of meaning of words is the problem of explaining how the hearer knows the object under the particular mode.

The mode of presentation of an object, according to Navya-nyāya, must be a property of the object. A property is anything which may be said to be *in* the object. According to Navya-nyāya, this is the case if and only if what is to be regarded as a property of an object is related to the object by *an occurrence-exacting relation.* Anything related to an object by such a relation will be its property.

The cognition produced by a word in a hearer is of an object which is a complex of three elements, a-R-b,[1] where R is an occurrence-exacting relation. The mode of presentation which is a property of the object is always the second term of the relation, R.

A. The structure of the cognition caused by hearing the word 'O' may be represented schematically as

Cognition of O-R-(O-ness)

where

(i) O is the qualificand of the cognition caused by 'O'
(ii) O-ness is the qualifier
(iii) R is the qualification of the cognition.

B. The structure of the meaning-relation may be represented by the following diagram:

Diagram 1[2]

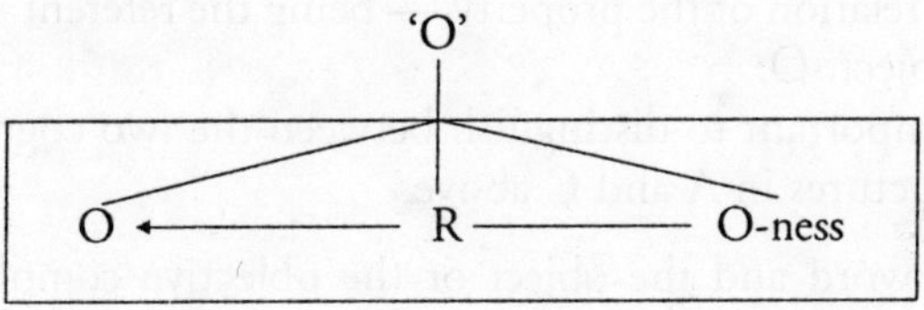

The word 'O' is related to the three elements in the objective complex by different relations which together constitute the meaning-relation which is one relation.

O-ness has the following three features:

(i) O-ness is a property of O
(ii) O-ness is the qualifier of the cognition produced by 'O'
(iii) O-ness is meant by 'O'.

Analysis of the Meaning-relation

The meaning-relation obtains between a word on the one hand and an objective complex on the other, in the sense that hearing a word one has the cognition of the object. Thus the semantic relation between word and object is explained in terms of the relation between the cognition and the object. A word is semantically related to an object if and only if the cognition generated by the word is of that object.

C. The structure of the cognition of the meaning-relation (M) may be represented as follows:

Cognition of: 'O' M O, or usually
Cognition of: O M 'O'

In this cognition of M
(i) O is the referent
(ii) O-ness is the limitor of the property — being the referent — resident in O
(iii) 'O' is the locus of M, i.e. has the semantical power to refer to O.

Thus, although in the knowledge of M, only O is the referent of 'O', still O-ness and R are known as the limiting property and limiting relation of the property — being the referent — resident in the object, O.

It is important to distinguish between the two cognitions and their structures in A and C above.

D. The word and the object or the objective complex are not related by any real or natural relation, but by will or convention. The general theory is that this convention is eternal or Divine Will.

The structure of this Divine Will is as follows:
(DW) O be the object of the cognition caused by 'O'

where

(i) O is the qualificand of DW
(ii) O-ness is the limitor of the property of being-the-qualificand-of-the-Divine-Will resident in O.
(iii) Being the object of the cognition caused by 'O' is the qualifier of the DW. Thus to be the referent of 'O' is to be the qualificand of the DW. To be the mode of representation of O, the referent of 'O', is to be the limitor of the property of being the referent. Thus O-ness is the mode of presentation

of O. O-ness as the limitor of being qualificand of DW resident in O which is the referent (ii), is called 'the limitor of the property of being the referent (referentness)'.

(iv) The mode of presentation of the referent being always a property of the referent must be related to the referent by R, an occurrence-exacting relation. This R which is a part of the meaning of 'O' must be related to the DW. This relation is due to the fact that R is the limiting relation of the property of being the qualificand of the DW resident in O.

Raghunātha, however, does not accept this usual Navya-nyāya theory. According to him, O-ness is (a) the limiting property of being the qualificand of the DW resident in O, and also (b) is the limiting property of being the qualificand of the cognition caused by 'O' resident in O. But this does not mean that O-ness has to be meant by 'O'. The point that Raghunātha is making here is that only the referent is the meaning of the word, and that, neither the mode of presentation of the referent, nor its relation (R) to the referent need be included in the meaning of the word. The fact that O-ness is the qualifier in the cognitive structure A, and is also the limitor of the property of being the qualificand of DW, does not mean that O-ness has to be part of the meaning of 'O'.

E. The mode of presentation of the referent which must be a property of the referent is also *the cause for the application of the word to the referent.* According to Navya-nyāya, if a word refers to any object, there must be a cause for the application of the word to the referent.

II GADĀDHARA'S THEORY OF ANAPHORA[3]

1. The problem:

S_1 — There is a jar over there, bring *it.*[4]
S_2 — There is a table over there, bring *it.*

In S_1 and also in S_2 the pronoun 'it' occurs. Grammatically, i.e. syntactically, in S_1 'it' refers to the noun 'jar', in S_2 to 'table'. The problem which Gadādhara raises here is semantical. What will be the meaning of 'it' in S_1 and in S_2?

It appears that 'it' does not have any one meaning, but has many meanings. In S_1 'it' means the jar cognized under the mode of jarness, in S_2 'it' means the table cognized under the mode of tableness. As a pronoun like 'it' can occur in various contexts, it has many meanings.

A word has many meanings if and only if it refers to different objects under different modes, i.e. when the word has different properties limiting the referentness of the word. The rule is that if the limiting properties of the referentness of a word be many, then the word has many referentnesses and so becomes equivocal. If, on the other hand, the referents of a word, like 'jar', are many, yet all referents are cognized under one mode, i.e. if the limiting property of referentness of word is one, then the word has only one meaning.[5]

The following diagrams will make this distinction clear:

Diagram 2: Univocation

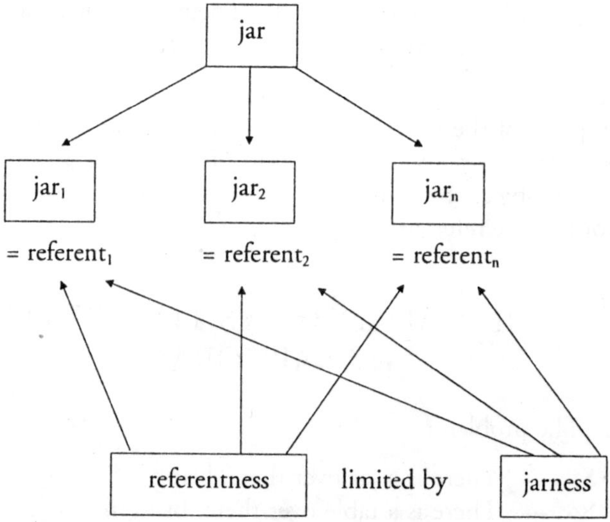

Diagram 3: Equivocation

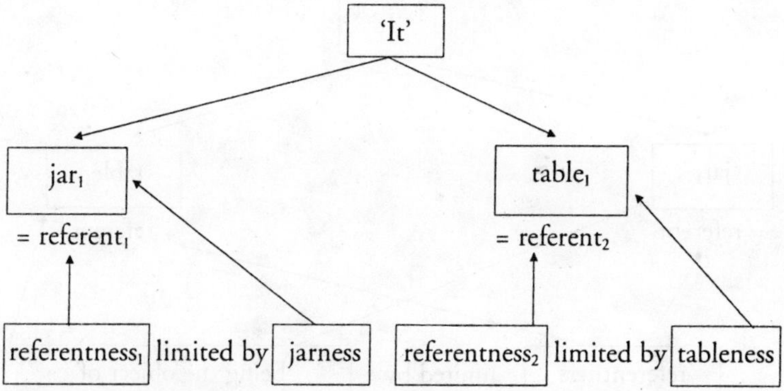

2. An attempt may be made to show that a pronoun like 'it' has only one meaning despite the fact that it means a jar in S_1 and table in S_2. For we may say that 'it' means the object which is the object of cognition of the speaker. This is the meaning of 'it' usually given in dictionaries. For example in *OED*, we get 'The *little one* is used anaphorically if it means "the little flower" or whatever it is that has just been mentioned'. In *Webster's Encyclopedic Unabridged Dictionary* we get, 'It, pron. 1. (used to represent an inanimate thing understood, primarily mentioned or about to be mentioned)'. So 'it' refers to the object 'understood or 'that has been just mentioned'. In Navya-nyāya terminology, 'it' refers to an object under the mode 'being an object of cognition of the knower' which becomes the limitor of referentness of 'it'. So 'it' has one meaning. Therefore, Diagram 3 has to be modified as in Diagram 3 M. In this interpretation 'it' becomes univocal.

3. Gadādhara finds this interpretation faulty, because when the speaker says S_1, the cognition which the hearer has is that he is to bring the jar cognized under the mode of jarness, and not under the mode of being the object of cognition of the speaker. So being the object of cognition of the speaker cannot be the limitor of referentness of 'it'. It will be jarness in S_1 and tableness in S_2. So 'it' is equivocal (Diagram 3).[6]

4. In reply to this objection of Gadādhara, the opponents modify their theory. They now distinguish between limitors of different

Diagram 3 M

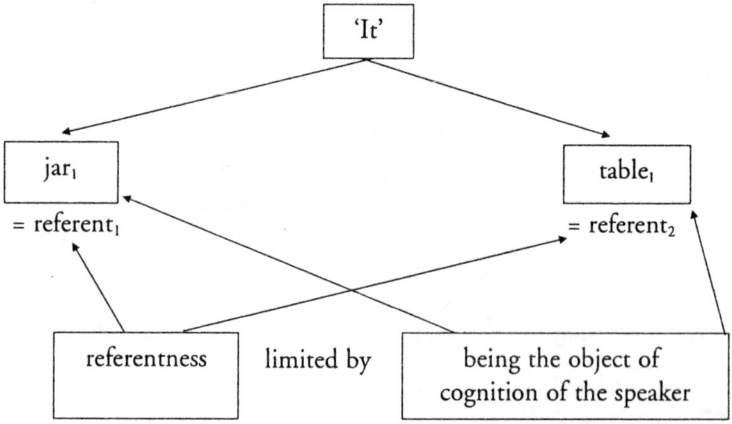

orders. The first order limitors of being the referent of 'it' are, indeed, many, jarness in S_1 and tableness in S_2 and so on. But if we rise to a second order limitor of being the first order limitor of being the referent of 'it', then we have only one property, namely being the object of cognition of the speaker. This may be explained by Diagram 4.

Now Gadādhara criticizes this theory, too. He points out that 'being the object of cognition of the speaker' which is the second order limitor is a qualifier of the first order limitors. Universals like jarness when unqualified and not referred to by any word are cognized in and through themselves. But if the universals are qualified then they are always cognized through their qualifiers. Now in S_1 the hearer cognizes that he is to bring the jar cognized through the mode of jarness; but if we accept the theory represented in Diagram 4, then we ought to cognize the jar in S_1 through the universal jarness which again is qualified by being the object of cognition of the speaker. But a hearer of S_1 cognizes the jar under jarness not further qualified by any qualifier. Hence this modified theory cannot explain how 'it' is univocal.

The opponents whom Gadādhara mentions here come up with an attempted solution to his objection, by distinguishing between a qualifier (*viśeṣaṇas*) and a pseudo-qualifier (*upalakṣaṇa*). A qualifier has been distinguished from a pseudo-qualifier on various grounds, of which only some will be explained.

Diagram 4

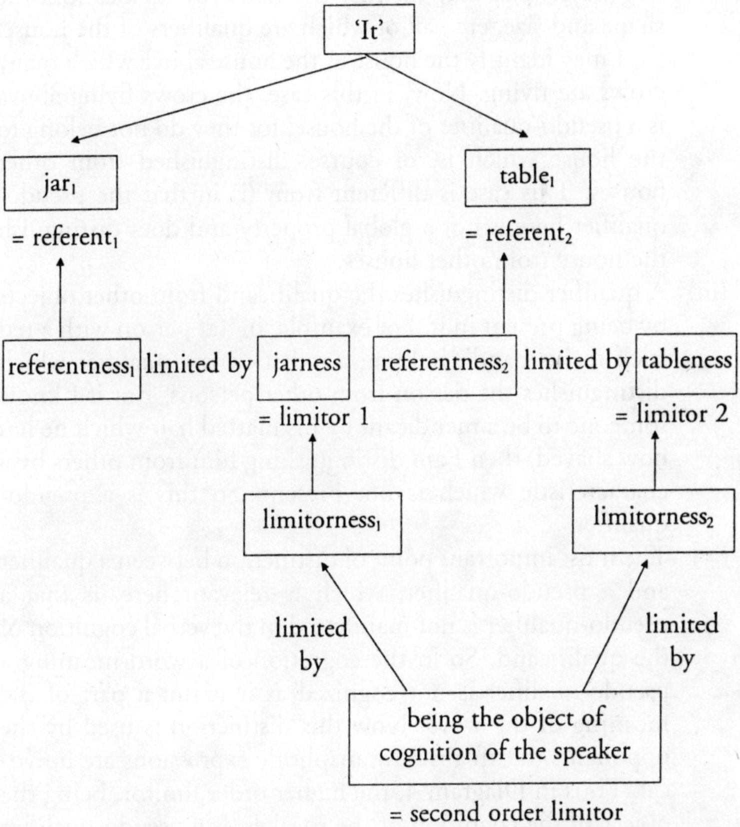

(i) A qualifier distinguishes a qualificand to which it belongs from all other objects. For example, in '(a) blue lotus', blue (colour) is a qualifier of the lotus for the colour distinguishes it from lotuses which are not blue. But in 'a lotus which is an object of true cognition' the grammatically adjectival clause 'which is an object of true cognition' means not a qualifier, but only a pseudo-qualifier. For, according to Nyāya, every object is an object of true cognition (at least of God who is omniscient); hence being an object of true cognition is a global property which cannot distinguish any qualificand from others.

(ii) If I want to identify Devadatta's house, I may proceed in two ways: (a) I may identify the house by its location, the shape and size, etc., all of which are qualifiers of the house. (b) I may identify the house as the house above which many crows are flying. Now, in this case, the crows flying above is a pseudo-qualifier of the house, for they do not belong to the house which is, of course, distinguished from other houses. This case is different from (i) in that the pseudo-qualifier here is not a global property and does distinguish the house from other houses.

(iii) A qualifier distinguishes the qualificand from other objects by being present in it. For example, in '(a) person with a red stick in his hand', holding a red stick is a qualifier which distinguishes the person from other persons. But if I know someone to be a mendicant by his matted hair which he has now shaved, then I am distinguishing him from others by a characteristic which is not present. So this is a pseudo-qualifier.

(iv) The most important point of distinction between a qualifier and a pseudo-qualifier, which is relevant here, is that a pseudo-qualifier is not manifested in the verbal cognition of the qualificand. So in the cognition of a word-meaning a pseudo-qualifier is not cognized as it is not a part of the meaning of the word. Now this distinction is used by the opponents to explain how anaphoric expressions are univocal. Thus in Diagram 4, the higher order limitor, being the object of the cognition of the speaker, is a pseudo-qualifier of jarness and tableness; hence this is not cognized in the meaning of 'it' in S_1 and S_2. The jar under the mode of jarness cognized as the meaning of 'it' in S_1; so also the table under the mode of tableness in S_2.

Gadādhara, however, criticizes this theory as giving no reason why the second order limitor should be a pseudo-qualifier. If this is recommended without any justification, then even the first order limitors may be regarded as pseudo-qualifiers. Then 'it' in S_1 will refer to the jar without any mode at all; when anyone hears it, he will have the cognition of the jar without any qualifier. But this is not possible; for according to Nyāya, no substance can be cognized except through a mode.

Thus the attempts of the opponent to explain how anaphoric expressions can be univocal are not successful. But we cannot also hold that they have many meanings. The theory that 'it' is not univocal, leads to insuperable difficulties. It will not be possible for anyone to learn all the meanings of a term like 'it'. For, in S_1 'it' means the jar, in S_2 'it' means the table, and there will be innumerable sentences in which 'it' will occur and in which it will mean innumerable objects different from one another. The point is that even though the speaker always cognizes the object referred to by 'it' under a proper mode, the hearer cannot always cognize it, for the hearer may not know that meaning of 'it'.[7] So to explain how anyone can learn the meanings of anaphorical expressions, they have to be regarded as univocal. The problem is how to justify this semantically.

Now I explain Gadādhara's solution. This is to explain why the second order limitor is a pseudo-qualifier. 'It' means, first, the limitor of being the object of cognition of the speaker. For example in S_1, the object of cognition of the speaker is the jar. The limitor of being the object of cognition is jarness. So jarness has the property limitorness of being the object of cognition of the speaker. But this property is a pseudo-qualifier of jarness, and the jar limited by jarness is the referent of 'it'.[8] This may be explained by Diagram 5. So when one hears S_1 or S_2, one has the verbal cognition of an object under a mode at 2, but not again under the mode at 3. Now Gadādhara explains the reason why 3 is a pseudo-qualifier, but not 2.

We have to remember here that a limitor of being the referent has three characteristics:-

O-ness is the limitor of being the referent O, if and only if,

(i) O-ness is a property of O
(ii) O-ness is the qualifier of the cognition generation by 'O'
(iii) O-ness is meant by 'O'.

Now the second order limitor in Diagram 4 is not meant by 'O' and hence it cannot be the qualifier of the cognition generated by 'O'. For the Divine Will is *not* of the form

DW1. 'It' produces the cognition of being-the-limitor (limitorness) of being the object of cognition.

Diagram 5

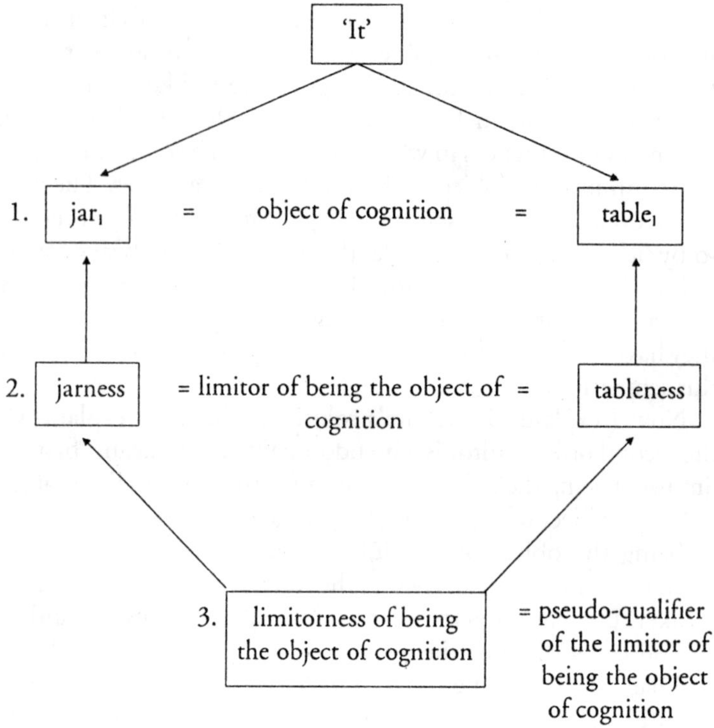

In the absence of DW in this form, being the limitor of being the object of cognition of the speaker cannot be a qualifier of the cognition generated by 'it'; hence 3 in Diagram 5 is not manifested in the cognition of it.

The Divine Will which is the semantic relation between 'it' and its meaning is of the following form:

DW2. 'It' produces the cognition of what is limited by what is indicated by a pseudo-qualifier which is the limitorness of being the object of cognition.

Note: A pseudo-qualifier indicates whatever of which it is the pseudo-qualifier. Thus indication is the relation between a pseudo-qualifier and that of which it is the pseudo-qualifier.

Now we explain DW2. 'It' in S_1 refers to the jar which is the object of cognition of the speaker. Jarness is the limitor of being the object of cognition of the speaker. So jarness has the property of being the limitor of being the object of cognition of the speaker. What this property — being the limitor of being the object of cognition — indicates is jarness. So this property is a pseudo-qualifier.

Diagram 6

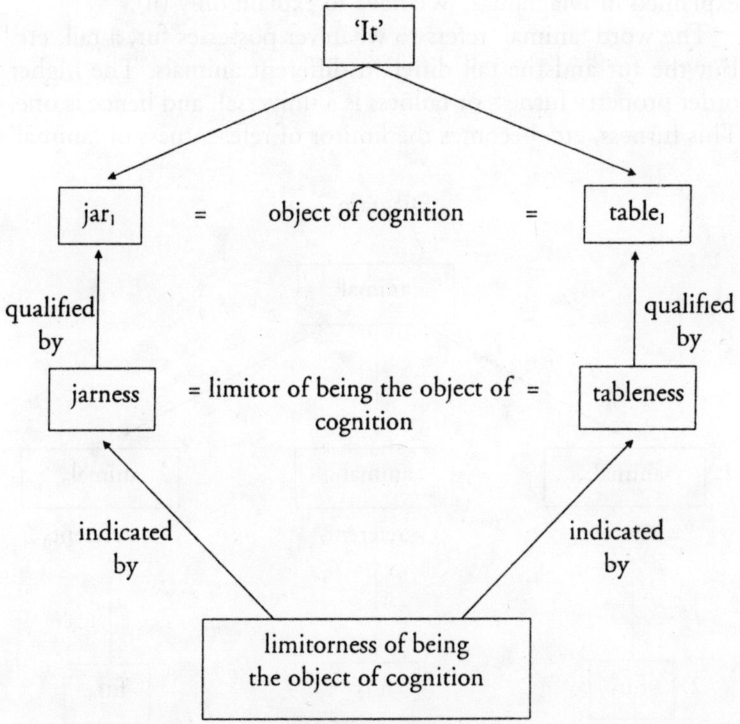

Now let us examine how this theory of Gadādhara can make anaphoric expressions univocal. It will be necessary to supplement the explanation of univocation and equivocation given on p. 21 (see Note 6). The full explanation is

A word is univocal if and only if (i) there is one limiting property of referentness; or (ii) there is a higher order property which becomes only indirectly a limitor of referentness; or, (iii) there is one pseudo-qualifier indicating all the limitors of referentness of a word. Otherwise, the word is equivocal.

Explanation

(iii) is explained in Diagram 6, which makes 'it' univocal. (i) is explained in Diagram 2. We have to explain only (ii).

The word 'animal' refers to whatever possesses fur, a tail, etc. But the fur and the tail differ in different animals. The higher order property furness or tailness is a universal, and hence is one. This furness, etc. becomes the limitor of referentness of 'animal'

Diagram 7

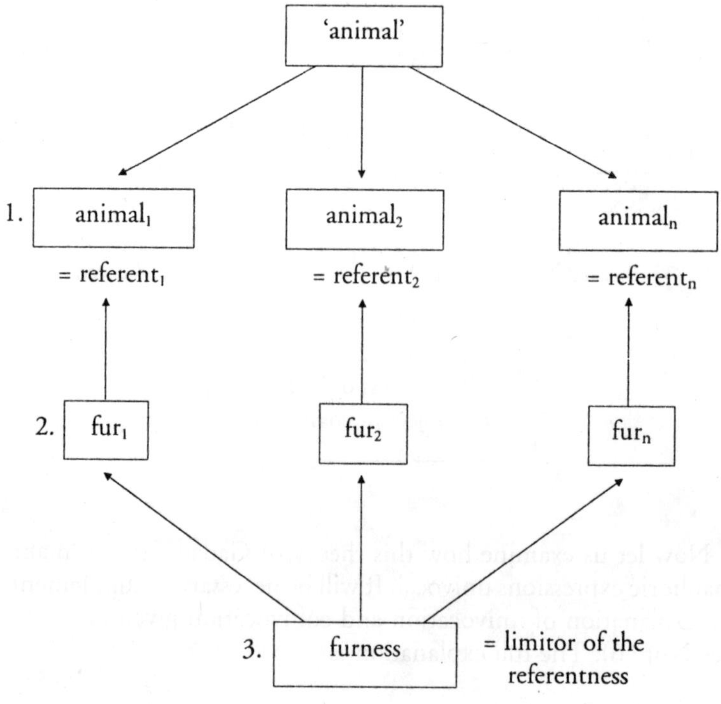

only indirectly via the particular furs, etc. of the animals.[9] This may be explained by Diagram 7. As this diagram makes it clear, furness becomes the limitor of referentness of 'animal' only indirectly via stage 2. But as this limiting property, though indirect, is one, 'animal' has one meaning.

Diagram 8: Equivocation

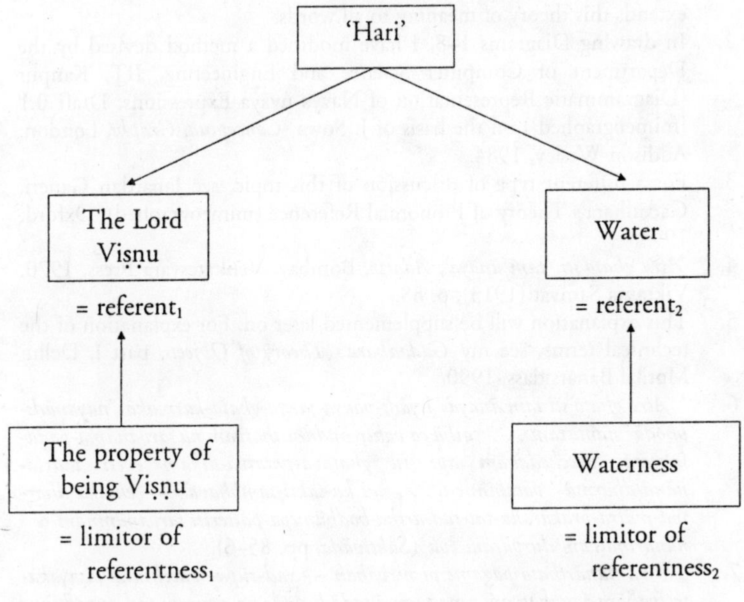

As there is no one higher order property, either as a qualifier or as a pseudo-qualifier which can unite the two referent-nesses, 'Hari' is ambiguous.

NOTES

* Gadādhara Bhaṭṭācārya (*c.* 1620–1700), has divided his major work on word-meaning, *Śaktivāda*, into three sections. In Section 1, he develops his general theory of word-meaning; in Section 2, he discusses meanings of some special words. His discussion of pronouns like 'it', 'that-which',

etc. belongs to this section. In Section 3, which is an appendix, he reopens issues of Section 1, and discusses them in much greater detail.

The Sanskrit term for pronouns is 'sarva-nāma', lit., 'names of everything', as pronouns can stand for any noun grammatically.

1. It is interesting to note that Mill in his theory of what he calls 'concrete general names', says: 'The word *man*, for example, denotes Peter, Jane, John, and an indefinite number of other individuals, of whom, taken as a class, it is the name. But it is applied to them, because they possess, and to signify that they possess, certain attributes.' J.S. Mill, *A System of Logic*, London: Longmans, Green and Co., 1956, p. 19. Only Gadādhara extends this theory of meaning to all words.
2. In drawing Diagrams 1–8, I have modified a method devised by the Department of Computer Science and Engineering, IIT, Kanpur (Diagrammatic Representation of Navya-nyāya Expressions, Draft 0.1 [mimeographed]) on the basis of J. Sowa, *Conceptual Graphs*, London: Addison-Wesley, 1984.
3. For a different type of discussion of this topic, see Janardan Ganeri, Gadādhara's Theory of Pronomial Reference (mimeographed), Oxford: 1993.
4. *'Atra ghaṭo' sti, tam ānaya'*, *Ādarśa*, Bombay: Venkatesvara Press, 1970, Vikrama Samvat (1913), p. 88.
5. This explanation will be supplemented later on. For explanation of the technical terms, see my *Gadadhara's Theory of Objects*, part I, Delhi: Motilal Banarsidass, 1990.
6. *' "Atra ghaṭo' sti tam ānaya" ityādi-vākya-janya-ghaṭa-karmakā "nayanādi-bhodā" nantaram . . . tathā ca vaktṛ-buddhisthatvaṁ na sarvanāma-pada-śakyatā 'vacchedakam api tu ghaṭava-paṭatvā-dikam eveti tādṛśa-nānā-dharmā-' vacchinneṣu . . . nai ka-śakti-saṁbhavaḥ — ekaika-dharma-mātra-prakāraka-tat-tad-artha-bodhānupa-patteriti sarvanāmnāṁ nānārthatvaṁ durvāram eva'* (*Śaktivāda*, pp. 85–6).
7. *'Na vā nānārthato pagame pi nirvāhaḥ — yad-rūpā-"vacchinna viṣayaka-vyavahāra-janakatvaṁ yena puṁsā tadādi-pade na gṛhitaṁ tad-avacchinne tena tac-caktyā' prakrāntasyā' pi tad-avicchinnasya tatpuruṣīya-tadādi-pada-janya-bodha-viṣayatā 'nupapatteḥ'* (*Śaktivāda*, p. 86).
8. *'Tat-padaṁ-buddhi-viṣayatā-'vacchedakatvo-palakṣita-dharmā-'vacchinnam bodhayatu' ityākārakasya vā bhagavat-saṁketasya svīkārāt tadādi-sarva-nāma-padeṣu buddhi-viṣayatā-'vacchedakatvo-palakṣita-dharmā-'vacchinna-nirūpitāyāḥ śakter aikyān na nānārthatā, buddhi-viṣayatā-'vacchedakatvo-palakṣita-dharmaśca kvacid ghaṭatavādikam kvacit paṭatvādikam ityanyad etat, teṣāṁ sarveṣāṁ buddhi-viṣayatā-'vacchedaktvo- palakṣitatvenā 'nugamaḥ. 'Tatra ghaṭosti tam ānaya' ityatra, buddhi-viṣayatā ghaṭe, buddhi-viṣayatā-'vaccedakaṁ ghaṭatvam, buddhi-viṣayatā-vacchedakatvam ghaṭatve, buddhi-viṣayatā-'vacchedatvo-palakṣita-dharmaśca ghaṭatvaṁ, tad-avacchinno ghaṭas tad-bodhakatvaṁ tat-pade-prāptam'* (*Ādarśa*, p. 85).

9. '*Paśvādi-sthale ca "paśu-padād lāṅgula-viśiṣṭo loma-lāṅgula-viśiṣṭo vā boddhavyaḥ" ityākāraka-bhagavat-saṁketa-svīkārāt paśu-vyaktau lāṅgulam viśeṣaṇaṁ, taccā 'nekam eva, tatra lāṅgulatvaṁ viśeṣaṇaṁ tacca ikam eva, tacca lāṅgulatvaṁ lāṅgula–dvārāi va śakyatāvacchedakaṁ iti paramparayā śakyatāvacchedakaṁ jātaṁ*' (*Ādarśa*, pp. 88–9).

REFERENCES

BHATTACHARYA, Gadādhara, *Śaktivāda*, edited by Sudarsanacarya Sastri with his own Commentary, *Ādarśa* (Venkatesvara Press, Bombay, 1970 [Vikrama Samvat, 1913]).

BHATTACHARYYA, Sibajiban, *Gadādhara's Theory of Objects*, part I (Motilal Banarsidass, New Delhi, 1990).

GANERI, Janardan, Gadādhara's Theory of Pronomial Reference (mimeographed) (Oxford, 1993).

Kanpur IIT, Diagrammatic Representation of Navya-nyāya Expressions, Draft 0.1 (mimeographed) (Department of Computer Science and Engineering, IIT, Kanpur).

MILL, J.S., *A System of Logic* (Longmans, Green and Co., London, 1956).

SOWA, J., *Conceptual Graphs* (Addison-Wesley, London, 1984).

Chapter 2

The Earliest Brahmanical Reference to Buddhism?

Richard Gombrich

In his famous work on Indian logic and epistemology, Bimal K. Matilal was much concerned with the interaction between Buddhist and brahmanical thinkers. I am not competent to write about the more technical side of Indian philosophy; but it may be appropriate if in his memory I draw attention to an early brahmanical reference to Buddhism which seems so far to have escaped attention. That is the main point of my essay; but I shall also use the opportunity to make a few observations about the treatment of world renunciation in the earliest law-books.[1]

I

The material I am dealing with forms part of the discussion in early brahmanical normative texts, *dharma-sūtras,* of ways of life (*āśrama*) available to brahmin males.[2] According to all such texts, a brahmin male has to be initiated into the study of the Veda and then spend several years as a celibate student pursuing that study. The widespread modern understanding of what went on in ancient India is based on the norms presented in the law-book of Manu: that the student in due course married and so became a householder (*gṛhastha*); that he later retired to become a forest-dweller (*vaikhānasa/vānaprastha*); and that he finally moved on from that to become a renouncer (for which there are various terms, but the one mainly used nowadays is *saṃnyāsin*). That a population should actually have lived by this rule has always strained credulity; yet I have seen a school textbook prescribed by the Indian government[3] which solemnly avers that in ancient India men spent the first 25 years of their lives as students, the next 25 as householders, at the age of 50 became forest-dwellers and at 75 renounced the world!

In an important article,[4] more than twenty years ago, Patrick Olivelle showed that the original theory of *āśrama* was not that they were four obligatory stages in a life cycle, but a classification of options facing a man at the end of his period of celibate Vedic study. At this point a man could either marry and enter the *āśrama* of a householder, or take to one of what were classified as three celibate life-styles: as a lifelong student; as a forest-dweller; or as an itinerant ascetic. Moreover, among the three law-books which scholars concur in regarding as the oldest, the *dharma-sūtras* of Gautama (G), Baudhāyana (B) and Āpastamba (A), the former two concur in condemning the celibate life-styles outright and demanding that brahmins marry and beget children; only A begins by presenting them as open options and reports arguments for their superiority, though he too concludes by deciding against them in favour of the householder's life.

While there must be some artificiality in the attempt to classify the lifelong celibates as leading three standardized and distinct ways of life, Olivelle's — obviously correct — reading of the texts makes them much more plausible as reflections of social reality. Obviously most brahmins did indeed marry, have children, and finally die as householders; the model presented by the earliest law-givers is not some wild cultural fantasy but closely related to what was going on around them.[5] This makes it less surprising that one of the texts should mention Buddhists, however much brahmins may have disapproved of them.

II

The text of B[6] was first presented to the West in the form of an English translation by Georg Bühler, published in 1882 in the *Sacred Books of the East* series. The Sanskrit original was printed for the first time just two years later by Bühler's pupil Hultzsch.

For his edition Hultzsch used the same materials as Bühler had used for his translation, with the addition of one manuscript.

In the Introduction to his translation, Bühler argued[7] that the fourth and final book (*praśna*) was later than the third, and the third later than the first two, and that even in them there were probable interpolations. His basic stratification and line of argument have been accepted by subsequent scholars and somewhat elaborated. There are in the second book two passages about world renunciation. The first, 2, 6, 11, 9–34[8] is the one on which I shall focus in this article. The second, much longer one, 2, 10, 17, 1–18, 27, forms the last part of book 2, and Olivelle has argued that it is a later addition, perhaps coeval with book 3. I find his argument completely convincing. He says[9] of the latter passage that it 'is the only text in all the extant Dharmaśāstras that deals with the procedure for becoming a renouncer', and also that it contradicts the earlier passage on a cardinal point, for it not only approves of renunciation, but regards the *āśrama* as sequential stages. In other words, the second passage has the same view of renunciation as does Manu, and must come from a later period than the passage with which I am concerned.

The text of this first passage appears in the manuscripts in two forms, so that one might even speak of two recensions. This must slightly complicate the picture. But I agree with Bühler and Hultzsch in regarding the longer version of the text as likely to be more authentic, so I shall use that as my basis. The variant version I shall discuss in Section VII.

To come straight to the point: I see a reference to Buddhists in ss.2, 6, 11, 26.[10] It reads:

apavidhya vaidikāni karmāṇy ubhayataḥ paricchinnā madhyamaṃ padaṃ saṃślisyāmaha iti vadantaḥ

Two translations are possible, and though the difference is slight and inconsequential I give both. The more probable one is:

> Saying, 'We reject Vedic rites and adhere to the middle path, being delimited to both sides.'

It is also possible that the first three words stand outside the speech, so that one should translate:

Having rejected Vedic rites, saying, 'We adhere to the middle path, being delimited to both sides.'

In the very next words, s.27, B declares against all the ways of life described in the previous sūtras (for a full translation see Section V below), so it is clear that the passage is not prescriptive. Buddhist ascetics (i.e. monks and nuns) do of course reject Vedic rites. Though they were recruited from all strata of society, there is strong evidence that among them brahmins were disproportionately well represented.[11] In condemning brahmins who had turned to Buddhism, B was thus not mentioning a merely theoretical category. Buddhist monks and nuns follow the middle path (Pali: *majjhimā paṭipadā*) described in the Buddha's first sermon as steering between the two extremes of indulging and mortifying the flesh.

Why has this reference to Buddhism remained so long unnoticed? Bühler's translation of the sūtra reads:

(Ascetics shall) say, 'Renouncing the works taught in the Veda, cut off from both (worlds), we attach ourselves to the central sphere (Brahman)'.

By supplying 'ascetics shall' he turns the sentence into a prescription. Other words supplied in brackets carry the main burden of the interpretation. I am not aware that Brahman is ever called 'the central sphere' in any other text.

Bühler did not invent the interpretation embodied in his translation. He took it straight from the only known extant commentary on B, by Govindasvāmin. The date of the latter is unknown, but Bühler and Kane agree in regarding him as recent.[12] I shall show, however, that his interpretation probably rests on an ancient tradition. To do so I must broaden the discussion to bring in a wider range of passages from the ancient *dharma-sūtra* literature.

III

P.V. Kane, the greatest authority on the history of the Sanskrit law-books, tentatively places, G, B and A in that chronological order; Olivelle follows him; and I am aware of no compelling reason to demur.[13] At the same time, one must take into account what it means to date oral literature. In my opinion, the earliest brahmanical text which looks as if it has always been a written work is Patañjali's *Mahābhāṣya*, which is generally dated to the middle of the second century BC.[14] The first dated evidence for writing in India (except perhaps for the extreme north-west) is Asoka's inscriptions of the third century BC. Vedic literature was composed and preserved orally for many centuries before that — hence the many years of studentship required of every young brahmin male — and there is no reason to think that the brahmins changed their customs by committing their texts to writing as soon as that invention became available. Whenever G, B and A were composed, we have not the slightest idea when they were written down; but we can be sure that they were preserved orally for a long time, probably at least a few centuries. On the other hand, these texts were of a comparatively low order of sanctity, which means that they were all the more liable to interpolation and other changes in the course of transmission.

Moreover, the three texts are related. They do not descend from anything like a single archetype, but from a single way of thinking about how life should be led. They were separately preserved by different Vedic ritual lineages (*śākhā*), which were also to a large extent clusters of patrilineages; but corporately they constituted a school of thought. This is obvious as soon as one looks at what they have to say about almost any of the topics they treat: the texts show signs of mutual influence both in their initial formulation and later; they borrowed from each other and interacted.

IV

I am dealing with the passages on the *āśrama* of renunciation in the earliest *dharma-sūtras*: G 3, 11–25; B 2, 6, 11, 16–34; A 2, 9, 21, 7–17. I have explained in Section II above that I am ignoring the long passage at the end of book 2 of B because it has no parallel in the other texts and is evidently a late addition. Moreover, there is no need to discuss ss.29–34 of our passage in B, because they simply support the conclusion that only the householder's life is correct by adducing Vedic quotations.

My discussion does not aim to be exhaustive. Its purpose is to throw light on the context in which B refers to Buddhists.[15]

The first parallelism between the passages, so obvious that one could overlook it, is that they are of similar length. The second is that all introduce the topic with a sūtra which just lists the four *āśrama*: G 3, 2; B 2, 6, 11, 12; A 2, 9, 21, 1. (The references are not contiguous to the passages I translate because some remarks on lifelong studentship and on other topics intervene.)

G calls the renouncer *bhikṣu*, B *parivrājaka*, A *parivrāja*. These are not terms of art but literally descriptive. G's name shows that this life-style is defined by begging for one's food, the other names show that it is defined by having no fixed abode (except in the rainy season). Those two characteristics are the constant and defining features of the life-style. Other features vary; the texts, in their attempts to categorise men who wish to escape from conventional categorise, are giving us a composite picture.

I have a bold, perhaps wild, suggestion about the change in name. The Buddha called his monks *bhikṣu*. G, I would suggest, used the word to name a brahmanical institution when he had not heard of Buddhist monks (perhaps because they did not yet exist). B knew of Buddhist *bhikṣu* and disapproved of them, so he avoided that name. A simply copied B (as I shall show below he did in other ways).

As I said above, the fact that G and B condemn renunciate life-styles out of hand means that they are describing rather than prescribing. How, then, should one translate the optative when it occurs in the sūtras? The normal translation is 'should', but that does not fit a description.[16] The English equivalent in this context is 'would'. A description of the world in terms of *dharma* is a long way from a purely objective ethnography — should such a thing be possible; it is an attempt to fit the world into a set of categories. Modern parallels would be such statements as 'A burglar would be equipped with a crowbar' or 'An adulterer would come in by the back door'. Should the reader find these unconvincing, one might look at the generalizations offered by prejudiced people, such as 'An X [substitute the name of an ethnic minority] would never give you a straight answer'. What this amounts to is that in these texts the difference between the indicative and the optative is of no account.

V

I turn to comparing G and B. G I translate as follows:

(11) The beggar (*bhikṣu*) keeps no possessions. (12) His semen is up (i.e. never shed). (13) In the rainy season he stays put. (14) For alms he would visit a village. (15) He would go there last and not return.[a] (16) He expects nothing (by way of food). (17) He is restrained in word, look and action. (18) He would wear clothing (just) to cover his private parts. (19) Some say he would wear (clothing) which has been abandoned, after washing it.[b] (20) He would not take any part of a plant or tree which is not (already) detached. (21) Outside the (rainy) season he would not stay in a village more than one night. (22) He is shaven-headed or wears a topknot. (23) He would avoid destroying seeds. (24) He is the same towards beings whether they harm or help him. (25) He undertakes nothing.[c]

Notes

[a] From other texts we can deduce that this means: he goes begging to people's homes after they have eaten and does not come back for a second helping.

[b] Or maybe: 'Some would wear . . . '

[c] My literal translation cannot convey the force of *anārambhī*. Perhaps because the verb *ā-rabh* was particularly used for undertaking a sacrifice, it has strong overtones of violence. In the vocabulary of renouncers, 'undertaking' (*ārambha*) implies aggression, in Jain Prakrit and Buddhist Pali as well as here in brahmanical Sanskrit.[17]

B I translate as follows:

(16) The itinerant, having left his family, would go forth (to the homeless life) without possessions according to rule. (17) Going to the forest (18) with his head shaven except for a topknot, (19) wearing a loincloth, (20) he stays in one place during the rainy season. (21) He wears yellow-stained clothing. (22) He would beg food when the pestle has been laid down, there are no (more) live embers, and the collecting of plates is over. (23) He has no hostility towards beings by aggression of word, thought or deed. (24) For purity he would carry a water-strainer, (25) performing water ritual with water drawn (from a well) and purified (with the strainer). (26) (They are) saying: 'We reject Vedic rites

and adhere to the middle path, being delimited to both sides.' (27) But the teacher/s (approve) having but one way of life, because the others have no children. (28) To that they adduce: There was an anti-god (*asura*) called Kapila, son of Prahlāda. Vying with the gods, he made these divisions. A wise man would take no account of them.

Let me consider the end of this first. At s.28 the text changes character; it breaks out of the style of sūtra texts and tells an aetiological myth, however brief, in the style of the ancient *Brāhmaṇa* texts. As mentioned above, the text goes on to cite Vedic verses. There is nothing like this heterogeneous passage in G.

The true conclusion is in s.27. The words giving the reason for the conclusion, 'because the others have no children', are also in G 3, 3. There they sit rather oddly as if they were part of the view to be refuted (*pūrvapakṣa*). G's refutation, at the end of the section in s.3, 36, runs: 'But the teacher/s (approve) having but one way of life, because being a householder is manifestly enjoined (in the Veda).' Thus B's conclusion combines the first half of this sūtra, the words *aikāśramyaṃ tv ācāryāḥ* (where it is not possible to say whether the plural form *ācāryāḥ* literally refers to several teachers or is an honorific form referring to one teacher), with the second half of G 3, 3. In this case it is B which seems more coherent than G, though I would not argue that the more coherent version of a text must be the older.[18]

For the rest, it is B which seems the less coherent of the two texts. Indeed, B bears in its language unmistakable indications of a composite character. The ascetic is described in the (generic) singular until s.26, where both 'saying' and the speech ('we adhere') are in the plural. The word 'saying', indeed, is a present participle suspended without a subject or main verb. S.25 too has a present participle,[19] this one in the singular, but it is possible to construe it with s.24. The absolutive *gatvā* in s.17, though easy to construe, also seems to hang without clear syntax.

The word 'saying' in s.26 suggests that the author is reporting on people alien to him. When these texts merely record an alternative view (or practice) they use *eke* ('some men') — see G s.19 above and A s.12 below. This fits my claim that s.26 is referring to Buddhists. S.21 may also be referring to Buddhists. The word *kaṣāya* means 'stained' and also refers to any colour in the yellow–orange–brown range; it is regularly used to describe the robes of Buddhist monks. The wearing of such robes may not have been peculiar to Buddhists — a thousand years later, the orders of *saṃnyāsin* founded by Śaṅkara adopted them too — but for Buddhist monks it was compulsory. Neither G nor A refers to such clothing, and I take this as further corroboration that B, and B alone, refers to Buddhists.

B's remarks on clothing seem inconsistent. S.19 says that the ascetic wears a loincloth, s.21 that he wears a yellowish robe. It is unlikely that these two statements can be taken in conjunction: Buddhist monks, and others, do wear an undergarment, but that is not referred to as a loincloth. G 18 uses the same word as B 19. The text of B is also untidy in that s.20 comes between the two sūtras on dress. Both G and B (and A — see below) seem to report on alternative possibilities for dress; but only B mentions the yellow-stained robe.

There is a similar relationship between G and B in the matter of hair-style. G s.22 gives two styles: the ascetic removes all his hair or leaves just a topknot. B s.18 uses the curious word *śikhā-muṇḍa*, which might possibly be a corruption of G's wording, but is more likely to be an independent, though influenced, version and to mean 'shaven but for the topknot'. Buddhist renouncers shaved off all their hair, Jains had to tear it out; both were *muṇḍa*. So far as I know, the retention of a topknot was a style peculiar to brahmins. When brahmin disciples of the Buddha complain to him that their former associates are mocking and abusing them for having joined his order, one of the insults they report is being called *muṇḍaka*, 'little baldies'.[20] So maybe B wanted to avoid that style, just as he avoided the term *bhikṣu*.

I have elsewhere shown[21] that B s.22 is echoed in a Buddhist canonical text when it describes brahmins. This of course does not show that B reciprocally knew the Buddhist text. But it is a striking coincidence that it is the same text as that in which the brahmin converts are insulted for having shaved their heads.

B differs from G in referring to the use of a water-strainer for purificatory rites. The archetypal picture of an ascetic renunciate in ancient India is of a complete drop-out who had left behind him all brahmanical rules for the conduct of life, and in particular all concern for purity and impurity. That brahmin renouncers very early moved back to a life of ritual regulation — if indeed there was ever a time when they wholly escaped it — can be ascertained by a glance at the passage on renunciation at the end of B book 2. Here ss.24 and 25 already represent a step in that direction. The use of a water-strainer *per se* does not imply purity rules: Jain ascetics can drink only strained water because they must avoid taking the lives of any insects in it, and if a strainer is one of the eight possessions of a Buddhist monk, no doubt it is for the same reason. However, the reference to purity (*śauca*) in s.24 and to 'water ritual', which must mean purification, in s.25, is another matter, and clearly a brahmanical matter. These two sūtras are wholly incongruous with the next one, in which men are rejecting Vedic ritual.

In sum, our passage of B is a hotchpotch. Two sūtras, 26 certainly and 21 very probably, refer to Buddhists. On the other hand, ss.24 and 25 seem alien to the picture of the institution painted by G and represent a gain for brahmin orthopraxy. While I see two positive references to Buddhists, I also conjecture two negative references: the avoidance of the name *bhikṣu* and of the fully shaven-headed style. However, such arguments from silence can be little better than guesswork; they are mere inferences made plausible, if at all, by the positive evidence. That B, in reacting to Buddhists on the one hand by mentioning them and on the other by not mentioning them, would be inconsistent in approach is neither here nor there, given the other manifest inconsistencies in the passage.

VI

I now translate A:

(7) Now the itinerant. (8) It is after this (i.e. studentship) that he goes forth (to the homeless life), celibate. (9) For him they prescribe: (10) He should be without a fire, without a home, without comfort, without resort, a silent sage (*muni*); uttering words only in Vedic recitation, acquiring enough to support life, he should live without (caring for) this world or the next. (11) For him is

prescribed a discarded (cloth for) covering. (12) Some (prescribe?) complete nakedness.[a] (13) He should abandon truth and falsehood, pleasure and pain, the Vedas, this world and the next, and should seek the Self. (14) In understanding it there is the attainment of security.[b] (15) That is contradicted by the teachings (*śāstra*). (16) If in understanding there is the attainment of security, one would not experience suffering (thereafter) in this very life. (17) For this reason what follows (that attainment) has been expounded.[c]

Notes

[a] For *parimokṣa* Monier-Williams' *Sanskrit–English Dictionary* records only meanings like 'release', both literal and metaphorical; that meaning would not fit here. Perhaps this release is from clothing.[22] As in G s.19, it is not clear what verb to understand with the subject 'some'.

[b] 'Security' (*kṣema*) is a term for salvation. The next sūtra rejects this view.

[c] Bühler, whose interpretation of this passage I follow throughout, explains in a footnote that what is expounded is yogic practices.

Just before this passage, at 2, 9, 21, 2, A says that security (salvation) can be attained by someone in any of the four *āśrama* — though in his conclusion he says that the householder is best. Given this view, and the reference to prescription, I have here translated the optatives as 'should'.

Ss.8–10 are repeated as ss.19–21 at the beginning of A's treatment of the forest-dweller's *āśrama.* Ss.9–10 immediately strike one as out of character. The first part of s.10 — up to my semicolon — is a line of *anuṣṭubh* verse, and so unlikely to be an original feature of a sūtra work. In fact s.10 is based on a verse (2, 10, 18, 22) in the passage about renunciation at the end of book 2 of B: the first line is the same, the latter half goes *bhaikṣārthī grāmam anvicchet svādhyāye vācam utsṛjet*: 'When he wants alms he should seek a village. He should use his voice in Vedic recitation (only).' Since this scans, we would know that it was original and A the borrower even if A s.9 did not identify it as a quotation.

It is a somewhat incongruous interpolation. On the one hand, it is the only sūtra in all our three passages to mention silence as a feature of this life-style. Moreover, to say that he should recite the Veda seems to contradict s.13. On the other hand, s.13 duplicates the clause about not caring for this world or the next.

With s.10, A moves the itinerant ascetic further towards Vedic orthopraxy. But it is a relatively late addition. Without that sūtra, A has little to say about how the ascetic lives. S.13 does say, however, that he abandons the Vedas.

I suggest that s.13 is a reminiscence of B s.26. Ignoring the participle 'saying', which indicates that B is describing some other group of people, A seems to be presenting his own understanding of that sūtra. He repeats the statement that the ascetic rejects the Vedas, but then nevertheless, presumably because he knows nothing of Buddhists,[23] turns the rest of the sentence in an orthodox (Vedantic) direction. What makes my suggestion plausible is that A would be interpreting — twisting — B s.26 exactly as Govindasvāmin has; in other words, the misunderstanding of Govindasvāmin and Bühler could be traced all the way back to A.

VII

It remains to look at the variant recension of B. Hultzsch prints it in an appendix. I ignore the variant readings within it; they can be consulted there. This variant recension goes from s.12 to s.26 both inclusive; I pick it up at the end of the account of the forest-dweller, i.e. at the point corresponding to s.16. This recension, however, is not divided into numbered sūtras by the editors and reads as connected prose — a point to which I shall return below. In my rendering I put semicolons where one might reasonably divide the text into sūtras.

Some men, renouncing rites, without fire, without a home; wearing a loincloth; staying in one place during the rainy season; performing water ritual with water drawn (from a well) and purified; begging food when the pestle has been laid down, there are no (more) live embers, and the collecting of plates is over; some (prescribe?) complete nakedness; saying, 'We reject Vedic rites and adhere to the middle path, being delimited to both sides.'

As Hultzsch points out, after the first few words (up to my first semicolon) this text consists of the following sūtras, in this order: 19, 20, 25, 22, A s.12, 26. I would add that 'without fire, without a home' corresponds to A s.10, though there the words are in the singular and here in the plural, agreeing with 'some men'. So we seem to have two borrowings from A, one of them from what we have shown to be a later interpolation in A.

In ss.19, 20, 25 and 22 the singulars of the other recension appear here as plurals, so that the masculine nominative plural appears throughout. Even so, the build-up of participles fails to culminate in a main verb, and with the intrusion of A s.12 it is not clear how many sentences we have here, so that the syntax of this recension is still unconvincing, even if it seems less broken up than the other. What really tells against the authenticity of this version is the fact that it has external sandhi throughout, a transparent effort to make it read like connected prose. But such connected prose is just what a text of this genre is not. In other words, a late editor has changed the phonetics and the grammar in the direction of classical Sanskrit prose, but in doing so has denatured the text.

One might leave it there. But I am intrigued by the insertion of *parimokṣam.* When translating A in Section VI above, I commented that no appropriate meaning for this word is attested in the dictionary. If someone was cobbling this passage together and borrowed from A a sūtra about clothing (or lack of it), it is odd that he did not put it next to the one about wearing a loincloth. I wonder whether this sūtra may not originally have belonged here after all, but meant something different from nakedness. It could be read with the next sūtra, 26, and would make sense if *parimokṣam* had its normal meaning of 'release': 'Some, (seeking?) complete release, reject Vedic rites, saying, "We adhere to the middle path, being delimited to both sides."' The problem with this suggestion is that it is almost too good: if the text was so straightforward why did it get corrupted? Another possibility is that originally the text had *prātimokṣa,* the name of the Buddhist monastic code. Since this word is unknown to brahminical Sanskrit, it would be extremely liable to corruption. The text might then have said that some kept entirely to the *prātimokṣa* instead of Vedic ritual injunctions — presumably the verb for 'keeping' was lost when it no longer seemed to make sense. Alas, the evidence is too scanty to allow for more than speculation. I hope that someone may yet shed more light on the text.

VIII

How old is this reference to Buddhists? And is it the earliest in a brahminical text?

To take the second question first: I can see no rivals. There are no certain references to Buddhism or Buddhists in Vedic or epic literature, or indeed in any Sanskrit text which seems to date from the oral period (see Section III above).

There is a *possible* reference to Buddhists in Pāṇini. In the section on the formation of *karmadhāraya* compounds, s.2, 1, 70 reads *kumāraḥ śramaṇādibhiḥ*. This tells one that one can form a word like *kumāraśramaṇā*, 'a virgin female ascetic'. A Buddhist nun could in general terms be referred to as a *śramaṇā*, but the term applies to any female ascetic in a non-brahmin religious group, and the Buddhists do not seem to have used it very commonly of themselves. (The corresponding Pali is *samaṇī*.) Neither Kātyāyana nor Patañjali has any comment on this sūtra.

In the first edition of his *History of Dharmaśāstra* Kane dated Baudhāyana 500–200 BC.[24] In his second edition he changed that to 600–300 BC, but added: 'All these dates are more or less tentative and there is no finality about them.'[25] Since I have dated the Buddha to the fifth century BC (approx. 490–410),[26] this allusion to Buddhists could not be older that the late fifth century. The fourth or third century BC would be more probable. I would therefore claim that this is by far the oldest clear reference to Buddhism in brahminical literature.

APPENDIX

Gautama 3, 11–25.

(11) anicayo bhikṣuḥ (12) ūrdhvaretāḥ (13) dhruvaśīlo varṣāsu (14) bhikṣārthaṃ grāmam iyāt (15) jaghanyaṃ anivṛttaṃ caret (16) nivṛttāśiḥ (17) vākcakṣuḥkarmasaṃyataḥ (18) kaupīnācchādanārthaṃ vāso bibhṛyāt (19) prahīṇam eke nirṇijya (20) nāviprayuktam oṣadhivanaspatīnām aṅgam upādadīta (21) na dvitīyām apartu rātriṃ grāme vaset (22) muṇḍaḥ śikhī vā (23) varjayed bījavadham (24) samo bhuteṣu hiṃsānugrahayoḥ (25) anārambhī.

Gautama 3, 36.

aikāśramyaṃ tv ācāryāḥ pratyakṣavidhānād gārhasthyasya.

Baudhāyana 2, 6, 11, 16–28.

(16) parivrājakaḥ parityajya bandhūn aparigrahaḥ pravrajed yathāvidhi (17) araṇyaṃ gatvā (18) śikhāmuṇḍaḥ (19) kaupīnācchādanaḥ (20) varṣāsv ekasthaḥ (21) kāṣāya-vāsāḥ (22) sannamusale vyaṅgāre nivṛttaśarāvasaṃpāte bhikṣeta (23) vāṅmanaḥkarmadaṇḍair bhūtānām adrohī (24) pavitraṃ bibhrec chaucārtham (25) uddhṛtaparipūtābhir adbhir apkāryaṃ kurvāṇaḥ (26) apavidhya vaidikāni karmāṇy ubhayataḥ paricchinnā madhyamaṃ padaṃ saṃśliṣyāmaha iti vadantaḥ (27) aikāśramyaṃ tv ācāryā aprajanatvād itareśām (28) tatrōdāharanti: Prāhlādir ha vai Kapilo nāmāsura āsa. sa etān bhedāṃś cakāra devaiḥ spardhamānas. tān manīṣī nādriyeta.

Baudhāyana 2, 6, 11, variant version of 16–26.

saṃnyasyāike karmāṇy* anagnayo 'niketanāḥ / kaupīnācchādanā+ / varṣāsv ekasthā+ / uddhṛtaparipūtābhir adbhir apkāryaṃ kurvāṇāḥ / sannamusale vyaṅgāre nivṛttaśarāvasaṃpāte bhikṣantaḥ / sarvataḥ parimokṣam eke+ / 'pavidhya vaidikāni karmāṇy ubhayataḥ paricchinnā madhyamaṃ padaṃ saṃśliṣyāmaha iti vadantaḥ.

* Hultzsch reads karmāṇy but reports karmāṇy as a variant.

\+ Here the MSS have external sandhi between what look like separate sūtras. But since the other cases are ambiguous, there could be sandhi throughout.

Āpastamba 2, 9, 21, 7–17.

(7) atha parivrājaḥ (8) ata eva brahmacaryavān pravrajati (9) tasyōpadiśanti (10) anagnir aniketaḥ syād aśarmāśaraṇo muniḥ svādhyāya evōtsṛjamāno vācaṃ prāṇavṛttiṃ pratilabhyāniho 'namutraś caret (11) tasya muktam ācchādanaṃ vihitam (12) sarvataḥ parimokṣam eke (13) satyānṛte sukhaduḥkhe vedān imaṃ lokam amuṃ ca parityajyātmānam anvicchet (14) buddhe kṣemaprāpaṇam (15) tac chāstrair vipratiṣiddham (16) buddhe cet kṣemaprāpaṇam ihāiva na duḥkham upalabheta (17) etena paraṃ vyākhyātam.

NOTES

1. In the hope that this article may interest some who know little or no Sanskrit, I have tried to keep the text accessible to them by using as few Sanskrit words as possible. For Sanskritists, the body of the textual material I have used is printed in the Appendix. I keep the established rendering 'law-book' for *dharma-śāstra.* The oldest law-books are called *dharma-sūtras*; 'aphorisms on norms' might be the closest translation. The numbered units of their contents are individual sūtras. I have kept the term sūtra in English and sometimes abbreviated it to s.
2. The texts claim to lay down norms for the top three estates (*varṇa*), for *kṣatriya* and *vaiśya* as well as brahmin; but there is ample reason to think that only on brahmin life did they ever have much effect, and indeed many of the prescriptions are only applicable to brahmins. I shall simply refer to brahmins throughout.
3. S.L. Kaeley, R.N. Khanna and V.K. Bhandari, *Indian History and Culture*, Delhi, revised edition, 1984, p. 30. The book is (or was when I saw it in 1986) prescribed by the Delhi board for ICSE class X.
4. Patrick Olivelle, 'The Notion of Āśrama in the Dharmasūtras', *Wiener Zeitschrift für die Kunde Südasiens*, XVIII, 1974.
5. In Clifford Geertz's famous phrase, the earliest law-books were 'models of and models for' reality. As society changed, the prescriptive literature tended to lag behind, and its authors sometimes venerated old texts without fully understanding them.
6. For the editions I have used and the translations I have consulted, see Bibliography.
7. Georg Bühler, trans., *The Sacred Laws of the Āryas. Part II: Vāsishtha and Baudhāyana* (*Sacred Book of the East*, vol. XIV), Oxford, 1882, pp. xxxiii–xxxv.
8. References to B misleadingly suggest a work subdivided at four levels. In fact, the second and third figures refer to paratactic divisions so that the second figure is logically redundant. The same is true of A. However, I have kept to standard usage in my references.
9. Olivelle, 'The Notion of Āśrama', p. 29.
10. I have briefly drawn attention to this in 'The Buddha's Book of Genesis?', where I say (p. 173) that 'it would deserve an article in itself'. This is that article.
11. B.G. Gokhale, 'Early Buddhism and the Brahmins', in A.K. Narain, ed., *Studies in the History of Buddhism*, Delhi, 1980.
12. Bühler, *Vāsishtha and Baudhāyana*, p. xlv, 'said to be a modern writer'; P.V. Kane, *History of Dharmaśāstra*, vol. I, Poona, 1930, p. 32, 'appears to be a very late writer'.
13. S.C. Banerji, *Dharma Sūtras: A Study in Their Origins and Development*, Calcutta, 1962, p. 49, argues that Āpastamba preceded Gautama, but his reasons seem flimsy.

14. Hartmut Scharfe, *Grammatical Literature* (*A History of Indian Literature*, vol. V, fasc. 2), Wiesbaden, 1977, p. 153.
15. To preclude misunderstanding, let me state that I am not suggesting that the Buddhists had any part in formulating the *āśrama* system. Its proponents were just as much brahmins as were their conservative opponents like G and B. But B may have been trying to smear his opponents by associating them with Buddhists.
16. Olivelle has however pointed out to me (personal communication) that one can interpret these passages as the hypothetical words of the opponent (*pūrva-pakṣa*), in which case the translation 'should' can stand.
17. Thus Bühler's translation, 'He shall not undertake (anything for his temporal or spiritual welfare)', rather misses the point.
18. Bühler, *Vāsishtha and Baudhāyana*, p. xi, mentions this as a 'not improbable' instance of B borrowing from G. (His reference to s.26 rather than s.27 is a slip.)
19. *kurvāṇaḥ*. The commentator reads the optative *kuryāt*, but that is probably his correction: 'Govinda has not rarely altered the text at his pleasure' (E. Hultzsch, ed., *The Baudhāyanadharmaśāstra*, Leipzig, 1884, p. viii).
20. *Aggañña Sutta*, *Dīgha Nikāya*, III, 81.
21. See Note 10.
22. Thus Bühler, following the (probably sixteenth century) commentator Haradatta; he adds in a footnote (Georg Bühler, trans., *The Sacred Laws of the Āryas. Part I: Āpastamba and Gautama* (*Sacred Book of the East*, vol. II, Oxford, 1879, p. 152) that Haradatta mentions an alternative interpretation: 'Some declare that he is freed from all injunctions and prohibitions'. Olivelle too (personal communication) finds the term obscure but, like Bühler, thinks there is a connection with the previous sūtra, in which he suggests that *mukta* might mean not 'discarded' but 'loose'. For a quite different line of thought see end of Section VII below.
23. In ss.14 and 16 the word for 'understand' is *buddha*, but the Buddhists had no monopoly in this word and I do not take it as a sign of Buddhist influence.
24. Kane, *History of Dharmaśāstra*, vol. I, 1930 edition, p. 30.
25. Kane, *History of Dharmaśāstra*, vol. I, 1968 edition, pp. 52–3.
26. 'Dating the Buddha: . . . ' This paper was written and circulated in 1989.

REFERENCES

Primary Sources

BÜHLER, Georg, trans., *The Sacred Laws of the Āryas. Part I: Āpastamba and Gautama* (*Sacred Books of the East*, vol. II), Oxford, Clarendon Press, 1879.

BÜHLER, George, *The Sacred Laws of the Āryas. Part II: Vāsishtha and Baudhāyana* (*Sacred Books of the East*, vol. XIV), Oxford, Clarendon Press, 1882.

—— ed., *Aphorisms on the Sacred Law of the Hindus by Āpastamba*, 2nd ed., revised, Bombay, Government Central Book Depot, 1892.

HULTZSCH, E., ed., *The Baudhāyanadharmaśāstra, Abhandlungen für die Kunde des Morgenlandes 8*, Leipzig, 1884.

STENZLER, A.F., ed., *The Institutes of Gautama*, London, Trübner, 1876.

Secondary Sources

BANERJI, S.C., *Dharma Sūtras: A Study in Their Origins and Development*, Calcutta, Panthi Pustak, 1962.

GOKHALE, B.G., 'Early Buddhism and the Brahmins', in A.K. Narain, ed., *Studies in the History of Buddhism*, Delhi, B.R. Publishing Co., 1980, pp. 68–80.

GOMBRICH, Richard, 'The Buddha's Book of Genesis?', *Indo-Iranian Journal*, 35, 1992, pp. 159–78.

—— 'Dating the Buddha: A Red Herring Revealed', in Heinz Bechert, ed., *The Dating of the Historical Buddha, Part 2*, Göttingen, Vandenhoeck & Ruprecht, 1992, pp. 237–59.

KAELEY, S.L., R.N. KHANNA and V.K. BHANDARI, *Indian History and Culture*, Delhi, revised edition, October 1984.

KANE, P.V., *History of Dharmaśāstra*, vol. I, 1st edition, Poona, Bhandarkar Oriental Research Institute, 1930, 2nd edition, Poona, 1968.

OLIVELLE, Patrick, 'The Notion of Āśrama in the Dharmasūtras', *Wiener Zeitschrift für die Kunde Südasiens*, XVIII, 1974, pp. 27–35.

SCHARFE, Hartmut, *Grammatical Literature* (*A History of Indian Literature*, vol. V, fasc. 2), Wiesbaden, Harrassowitz, 1977.

Chapter 3

Scepticism Revisited: Nāgārjuna and Nyāya via Matilal

D.P. Chattopadhyaya

I

When Professor Bimal Krishna Matilal tells us that '[a] philosopher has to learn to live with the sceptic . . . [because the latter] shares the same concern for truth with the philosopher . . . [and] is first and foremost an "inquirer" ', I am inclined to agree with him. I also endorse his view that both the philosopher and the sceptic 'persist in seeking [truth] and probing [into it]'. Further, when referring to Kumārila's criticism of the hyper-sceptic and Udayana's dismissive attitude towards the sceptic, Matilal disapproves of 'the frequent jokes and insults that are normally heaped upon the sceptic' on the grounds that they are 'wide of the mark', I think Matilal betrays, at least to start with, a constructive attitude toward scepticism.[1]

It is true that empiricism is generally inclined towards scepticism. But all forms of scepticism are not 'impractical' or 'scandalous'. On the contrary, to denounce scepticism without specifying its form is highly misleading and academically unfair. It has been rightly pointed out by Matilal that philosophers, being concerned as they are with truth, are akin to the sceptic who is also engaged in the inquiry into or search for truth. That every philosopher, rationalists like Descartes and Hegel or empiricists like Russell and Nāgārjuna, has to take scepticism seriously and is obliged to explain it at length shows its force and significance.

In Indian tradition the Buddhist position is often taken to be the paradigm of scepticism and its attending weakness. The point to be remembered here is that the different schools in Buddhism do not agree in their formulations of what is called scepticism. For example, according to the Vaibhāṣikas and early schools, both (empirical) *saṃsāra* and (transcendental) *nirvāṇa* are real. The *Sautrāntikas* maintain that *saṃsāra* is real but *nirvāṇa* unreal. *Per contra* the Yogācāra or the Vijñānavāda holds that *saṃsāra* is unreal but *nirvāṇa* real. The Mādhyamika defends the most radical 'view' which, to it, is the truth, viz. both *samsāra* and *nirvāṇa* are unreal, separately unreal. Among the leading Buddhist philosophers, the one who is most frequently singled out and critically referred to for his sceptical position is Nāgārjuna of the Mādhyamika persuasion.

Matilal himself has paid considerable attention to Nāgārjuna's critique of *pramā* (valid knowledge) and *pramāṇas* (means of attaining *pramā*).

It has been said time and again that Nāgārjuna's scepticism regarding all philosophical positions, comprising different forms of knowledge and their proofs, rests on his denial of what is called *bhāva* or existence. If every view, arguments for and against it, proofs and disproofs of it, are said to be without *svabhāva*, lacking in 'own-being', then the critic of Nāgārjuna is perhaps within his rights to point out that his sweeping generalization about *svabhāva-śūnyatā*, essencelessness, of all *pramā* and *pramāṇas*, renders his own sceptical thesis or theory totally inefficacious. What is worse, an apparent air of paradoxicality seems to have taken away all the deemed force of Nāgārjuna's critique of *pramā* and *pramāṇas*. In fact in Vātsyāyana's *Nyāyasūtra* precisely this line of argument has been carefully explicated (ii. 1.8–ii. 1.32).[2] But before one dismisses Nāgārjuna's critique of *pramā* and *pramāṇas* on the related grounds of alleged essencelessness/ meaninglessness and emptiness, one has to be more careful about the exact formulation of Nāgārjuna's own view, setting aside his detractors' interpretations put on his view.

In this connection the critic would be well advised to look into the very first sūtra of *Śūnyatāsaptati-Kārikā (SSK)*.

> Though the Buddhas have spoken of duration (*sthiti*), origination (*utpāda*), destruction (*bhaṅga*), being (*sat*), non-being (*asat*), low (*hīna*), moderate (*sama*) and excellent (*viśiṣṭa*) by force of worldly convention

(*lokavyavahāravaśāt*), [they] have not done [so] in an absolute sense (*tattvavaśāt*).[3]

Many of the objections raised against Nāgārjuna's so-called defence of scepticism rests on misunderstanding of such key concepts as *bhāva, svabhāva, śūnya, sat* (being), *asat* (non-being). The second *sūtra* of *SSK* suggests that many of our ideas related to emptiness are conventional in character and not to be taken in an absolute sense. Nāgārjuna also indicates the relativity and inadequacy of such linguistic designation as 'self' ('*ātman*'), 'non-self' ('*anātma*') and 'self-non-self' ('*ātmānātman*'). These designata, like '*nirvāṇa*', because of their emptiness (*svabhāvaśūnyatā*), are linguistically inexpressible.

All things, taken together, are said to be without substance (*svabhāva-śūnyatā*), causally or conditionally or totally or separately, and 'therefore', says Nāgārjuna, 'they are empty (*śūnya*)'. The points to be noted here are two. First, here *svabhāva-śūnyatā* 'stands for' unsubstantiality. Secondly, intriguing is the use of the word, 'therefore'. Are we to understand that emptiness (*śūnya*) is due to lack of substance? Moreover, to be noted is the inexpressibility of all 'expressible things' (*abhidheyabhāva*).

But from these views of Nāgārjuna one must not think that he is a nihilist in the received sense of the term about being (*sat*). He 'says' it *exists*. But the same cannot be said of non-being (*asat*) or being-and-non-being. Consequently, though non-being and being-and-non-being do not endure or vanish, being does exist. In fact it seems from Nāgārjuna's line of argument in *SSK* that he is more interested in expounding the view of dependent origination (*pratītyasamutpāda*) and the resulting indeterminable character of things. But nothing that Nāgārjuna says in this work, should be taken to suggest that being does not exist. True, he says that it 'does not arise'. But the point to be noted is what is *sat* can *exist* without being due to something else or without being productive of something else. Therefore, one might conclude that the main thrust of Nāgārjuna's argument is to indicate the inexpressibility and indeterminability of things. And this follows from the very basic Buddhist thesis of the momentariness or the *madhyamā* character of what exists.

To the objection that if things (*bhāva*) were empty (*śūnya*), cessation (*nirodha*) and origination (*utpāda*) cannot occur, Nāgārjuna's response is: being (*bhāva*) and non-being (*abhāva*) are successive and relative to each other but not simultaneous. Truly speaking, being (*bhāva*) must not be accepted. If *bhāva* as *sat* were there, we would be sure of permanence; if non-being (*asat*) were there, we would always have witnessed annihilation or extinction. But as a matter of fact we do not witness either permanence or perpetual annihilation. From these considerations Nāgārjuna concludes that there is neither being nor non-being.[4] Again he reminds us that true being (*sadbhāva*) cannot vanish or go out of existence. About *nirvāṇa* it is said that it is neither *bhāva* nor *abhāva*, neither due to destruction nor permanent (*śāśvata*). In effect Nāgārjuna discounts all that is conditional (*saṁskṛta*), — origination (*utpāda*), duration (*sthiti*) and cessation (*bhaṅga*). By discounting all these *saṃskṛtalakṣaṇa*, however, he is not committing himself to the existence of the *saṃskṛta* phenomenon.

From this dialectic of Nāgārjuna the discerning mind is not at all thrown into a state of despair. On the contrary, he realizes the inexpressible character of what is truly there. A similar conclusion becomes inescapable when one carefully goes through his views on the nature of *karma* (action) and *kāraka* (agent). In a sense both *karma* and *kāraka* are *śūnya* or phantom (*nirmitaka*). Therefore, the talks of *karmaphala* (results of action), enjoyment and suffering are all idle from the ultimate point of view. Also seem pointless the talks of *citra*, the manifold world, and its forms (*rūpa*).

Though by his dialectic Nāgārjuna tries to deny an ultimate or durable ontological status to the empirical world and the human agent, his actions and results thereof, that does not prevent him from explaining how *karma* (action) is caused by passion and how the human form, the result of karma.

Also his accounts of colour-perception and sense-fields give one the distinct impression that he is not at all oblivious of the common-sense world we talk of. It is equally significant to note that Nāgārjuna speaks of *māyā* in the context of *karma*-formations. Simply because our acts of consciousness are episodic and not propositional (in the modern sense), we cannot say that it is not on account of consciousness at all.

Nāgārjuna's discourse on consciousness (*vijñāna*), discernible object of consciousness (*vijñeya*) and how the embodied agent/knower is formed, makes it plain that he is trying to offer us a purely phenomenal account of all that with which we are concerned in our life. His dialectic is intended to free human minds of the trappings of the sense-bound body living within the framework of passions and habits and enmarked by the indefinite and alternative ways of viewing and reviewing the world. The true seeker is required to overcome the phenomenal duality of existence and non-existence and thus become calm (*śānta*). When Nāgārjuna clearly tells one that if one can tear off the nets of false views (*kudṛṣṭijāla*), one obtains 'nirvāṇa by abandoning desire (*rāga*), delusion (*moha*) and hatred (*dveṣa*)', we cannot doubt the existence of unsullied (*alipta*) persons only about whom the talks of *nirvāṇa* and the ways of obtaining that *nirvāṇa* make sense.[5]

Not only in *SSK* but also in his other works like *Catuḥstava (CS)* and *Bodhicittavivaraṇa (BV)*, Nāgārjuna sarcastically refers to 'the dogmatists' who get terrified by the lion's roar of emptiness '(*śūnyatāsiṃhanāda*)'. There are four main aspects of Nāgārjuna's dialectic. Ontologically speaking, it tries to show that all phenomena, because of their essencelessness and mutual dependence, are empty. Epistemologically speaking, the ultimate truth, the undeniable object of cognition, is not really objective in the ordinary sense of the term. It is so only metaphorically (*upadaya prajñapti*). Psychosomatically speaking, the ultimate truth, *nirvāṇa*, is the abolishment of all passions, cravings, hatred and delusion. Ethically speaking, the ultimate end of life is freedom from the bonds of *karma* and subjection to the inner imperatives of compassion, non-violence and love. To posit the ultimate aim of life and to indicate therewith *what* is really real, *how* that reality can be grasped in a non-discursive intuition (*prajñā*) and *what* are its psychosomatic accompaniments, normally show that Nāgārjuna and philosophers of his persuasion are not nihilist or sceptic as we understand the terms.

II

The main logical controversy between the Mādhyamika and the Naiyāyika centres round the question whether *pramāṇa-prameya* relation is or is not interdependent or circular, Nāgārjuna tries to show that *pramāṇa* theorists fail to show that *pramāṇa* is not a *prameya* and, further, that a *pramāṇa* itself is a *prameya*, i.e. in need of being proved. But the Naiyāyika claims to have established that a knowledge-claim or *pramā*-claim, despite its *prameya* character (in principle), is really knowledge. Otherwise the whole doctrine of *pramā* becomes hollow. Nyāya's initial *pratijñā* or proposition that *pramāṇa* and *prameya* are inseparable (*miśra*) provides Nāgārjuna a plank of attack on the former's position. Their inseparability, it is argued, indicates that neither is self-established. This dependence of one upon the other is said to be indicative of the inefficaciousness of both.

Understandably, the Naiyāyika rejects this interpretation of his view by the Mādhyamika. According to the former, *pramāṇa* and *prameya* as *padārthas*, though co-relative, i.e. in a sense circular, are not viciously so. On the contrary, they can establish each other or be mutually supportive. *Pramāṇa*, like light, is self-established or self-proved. But in case the correctness of a *pramāṇa* is questioned, it is possible to remove it by evidence, argument, example, etc. In this limited sense *pramāṇa* or what is said to be *pramāṇa* at one stage may well be regarded as *prameya* at another. That does not compromise their *padārtha* character in any way.

But this argument is not acceptable to the Mādhyamika. He introduces his well-known dialectic at this stage. If *pramāṇas* are said to be valid on their own account, the question of their being dependent upon anything else and at any stage does not arise. Besides, the Naiyāyika, having stated initially (*pratijñā*) that *pramāṇa* and *prameya* are inter-related, cannot logically claim that *pramāṇas* are self-established. For that amounts to *pratijñahāni*, retreat from the initial position. Additionally, it is pointed out that if *pramāṇas* themselves need something else for their self-establishment, it creates a sort of *anavasthā*, infinite regress. If the proofs cannot prove the provables, how can regress be stopped?

One of the main planks of the Buddhist form of scepticism is the alleged unreliability of perception as a means of knowledge or proof (*pramāṇa*). The objection against perception may be formulated in very many ways. But I propose to look into the Nyāya formulation of a possible objection against perception and how it could be met.

Perception is said to be due to sense–object contact, direct or indirect, proximal or distal. But the (possible) objector or critic can argue to this effect: perception can neither be prior to, nor posterior to, nor simultaneous with the objects of sense. First, perception cannot be prior to the object to be perceived, the colour of a flower, for example. How can we perceive a colour which is yet to come into existence? Secondly, perception (of the fragrance of a flower, for example) cannot be posterior to its (possible) existence. How can one smell a flower's fragrance when it is no longer there? Positively speaking, one cannot perceive the object which has gone out of existence in *every* sense of the term. When the very *objective* source of sense-perception goes out of existence, it cannot be logically credited to have sent a signal which could be perceived later on (by a percipient). However, the Nyāya formulation of the 'impossibility' of posterior perception has not been formulated in a very rigorous way, keeping in view, for example, perception of the very distant astronomical bodies which have gone out of existence but can be perceived later on for a definite length of time, depending upon the time-distance between the percipient and the body in question. Apparently, the reasons of these non-rigorous formulations are twofold. The Naiyāyika has in his mind the well-known Buddhist doctrine of the *momentariness* of time which denies both its pastness and futurity. Further, the Naiyāyika's own view of time, defined in terms of our relation to the sun, is not perhaps efficient enough to deal with our contemporary astronomical knowledge.

Thirdly, the critic also discounts the possibility of simultaneous perception, i.e. simultaneous existence of the percipient and the objects of perception, colour and smell of a flower, for example. According to the received view, perceptions are successive. One who perceives colour cannot at the same time perceive smell. Two perceptual cognitions are destined to be distinct and successive. But since temporal orders of succession between the sense and the specific object, between *pramāṇa* and *prameya*, are not definitely ascertainable, the outcome of perception remains doubtful. Alternatively, the critic's objection may be construed in this way. Colour-perception and smell-perception are not of the same (perceptual) kind. Inferential remembrance of colour may appear to be simultaneous with the smelling of fragrance. Since perception and inference are two different means of knowledge, they cannot be simultaneous. Besides, mind, believed to be atomic in substance and necessary for production of knowledge, can be 'in touch' only with one kind of knowledge *at a time*, perception *or* inference, but *not* both.

If these objections of the critic are valid, the Mādhyamika scepticism stands vindicated and the Nyāya position is put into jeopardy. Naturally, the Naiyāyika, in this case Vātsyāyana commenting on Gautama's *Nyāyasūtras*, tries to show the weakness of the above objections.

Unless the precise intention underlying the objections raised against perception as a means of knowledge is understood, the response can hardly be appropriate. Does the critic mean to deny the very existence or possibility of *pratyakṣa* (perception) as a *pramāṇa* (a means of knowledge) or does he want merely to question only its efficaciousness in specific cases because of the attending difficulties or obstructions or non-obtainability of the necessary and sufficient conditions? The very attempt to deny the existence of perception, it is argued, commits the Buddhist to the view that there is perception (which may be refuted). On the other hand, if perception *gua pramāṇa can* be refuted, the Naiyāyika tells the Buddhist, it is (at least) established that there is a way (*pramāṇa*) of refuting a *pramāṇa's* claim that it is a *pramāṇa.* By implication the sceptic is then obliged to concede that there *is pramāṇa*, perceptual or otherwise. This concession takes away the proclaimed radical thrust of the critics like Nāgārjuna who in their dialectic question the very existence or possibility of perceptual knowledge.

Incidentally, the realist Naiyāyika tries to meet the sceptic's objection that *pramāṇa* (proof), being itself of the nature of a *prameya* (provable), is not of much epistemic or logical consequence. He points out that *prameya* is not a very weak concept because what is *prameya* or provable is not arbitrarily determinable. It is only within a definite system of principles that a claim or proposition or theorem can be shown to be *provable. Pramāṇa*, like a measuring instrument, is itself measurable. The eyes which ('as subject') enable us to see (objects) are themselves objects of perception. *Pramāṇa* and *prameya*, as stated earlier, are interdependent, but their relation is not viciously circular.

Prameya is the object of knowledge and *pramāṇa* is the cause of that knowledge. There is no fixed order of anteriority or posteriority or simultaneity between the two. In some cases the causes of knowledge precede the objects of knowledge. For example, the sunlight as cause remains operative before the objects (to be) seen by it come into being. In some cases the objects of knowledge are there before the cause of knowing them is made available to the percipient. For example, the table, chairs and books of my study room were there before I put on the light and saw the same. Again, in some cases (smoke and fire, for example) the cause and the object of knowledge are found to be coexistent.[6]

The Nāgārjunite critique of perception as a means of knowledge, at least in its manifest form, may be met without much difficulty. But once we take seriously the Naiyāyika's point that no (purported) refutation of a *pramāṇa* is possible without prior acceptance of some or other *pramāṇa*, the Buddhist's criticism of perception loses much of its initial thrust. This is not to affirm that perceptual scepticism is being totally rejected. The chief exponents of the Mādhyamika school like Nāgārjuna, Āryadeva, Buddhapālita, Candrakīrti and Śāntideva, somewhat like the Śāṅkara Vedāntin, try to show how our discursive knowledge of the phenomenal world is beset with contradictions, antinomies and sublation.[7] Only when we have at our disposal some non-discursive mode of knowledge which is disclosive of and affine to the very nature of reality to be known, we get glimpses of knowable objects even at the level of perceptual judgement. The ultimate reality or things-in-themselves, whatever they may be, lend themselves to alternative (*vaikalpika*) modes of apprehension. What is there is *not* exactly graspable. The unavoidable gap between *what* is there (to be known) and *who* is here to know it is bound to subdue our praise for perceptual statements. Perceptual identification of the object always leaves an element of indefiniteness attending or around both the perceiving mind and the perceived object in question. Reidentifiability of the object does not improve the situation significantly or can hardly remove the said indefiniteness. Lingering effects of previous perceptions, force of habits, undercurrents of the web of beliefs, etc. are always there to make the picture more or less opaque, despite its *objectivity*.

Not only perception but also inference as a means of knowledge has been questioned. The forms of questioning are numerous and naturally reflective of the questioners' diverse points of view. The Nāgārjunite criticizes it as a part of his project of critique of all sorts of *pramāṇas.* The Cārvāka the materialist-empiricist who recognizes only perception as means of knowledge rejects the claim of inference as an independent means of knowledge. In recent times the validity of inference has often been questioned on (i) the alleged arbitrary and purely conventional character of the rules of inference, (ii) the untenability of the proclaimed self-justifying nature of the rules of inference, and (iii) diverse interpretations of logical constants.

Some of the objections raised against inference may be briefly indicated here. Generally speaking, the universal relation (*vyāpti*) believed to be obtained between probanda and probans is at times misperceived for some reason (partial analogy, for example) or other. If ants are seen to be carrying off their eggs or the peacocks are heard to be screaming, we infer the gathering of rainclouds and rainfall. A swollen river induces us to infer that there has been rain. But all these inferences may be incorrect. A river may overflow because of its embankments. Ants may carry off their eggs because their anthills have been damaged. The so-called screaming of a peacock may be nothing but mimicking by an expert.[8]

However, the Naiyāyika refuses to recognize these alleged irregularities or counter-examples as invalidating inference (as a means of knowledge). On the contrary, he holds that the very identity of the probans, the true nature of 'the overflowing river', 'the (frightened) egg-carrying ants' and 'the screaming peacocks', has not been correctly grasped by the objector (against inference) and therefore his formulation of (the examples of objection) is faulty, i.e. fails to show that the inference is irregular.

The objection against inference may be reformulated in other ways, focusing attention, for instance, on the fallible perceptual component of the universal relation (*vyāptisambandha*), viz. *a priori* (from cause to effect, from clouds to rain), a posterior (from effect to cause, from swollen river to rain) and '[un]commonly seen' traits (from horns of beast to its having a tail).[9] Only the probabilistic or statistical forms of inference having no obvious perceptual component in them can perhaps avoid this objection.[10]

But in the Indian tradition of logic-cum-epistemology one does not come across what may be called *purely formal structure* of inference completely devoid of perceptual reference or relevance. Matilal tells us, 'Nyāya gives an account of inference as a sequence of psychological events, the final event being held to be causally connected with the immediately preceding'. But he also tries to show that those psychological events do have the 'intentional structures or *formal* properties' of the objects they purport to grasp and express in language.[11] Also in his analysis of the Indian views of knowledge or *pramā* he exhibits a systematic ambivalence. On the one hand, he recognizes the psychological/episodic character of judgement and inference and on the other hand, taking cues from the anti-psychologism of Frege and Husserl, he tries to form 'abstract structures' from judgemental and inferential episodes. For this purpose, Matilal relies more on the syntactic than on the semantic aspects of linguistic expressibility of cognitive episodes. Apparently one of his aims as evident in this exercise is to save his version of 'direct realism' from the attack of psychologism/empiricism. He often speaks of cognition as representation of objective states of affairs in substantial disregard of the intervening factors. One wonders if the rules of inference can be defensibly described as representative.

III

The position of the dialectician, whether he is a Mādhyamika or a Vedāntin, needs to be treated with circumspection. It seems to have two aspects geared to two related aims, negative and positive. The negative or logico-analytic aspect is purported to point out the vulnerability of the stands taken by the opponents and, if possible, demolish them. Since this analytic-critical aspect is often elaborated in detail, it draws close attention and is widely known. This so-called negative dialectic is neither entirely negative nor itself in the nature of a refutable theory or position. One is advised to avoid such *relative* terms like 'position' (as distinguished from 'opposition') and 'theory' (as distinguished from 'practice') because both aspects of dialectic, separately as well as jointly, are equally vulnerable. Strictly speaking, dialectic, like dialogue, is not entirely negative. It is disclosive or, to use Hegel's favourite term, sublative. The entire richness, complexity and nuances of our experience cannot be neatly captured, without residue, by position and/or opposition. *Via negativa*, following the *neti mārga*, the 'position', the *iti*, that is arrived at is *not* relative. Whatever it is, (Buddhist's) *śūnya* or (Vedāntin's) Brahman, it is *absolute*. It is not conceptualizable. Perhaps it is not even linguistically expressible. Does not one by 'knowing' what *śūnya* or Brahman get in any way cognitively closer to it?

Under one interpretation, by knowing that the phenomenal world is not real (*tattva*) one realizes the 'futility' of *positivism*. The 'futility' of positivism also brings to light the inaccessibility of our sense-mind to what is *tattva* or *śūnya*. But *śūnya*, in whatever way we understand it — negation, non-being, non-existence or essencelessness, need not cause despair in us. It may well be construed as the dawning of intellectual intuition (*prajñā*). *Śūnyatā*, inaccessible to discursive judgement (*dṛṣṭi*), is comprehensible only by *prajñā*. To be more precise, *śūnyatā* and *prajñā* are identical. *Śūnyatā* or *prajñā* is negation of negation, absence of absence. In brief, it is the differenceless identity of *jñeya* and *jñāna*.

The confirmation of this interpretation is found in Nāgārjuna's observations on the nature of *nirvāṇa*, *saṃsāra* and their relation. 'There is no difference at all between *nirvāṇa* and *saṃsāra* . . . the bliss consists in the cessation of all thought, in the quiescence of plurality.' And Buddha is said to have preached no other separate reality.[12] The same view is reformulated when it is said that the causally co-ordinated plurality of things known as phenomena is identical with *nirvāṇa* or *śūnya* when viewed without causality, without co-ordination.

Under another interpretation, the negative one, Mādhyamika dialectic causes despair in us and makes us think that all that we experience in this world is momentary, discrete and unreliable. Since I have already discussed this interpretation, I do not wish to repeat it here. But by referring to these two interpretations what I want to highlight is this. These two interpretations need not be viewed as contradictory. Rather they are complementary. What is or seems to be phenomenally unreal is transcendentally real. This double-aspect view of reality takes cognizance of what seems to be empirically unreal and also gives an account of why it does so appear.

IV

The scepticism about the external world centres mainly round the arguments for and against the efficaciousness of the *pramāṇas*, in general, and perception, in particular. It is true that the Buddhists like Dharmakīrti and Dharmottara recognize perception and inference as *pramāṇa.* But other Buddhists like Diṅnāga maintain that perception may be viewed under *samyagjñāna* or valid knowledge and need not be recognized separately. Somewhat like Diṅnāga, Śāntarakṣita does not speak of knowledge in his definition of perception. However, this is not to suggest that they do not recognize perception as a sort of knowledge. The only point to be mentioned generally is that their treatment of perception gives one the impression that perceptual knowledge is not capable of providing us *direct* knowledge of the external world. The sort of knowledge that we have due to perception is said to be more or less indefinite. An element of construction or *kalpanā* gets mixed up in our perception of the external world. From Dharmakīrti's account of perception found in *Nyāyabindu* it is clear that he is trying to defend a sort of perceptual representationalism. Perceptual knowledge 'corresponds to' or 'agrees with' the world outside us. And it may be confirmed by our practices and fulfilment of our aims. But he does not rule out the role of some intermediate psychological states between (the percipient's) perception and the world perceived.[13]

This account of perception may not appear to be convincingly realist or even objective. Some philosophers like Diṅnāga defend perception as that form of knowledge which is free from *kalpanā* or conceptual construction. In inference the role of *kalpanā* remains always in some form or other. To make his view of perception very rigorous Diṅnāga qualifies it by '*abhrānta*', free from errors and illusions. For he apprehends that, given the room for illusions and errors, the claim of inference to be in agreement with what is objectively real gets weakened.

The talk one often hears of scepticism about the external world and objects thereof is basically due to a mix-up of ontological and linguistic issues. Certainly in a sense the realist is justified in asserting that the apple I see is there independently of its being seen by me; its shape, colour and weight (if measured) are not due to my interpretation (of what I see of the apple in question). But can I justifiably assert merely on the basis of my visual *perception* of it that I *know* it? What about its weight? Without a weighing machine (and that too must be reliable) one cannot be sure of its weight. What about the other side of the apple not visible to me? Either I have to infer it or I have to go to that side from where its hitherto invisible side can be seen. But then I cease to see *this* (previously seen) side and about it I have to rely on my memory or fall back again upon inference. This line of analysis may be further refined and many hidden complexities brought out indicating (a) inadequacy of direct/naive realism and (b) the ineliminable role of construction not assignable to this or that specific perceptual capacity.

Matilal's discussion of the Abhidharma phenomenalism of Nāgārjuna and Vasubandhu brings out this point clearly. Not only the *dharmas* perceptible by the five human sense organs but also those graspable by mind can be shown to be substantively there as existent (*dravya-sat*). But that is not recognized by Vasubandhu as good enough ground to deny what is empirically/phenomenally there (*saṃsāra-/saṃvṛti-sat*). In a Heideggerian vein one may point out that both *saṃsāra* and *saṃvṛta* etymologically mean what is partially hidden or enveloped. More relevant to my discussion is the co-present or complementary distinction between what is empirically real and what is substantive or existentially real.

Even to be a minimal realist the Buddhist is required to posit a weak causal link between what is empirically available and what is hiddenly or representationally so. But since this link itself, on the received Buddhist accounts, is perishable, the question of reduction of 'empirical real' to 'representationally real', even if pressed, cannot be satisfactorily settled one way or the other. Both Locke and Kant grappled with the issue in their own ways and, caught in the cross-currents of representationalism, realism and phenomenalism, had to remain content with a sort of dualism, a curious mix of realism and empiricism, far from execution of any reductive programme. Matilal's rejection of Dummett's implied (?) 'God's-eye view' of the (logically determinate) universe is understandable. Equally understandable is his reservation about the so-called 'humanistic' rehabilitation of Buddhist Abhidharma (phenomenalism and representationalism) and the Mādhyamika *śūnyavāda*. After all his aim is to defend direct realism mainly on the basis of objective universals (*jāti*) and nominal universals (*upādhi*). In this respect, while he follows the insights of the Naiyāyikas like Uddyotakara and Udayana, he keeps well in view the semantic–epistemic cues provided by the Navya-nyāya and the Euro-American analytic thinkers.

The steps Matilal takes to establish his thesis of direct realism and, in the process, stave off scepticism are in brief, as follows.

First, we cannot logically recognize particulars like substance, quality and action unless we recognize *real* universals like substancehood (*dravyatva*), qualityhood (*guṇatva*) and actionhood (*karmatva*). The relation between universals and particulars may be understood in two different ways, pro-reductionist and anti-reductionist. A universal *F* may be true of *a, b* and *c*. *F*-ness to be true (as predicate/universal) of logical subjects *a, b* and *c* need only to be simple property of *a, b* and *c*. *F*-ness can be said to be in existence if it is/can be property of the logical subjects *a, b* and *c*. In case *F*-ness is simple and unanalysable it may be regarded as 'natural kind' or 'metaphysical kind' and accorded a place in the core or 'inner circle' of Matilal's conceptual (rather, ontological) scheme. When *F*-ness satisfies this *necessary* condition of existence, i.e. is expressible in and through *a, b* and *c* or some observable features of objects, it is being understood in a pro-reductionist manner. According to Matilal, Nyāya admits that most of the universals or abstract entities are to be analysed observationally.

Cowhood of cow and cow may be perceived directly or, what Gangeśa calls, non-constructively. Matilal observes: 'The universals and basic properties are directly grasped by our awareness as ultimates.' But this awareness is non-qualificative and pre-linguistic, i.e. inexpressible. *Jāti* and *akhaṇḍa upādhi*, though said to be cognitively graspable, are not representable. Besides real universals and concrete properties, the only relational universal that is recognized as real and objective by Matilal, following the Nyāya tradition, is inherence (*samavāya*). Other relations like *samyoga* (physical conjunction) and *svarūpa-sambandhas* (identity relations), though objective and useful for analytic and communicative purposes, are not admitted into the 'inner circle' of Matilal's conceptual (or ontological) scheme.

V

To think, as Matilal does, that the real universals and basic properties of the core area of his ontological scheme are self-contained and whatever falls outside the core area needs the core contents for its understanding, seems to be flawed on several counts.

First, the implied core/periphery or existent/subsistent dichotomy is unclear and needs elucidation. Merely to speak of their asymmetry is not enough. What happens to the core contents if change takes place in the non-core area? Must we think that the asymmetrical relation between the above two areas is fixed and static? If the question is answered in the affirmative the ontological scheme fails to accommodate growth of knowledge or history of science.

Secondly, if the question is answered in the negative, Matilal or his defender is expected to justify the privileged status accorded to the 'inner circle' consisting of genuine universals and properties. How can the possibility, if not necessity, of interanimation and overlap between the two 'circles' be rationally ruled out? Related to this question is another equally pertinent question: If universals and properties, the two main species of the core (ontological) population, essentially linguistic in *our* (human) understanding, can be credited to have pre-linguistic existence, how can their scope and identity possibly be presented to other selves and even to ourselves? If the scope and identification conditions remain undetermined, how can the line of demarcation, if any, between 'the inner circle' and 'the outer circle' be drawn by humans (who are without God's all-pervading view)? Here, from the human standpoint, the talk of distinction between the ontological discourse and the epistemological one makes little or no sense and the attempt based thereupon to salvage direct realism is destined to collapse.

Thirdly, Matilal's difficulty is symbolic of all those who simultaneously profess empiricism in epistemology and realism in ontology without allowing God to play any role in the game. Dummett has a point when he states that without invoking God, rather the paradigm of the Berkeleyan God's knowledge, it is difficult to vindicate realism and to contain scepticism.[14] Since Matilal, like Dummett,[15] recognizes the importance of the (truth-value-wise) determinate character of cognitive statements, he is required to pay added attention to this problem. Otherwise, one feels his steps to direct realism are rather hasty and not promising.

Fourthly, apparently in recognition of the importance of the role of language in ascertaining the determinate (or otherwise) character of knowledge-claiming statements, Matilal takes up this issue once again in his monograph, *The Word and the World.* Referring to the two views on the word–object relationship found in the Nyāya-Vaiśeṣika school, viz. (i) it is fixed conventionally and (ii) it is fixed by God, he concedes 'both . . . contain some grains of truth'. In order to eliminate or minimize the effects of (i) 'conventionalism', relativism due to language-user's whim and those of (ii) 'eternalism' or counter-intuitive fixedness of word–object relationship, Matilal pleads 'for accepting a theory like the *sphoṭa* theory of language'.[16] For the ontologically warranted assumption is that in the mind of the competent speaker and the competent hearer the words and their meanings remain identical at all times and places. But is not this highly idealized assumption contrary to our bilingual/multilingual experience of expression, exchange and translation? I would like to hold that, following Bhartṛhari, *sphoṭavāda* leaves room for variation in the word–object relationship without causing any breakdown of communication. It is to be remembered here that Bhartṛhari's theory of language was not merely logico-epistemic but also aligned to rhetoric and poetics.[17]

Fifthly, close examination of Matilal's arguments, examples and references shows an element of ambivalence in his basic approach to semantics, ontology and epistemology, despite his avowed liking for direct realism. It is difficult to accept the external realist's claim that a particular object together with its underlying universal or the property of which it is an instantiation can be directly and unerringly grasped. Is our body a mute and passive spectator of the enterprise called knowing that 'takes place' in it? That our body and, particularly, brain are specially equipped to gather, categorize and retain information *about* our inseparably related 'inner' and 'outer' worlds is evident from the (more or less) success of the action performed in the light of those informations. The structured awareness of our body to which different informations, different aspects of the same object as information-source, along with their changing and stable relations are available, while the body itself by its active and reflective capacity can generate new information is often said to be human self.[18] When these facts about our psychosomatic system are rightly kept in view, the direct realist's image of knowledge appears simplistic, if not opaque and misleading.

Finally, *saṃśaya* (initial doubt) and *jijñāsā* (questioning) are not at all incompatible with *pramā* or *jñāna*, certainly not in the Indian tradition of philosophy.[19] *Saṃśaya* is a *cognitive* state of rest or arrest. Of the similar sense of cognitive-cum-explorative rest Hegel speaks in his *Phenomenology of Mind* in the context of scepticism. *Skepsis*, the root word, by its very nature means the movement between the truth that is 'already' (naturally) there in consciousness and 'not yet' (transcendentally) present in consciousness. Comparable in sense is *Śaṅkara's jijñāsā*. It is questioning of the self on the basis of inarticulate knowledge of the Self for its articulate realization. Neither in *saṃśaya* nor in *jijñāsā* the vaticination of knowledge is absent. With knowledge is present the world of objects, both 'inner' and 'outer'.

NOTES

1. Bimal Krishna Matilal, *Perception: An Essay on Classical Indian Theories of Knowledge*, Oxford: Clarendon Press, 1986, pp. 46–7.
2. *Nyāya*, Gautama's *Nyāya-Sūtra* with *Vātsyāyana's* Commentary, tr. Mrinal Kanti Gangopadhyaya, Indian Studies: Past and Present, Calcutta, 1982, pp. 77–98.
3. *Nāgārjuniana: Studies in the Writings and Philosophy of Nāgārjuna*, by Chr. Lindtner, Delhi: Motilal Banarsidass, 1987, pp. 34–7.
4. *SSK*, ss.20–1.
5. *SSK*, 73.
6. *Nyāya-Sūtra of Gotama* by Satish Chandra Vidyabhusana, New Delhi: Oriental Books Reprint Corporation, 1975, pp. 22–42.
7. T.R.V. Murti, *The Central Philosophy of Buddhism*, London: Unwin Paperbacks, 1980.
8. *Nyāya-Sūtra*, II, i.37.
9. Ibid., I, i.5.
10. D.P. Chattopadhyaya, *Induction, Probability and Skepticism*, Albany:

SUNY Press, 1991, p. 45; see also, R. Carnap, *Logical Foundations of Probability*, Chicago: Chicago University Press, 1950, pp. 205–8.

11. B.K. Matilal, *Perception*, Oxford: Clarendon Press, 1986, p. 124; see also, his *The Word and the World: India's Contribution to the Study of Language*, Delhi: Oxford University Press, 1990, pp. 49–74.
12. Quoted from Stcherbatsky's *Conception of Buddhist Nirvāṇa*, Varanasi: Bharatiya Vidyaprakashan, 1975, pp. 77–8.
13. Satkari Mookerjee, *The Buddhist Philosophy of Flux*, Delhi: Motilal Banarsidass, 1980, pp. 273–81.
14. Matilal, *Perception*, 1986, p. 244.
15. Michael Dummett, *Truth and Other Enigmas*, London: Duckworth, 1978, p. xxxix.
16. Matilal, *The Word and the World*, pp. 29–30.
17. 'Sri Aurobindo on Knowledge and Language' (ch. 11) and 'Enlightenment Communication and Silence' (ch. 15) in D.P. Chattopadhyaya, *Knowledge Freedom and Language: An Interwoven Fabric of Man, Time and World*, Delhi: Motilal Banarsidass, 1989.
18. Jaakko Hintikka and Patrick Suppes, eds, *Information and Inference*, Dordrecht-Holland: D. Reidel Publishing Co., 1970, pp. 3–27, 263–97.
19. Chattopadhyaya, *Induction Probability and Skepticism*, pp. 259–61, 282–4, 321.

Chapter 4

Matilal on Nāgārjuna

Mark Siderits

There are deep affinities between Madhyamaka and scepticism, both Pyrrhonian and post-Cartesian. There are likewise deep affinities between Madhyamaka and the sort of mysticism characteristic of Advaita Vedānta, Plotinus and Eckhart. While Madhyamaka was never the primary focus of Bimal K. Matilal's explorations of the Indian philosophical tradition, he clearly recognized its significance to that tradition. And on those occasions on which he sought to explain it to those outside the tradition, he exploited the affinities just mentioned—to scepticism and to mysticism. He always did so cautiously, for he was well aware of the Mādhyamika's tendency to resist labels or glib comparisons.[1] And his purpose was clearly always just to facilitate comprehension through accessible comparisons so as to foster dialogue between traditions. Still I think that the particular affinities he exploited are less illuminating and less robust than his writings on Madhyamaka would suggest. I shall examine Matilal's two characterizations of Nāgārjuna, the founder of Madhyamaka —as sceptic and as mystic.

I

Matilal's characterization of Nāgārjuna as sceptic occurs in his examination of the *Vigrahavyāvartanī (VV)* attack on the Nyāya theory of the means of knowledge or *pramāṇas.*[2] In *VV* vv.31–51, Nāgārjuna deploys what can be called the trilemma argument against the possibility of showing that certain classes of cognition may be accepted as reliable. Matilal takes the point of this argument to be to try to show that we can never attain knowledge: since we can never know that any of the means of belief-formation that we employ are in fact reliable, we can never be said to know of any belief that it correctly represents the world. Of course this argument succeeds in establishing a sceptical conclusion only if we presuppose the KK thesis, that knowledge requires that one know that one knows. Thus Matilal explicitly attributes the KK thesis to Nāgārjuna.[3]

In the Western tradition, we may distinguish between the radical scepticism of the Cartesian project, and the Pyrrhonian scepticism of the Hellenistic era. In the modern, radical form of scepticism, we find a tendency to focus on such issues as the problem of our knowledge of the external world,[4] or of other minds, or of the past, or the problem of induction. Ancient scepticism is not similarly focused on any particular class of knowledge-claims. We may then note that radical scepticism differs from its Pyrrhonian predecessor in two important respects. First, there is the tendency in radical scepticism to partition propositions into two broad classes, one of which is deemed epistemically privileged relative to the other. Thus Descartes takes propositions about the subject's mental states to be epistemically prior to propositions about the external world. Likewise the Humean problem of induction is grounded in the distinction between, on the one hand, propositions about the present and the past, and, on the other, propositions about the future. In each case we find the claim that the first class of propositions is better known or more certain than the second. We are then invited to conclude that the first class must serve as evidence for any claims we make concerning the second class. A second feature of radical scepticism, one that derives significant support from the first, is its employment of an internalist conception of justification: the knowing subject's being justified is here understood as a matter of her having justification. This is why, in order to be justified in believing some proposition about, e.g., the external world, I must be able to rule out such sceptical hypotheses as that I am dreaming. Otherwise I have no justification for the employment of those bridge principles that I presumably employ in passing, e.g. from propositions about my mental states to propositions about the external world. Hence the KK thesis: in order to know, one must know that one knows.

I want to claim that this radical form of scepticism does not play a significant role in the Indian tradition. This claim will initially strike many as implausible, given that the Yogācārins espoused a form of subjective idealism every bit as radical as Berkeley's. Surely, it will be objected, the Yogācāra thesis of consciousness-only (*vijñapti-mātratā*) is just as much an expression of external world scepticism as anything to be found in Descartes or Hume. I would suggest, though, that this claim does not stand up under closer scrutiny. When we look, for instance, at Vasubandhu's *Vijñaptimātratāsiddhi*, we find none of the epistemological considerations that figure so prominently in the post-Cartesian external-world problematic. It is true that Vasubandhu begins the work with an appeal to the illusion of hairs on the moon seen by those with cataracts. But nowhere does Vasubandhu suggest that we are somehow unable to prove ourselves to be in any better situation with respect to our own judgements concerning external objects. The point of the hairs-on-the-moon example is rather to induce the realist opponent to explicitly espouse representationalism rather than direct realism as the preferred account of perceptual cognition. (And here it is significant that the Sautrāntika school originally embraced representationalism on essentially metaphysical grounds, namely those considerations expressed in the time-lag argument.) This establishes the dialectical situation in which Vasubandhu can employ the metaphysical principle of lightness (Occam's Razor) in defence of consciousness-only. Notice, as well, how dreaming gets used in this defence based on lightness: not to make the point that we can never be certain we are awake (Vasubandhu accepts the distinction between dreams and waking experience), but just to block one representationalist strategy for showing that external objects cause our perceptual cognitions. Finally, we must note the centrality, to the Yogācārins' case for consciousness-only, of their arguments against the various realist accounts of the physical object (most famously their argument against atomism). Much of the debate between Yogācārin and realist is given over to discussion of how one or another of the realist theories about the ultimate physical particulars might yield an adequate account of our sensory experience. To the post-Cartesian external-world sceptic, such discussion must seem utterly besides the point: how could demonstrating the mere possibility of such an account serve to vindicate realism, so long as I am still unable to rule out such sceptical hypotheses as that I am dreaming? Unless we suppose that the Indian philosophers who engaged in this debate were quite uniformly poor at seeing where their arguments led, we must conclude that the epistemological considerations that motivate radical scepticism did not play a role in the Yogācāra theory of consciousness-only. Still, all that this might seem to show is that radical scepticism did not play the role in the Indian formulation of subjective idealism that we might have expected from the example of the post-Cartesian tradition. What I claim is that radical scepticism is wholly absent from the Indian tradition. I want to suggest that this is so for the reason that none of the parties to Indian epistemological theorizing held a version of internalism. When all competing theories of knowledge are externalist in their account of justification, it is difficult for the peculiarly radical forms of sceptical questioning that characterize the post-Cartesian project to gain a purchase. Even if the KK thesis is removed, one may still raise sceptical questions about particular knowledge-claims, but it is difficult to call into question all our knowledge of the external world, of other minds, etc.

The controversy in Indian epistemology over whether knowledgehood (*prāmāṇyatva*) is intrinsic or extrinsic (*svataḥ/parataḥ*) would appear to suggest otherwise;[5] it is tempting to view this as a debate over the KK thesis. I will try to show that it is not, that all parties to this debate hold externalist views. But if internalism was held by anyone in the Indian epistemological tradition, this debate is just the context in which we would expect it to be visible and prominent. So its absence here would strongly suggest its absence from the tradition as a whole.

Let us look more closely at the views of those who hold that knowledgehood is intrinsic to those cognitions that are instances of knowledge. Such theorists make three claims about such cognitions: (1) those factors that serve as the cause of the cognition also serve as the cause of the inherence of knowledgehood in that cognition (i.e. it was not some additional, extrinsic factor that brought about the knowledgehood of this cognition); (2) to have a cognition is to be aware not only of the content of that cognition but also of that very cognition itself (i.e. my ability to report that

I am aware of the object is not the result of an inference or act of introspection occurring subsequent to the awareness of the object); (3) whenever one has a cognition that is an instance of knowledge, one is aware of the knowledgehood of that cognition. Now thesis (3) certainly looks, to the eyes of one versed in post-Cartesian epistemology, like a strong affirmation of the KK thesis. But the fact that those who held (3) also held thesis (1) makes this interpretation seem somewhat less straightforward. For it would be odd for an internalist to conjoin an affirmation of the KK thesis with any claims about the causal genesis of veridical cognitions. Those are the sorts of claims that one expects the externalist to make.

There is, however, nothing odd about the conjunction of (1) through (3) if we read them in the following way: For any cognition that is an instance of knowledge, because (1) the cause of the veridicality of that cognition is intrinsic to the cause of the cognition itself, the knowledgehood of that cognition is intrinsic to the cognition. (Compare: because the cause of the materiality of my desk is an essential feature of those causes that brought about the existence of my desk, materiality is an essential property of my desk.) But (2) when one cognizes, one is aware not only of the content of one's cognition but also of that very cognition itself. Hence (3) when one has a veridical cognition one must be aware of the knowledgehood of that cognition. For to be aware of the cognition itself is to be aware of those properties that are intrinsic to the cognition, and knowledgehood is intrinsic to veridical cognitions. Now (3) does claim that when I am justified in accepting a cognition, then I have justification that is in some sense accessible to me 'internally'. Hence the appearance of internalism. The difficulty here is to get clear on the sense of accessibility involved. That it is not the sense involved in giving justifications is clear from the fact that those who hold these three theses also hold: (4) when a non-veridical cognition occurs, one may be aware of its non-knowledgehood (*apramāṇyatva*) only by means of a distinct cognition. That this claim is required to make the overall view coherent is clear from the fact that otherwise it would amount to the implausible position that whenever our cognitions are veridically produced, we always know, immediately and non-inferentially, that we have knowledge. Rather, what the proponents of the intrinsic knowledgehood thesis were seeking to do is give theoretical articulation to the fact that our ordinary attitude toward cognitions is one of acceptance; it is usually only upon reflection that we inquire into the credentials of our cognitions for representing the facts.[6]

I claim, then, that the KK thesis plays no role in Indian epistemology. To attribute knowledge to a subject is to remark on the causal history of that subject's cognition. It is not to attribute to that subject the ability to grasp whatever evidence would be required to rule out sceptical hypotheses. From the fact that, for all I now know, I might be dreaming, it does not follow that I am not justified in believing that there is a desk before me. Given this view of knowledge, radical scepticism does not arise. And Matilal was simply wrong to attribute the KK thesis to Nāgārjuna. We must look elsewhere to see the point of Nāgārjuna's asking how the means of knowledge are established.

As I pointed out above, radical scepticism is not the only form of scepticism. Pyrrhonian scepticism seems in many respects much closer in spirit to those elements of the Indian tradition that call into question our ordinary grasp of the world. Unlike radical scepticism, the Pyrrhonian variety is put forward (or rather, described—Pyrrhonians do not make recommendations, lest they appear dogmatic) as a way of life. While inquiring minds might want to know, knowledge is not to be attained on any of life's vexing questions. Tranquillity is, however, gained through the suspension of belief. Thus Sextus Empiricus presents us with a bewildering variety of arguments against the views of the 'dogmatists', those who purport to have arrived at the truth in some area of human inquiry. The point of such arguments is to help us attain tranquillity by enabling us to see that for any view for which reasons might be advanced, there may be found equally good reasons supporting the opposing view. When we recall that Nāgārjuna seeks to help us attain *nirvāṇa* by refuting all available metaphysical theories (*dṛṣṭi*), we can see how a case might be made for identifying Nāgārjuna as this sort of sceptic.[7]

What makes the case for such an identification particularly compelling in the present context is the fact that Sextus employs just the trilemma that Nāgārjuna uses against the claim that a particular procedure counts as a means of knowledge: such a claim is merely dogmatically asserted, or else its defence is vitiated by circularity, or its defence results in an infinite regress.[8] Now I wish to argue that it would ultimately be a mistake to think of Madhyamaka as a kind of Pyrrhonian scepticism. But before doing so, I shall use the significant similarities between Sextus and Nāgārjuna to make an important point about the latter's use of the trilemma.

Because the trilemma looks to be an extremely powerful sceptical engine, it is not difficult to understand how Matilal came to view Nāgārjuna as a universal sceptic. The trilemma is used to show that the epistemologist can ultimately give us no satisfactory justification for the claim that cognitions resulting from a certain procedure are instances of knowledge. Thus we seem forced to conclude that we can never have knowledge, for we can never know of any of our cognitions that it is the product of a reliable process. But notice that if the argument is formulated in this way, it has the KK thesis as an implicit premise. It would thus be peculiar to attribute such an argument to the Pyrrhonian, to whom the KK thesis would seem like just another piece of dogmatic theorizing. Moreover, if this were the argument that Sextus had in mind when he used the trilemma, how are we then to account for his incredible prolixity? The argument may be put very succinctly—we just did. Why, then, would Sextus feel compelled to show us how to confute the dogmatist in such areas as physics, grammar and ethics, if the impossibility of knowledge could be demonstrated once and for all just by the use of the trilemma? Was Sextus too dim-witted to see the force of his own argument? Was he merely a compiler of the arguments of others, arguments the respective powers of which he could not himself grasp?

All these questions are easily answered if we view Sextus's deployment of the trilemma as a product of the dialectical situation, and not as expressive of any theoretical commitment. This is the reading recommended by Williams, who writes,

> The Pyrrhonian does not offer sceptical arguments as freestanding theoretical problems or theses. He deploys them reactively. The problem of the criterion arises because there are actual conflicts of opinion. It becomes of such dominant interest because these conflicts move dogmatists to formulate theories, such as a theory of truth.[9]

It is the usual practice of the Pyrrhonian to seek to counterpose the conflicting views of opposing theorists on some one or another subject. In such agonistic contexts, it is not uncommon for theorists to seek to go epistemological—to bring their preferred account of knowledge to bear in support of their theory. In this case, the Pyrrhonian may deploy the trilemma as a way of revealing the resort to epistemology as just one more piece of dogmatic theorizing. For the Pyrrhonian, the trilemma does not by itself induce the tranquillity that comes from abandoning the pursuit of knowledge; it could function as a freestanding argument for scepticism only if we held certain substantive epistemological views. It is rather just one more tool in the sceptic's toolbox, to be deployed when the dialectical situation warrants its use, namely when the debate threatens to go epistemological.

Now I do not think that Nāgārjuna should be thought of as a Pyrrhonian sceptic. Nonetheless I believe we can learn something about Nāgārjuna's use of the trilemma by reflecting on the role that it plays in Pyrrhonism. To see Nāgārjuna's use of the trilemma as sceptically motivated is to take him as viewing epistemology as First Philosophy (with philosophy itself seen as the queen of the sciences). That this was not Nāgārjuna's view is clear from the overall dialectical structure of his writings. Epistemological issues are conspicuous by their virtually total absence from *Mūlamadhyamakakārikas*, the whole point of which is just to show the impossibility of constructing a coherent metaphysics given the requirement that a real be a *svabhāva*, i.e. something that bears its own essential nature. The locus of the trilemma, *Vigrahavyāvartanī*, is given over to objections that arise only after the arguments against *svabhāva* have been set forth. It is deployed in response to the Nyāya objection that the emptiness of all existing things could be proven only through the use, by a really existing subject, of a really existing means of knowledge that apprehends the entities deemed empty. The opponent thus takes this objection to show that there must be at least one *svabhāva*, namely whatever means of knowledge the Mādhyamika employs in his proof of emptiness. And this in turn is taken to show that there must also be really existent objects of knowledge. Nāgārjuna starts his response by claiming that since he does not prove the emptiness of all things by apprehending objects through some means of knowledge and thus ascertaining their emptiness, his procedure is not self-undermining in the way alleged by the opponent. He then deploys the trilemma in order to demonstrate that no theory of the means of knowledge can be shown to be ultimately true. This in turn is taken to establish that we cannot infer, from our ability to have knowledge of emptiness, that there must be things that are *svabhāvas.*

When Nāgārjuna begins his response by denying that he needs a real means of knowledge to prove emptiness, one might suspect that he is taking up a sceptical stance. His use of the trilemma might seem to confirm this suspicion. But we must pay careful attention to what he thinks he has shown, namely that if there are means of knowledge, then they too are empty, and thus that the Mādhyamika's proof of emptiness is consistent. There may be epistemological consequences in all this. But the point of the reply is just the usual one for Nāgārjuna—that all things are devoid of *svabhāva.*

It should now be clear why I think Nāgārjuna ought not be thought of as a Pyrrhonian sceptic. He thinks we can know that all things are empty. His use of the trilemma is just meant to show why this does not have the metaphysical realist consequences alleged by his Naiyāyika opponent. But if more evidence is needed, consider the treatment of doubt in *Vaidalyaprakaraṇa.*[10] In that work we find the trilemma once again deployed to refute the Nyāya claim that there are ultimately real means and objects of knowledge. Given the result that there are ultimately no means of knowledge, the question is raised whether this does not lead to doubt. The response is that doubt too is empty. On the Nyāya account, doubt arises when the object is apprehended as characterized by two distinct and incompatible properties, only one of which actually characterizes the object. But how can an unreal property give rise to an apprehension? Perhaps the view is that the object is characterized by indeterminacy with respect to these two properties. But this should give rise to the determinate apprehension of indeterminacy, not to doubt. Thus truly existent doubt is impossible. Now one might have reservations about this argument. (For my part, I think it succeeds at undermining a certain realistic view of psychological states.) The point is just that here doubt enjoys no privileged status, it falls into the same boat as knowledge (namely Neurath's boat). The Mādhyamika is no kind of sceptic.

Matilal has another characterization of Nāgārjuna—as a mystic. But before we examine that characterization, we should briefly investigate some of the epistemological consequences of the Madhyamaka trilemma, since these will turn out to have some bearing on the question of mysticism.[11] There is one great irony in this dispute between Madhyamaka and Nyāya over the possibility of a theory of the means of knowledge. This is that the Nyāya response to Nāgārjuna's use of the trilemma would be effective against the radical sceptic; but in answering the sceptic, this response only helps confirm the anti-essentialist thrust of the Mādhyamika. The Nyāya response makes critical use of both their 'extrinsic knowledge-hood' version of externalism and their extensionalism: because the theoretical claims of Nyāya epistemology are wholly reducible to their particular instances, and because the knowledgehood of a particular cognition may only be known through an examination of the factors from which it arose, their account of the means of knowledge may only be validated piecemeal. This makes it impossible for the sceptic to attempt to assess all our knowledge of the world all at once, a move that is critical to the radical sceptic's agenda. But this also leaves Nyāya unwittingly making just Nāgārjuna's anti-essentialist point. For now it emerges that something can be a means of knowledge only by virtue of standing in certain complex relations with certain objects of knowledge and certain other means of knowledge. Nothing bears the essential nature of a means of knowledge. All means of knowledge are empty.

The irony in all this is brought home most clearly when Candrakīrti—the Mādhyamika whose views are closest in spirit to radical scepticism—explicitly endorses the Nyāya account of the means of knowledge.[12] It should, of course, be pointed out that this endorsement is valid only at the conventional level of truth, not at the ultimate level. And this might lead us to suspect that the Naiyāyika has not, after all, unwittingly fallen in with the Mādhyamika. To see whether this is right, we will need to look more closely at the Madhyamaka distinction between the two truths. And the best way to approach this is through Matilal's account of Nāgārjuna as mystic.

II

Just as his characterization of Nāgārjuna as a sceptic grows out of consideration of the use of the trilemma in *Vigrahavyāvartanī*, so Matilal's discussions of Nāgārjuna as mystic tend to focus on that passage (*VV* v.29) in the same work where Nāgārjuna states that he has no thesis (*pratijñā*) to assert or deny. Matilal forges the connection by way of the thesis that mystical experience is strictly ineffable (the IME thesis).[13] The linkage is as follows. Faced with the objection that his claim that all things are empty must itself be empty, and is thus incapable of demonstrating that all things are empty, Nāgārjuna replies that he has no thesis. This is tantamount to the claim that the statement 'All metaphysical statements are false' is itself neither true nor false, hence strictly inexpressible. But the Mādhyamika also holds that the insight that leads to liberation comes about through comprehending that all metaphysical statements are false. Since mystics generally claim just such transformative powers for what they describe as mystical experience, the claim that all metaphysical statements are false must be an (indirect) expression of mystical experience. Hence Nāgārjuna subscribes to the IME thesis. He employs the semantic paradox that is generated by the claim 'All metaphysical statements are false' in order to *show* the soteriological insight that cannot be *said* (or whistled either).[14]

When Putnam gave his famous 'brains-in-a-vat' argument[15] for the conclusion that the statement 'We are brains in a vat' is necessarily false, this result was widely viewed as paradoxical. If it is indeed impossible for us to imagine that we are brains in a vat, how did we succeed in grasping the premises of the argument, which have us envisioning the situation of brains in a vat? What such interpreters of the argument uniformly overlooked was the possibility that it was intended as a *reductio* on the metaphysical realism underlying construction of brains-in-a-vat scenarios.[16] I want to suggest that something similar is afoot here. To view Nāgārjuna's 'I have no thesis' as expressive of a semantic paradox is to overlook the possibility that there is a similar sort of *reductio* somewhere in the vicinity.

Now *reductio* arguments are typically Nāgārjuna's preferred device for demonstrating the falsity of various metaphysical theories. And as Matilal has always been careful to point out,[17] the negation employed in the conclusions of these arguments is of the 'verbally bound' (*prasajya*) variety, rejecting the metaphysical thesis in question in such a way as to avoid commitment to any other thesis. It is by employing this sort of negation that Nāgārjuna can, for instance, deny that originating entities are produced from themselves without thereby committing himself to the thesis that they are produced from other things (see *MMK* I.2). The question is how long one can get away with this. We seem willing to tolerate the resultant failure of bivalence when it is locally confined, as in 'The son of a barren woman is neither dark nor non-dark'. But when bivalence failure seems apt to go global, incoherence appears to threaten. And the Mādhyamika seems to want to take bivalence failure global: the conclusion we are invited to draw from the refutation of a wide array of competing metaphysical theories is that all metaphysical statements are false (this being the equivalent in the formal mode of the material mode 'All existents are empty'). That is, not only do we see that these and those particular opposed metaphysical theses are false, we are supposed to see from a sufficient sample of particular failures that all such theses are to be rejected.[18] But if this rejection is truly universal, and the negation involved in the universal rejection is the commitmentless variety, then the assertion that all things are empty is incoherent. Yet the Mādhyamika asserts it nonetheless, boldly claiming all the while not to be making any assertion. What is going on here? How could my envisioning the plight of brains in a vat lead to the conclusion that I cannot imagine being a brain in a vat?

The beginnings of an answer are to be found in Bhāvaviveka's practice in his commentary to *MMK, Prajñāpradīpa.* There the Madhyamaka *reductio* is routinely formulated in such a way as to make clear that the metaphysical thesis under consideration is denied ultimately, i.e. is to be rejected as an expression of the ultimate truth. To see the bearing of this on the issue at hand, we need a clear formulation of the distinction between conventional and ultimate truth. That distinction was first drawn within the Abhidharma schools of Buddhism, but we may begin with a more generic characterization:

C: A statement is conventionally true if and only if it is commonsensically assertible.

U: A statement is ultimately true if and only if it corresponds to ultimately existent states of affairs, i.e. names only ultimately existent entities, and depicts those entities it names in a way that correctly pictures how they in fact stand in the world.

Given the nominalist ontology of Abhidharma, which viewed such complex entities as chariots, forests and persons as mere conceptual fictions posited only for the sake of discursive economy, we can then define an Abhidharma version of U as follows:

U_A: A statement is ultimately true if and only if it corresponds to reality and neither asserts nor entails that wholes exist.

Similar conceptions of ultimate truth could be formulated to reflect the preferred ontologies of other schools such as Nyāya. Now Nāgārjuna's chief task in *MMK* was to demonstrate that no statement is U_A true. In his other works, and in the works of later Mādhyamikas, those results were extended to other conceptions of U. But it is important to the Abhidharma project that conventional truth be in some sense reducible to ultimate truth: the practical efficacy of our common-sense warranted beliefs is to be explained by the fact that, while these beliefs are for the most part ultimately false, they do allow of reductive translation into statements that are ultimately true. The question at hand is what we are to say when it turns out that there are no statements that are ultimately true. Bhāvaviveka's point in prefixing 'ultimately' to the conclusions of the specific *reductios* in the Madhyamaka arsenal is that we should say there is only conventional truth. While the negations of the specific *reductio* arguments are of the *prasajya* (commitmentless) variety, the negation involved in the universal rejection of metaphysical theories is not. That bivalence fails in the domain of ultimate truth should be taken as indicative of something deeply problematic in the very concept of ultimate truth. The practical efficacy of conventional truth requires no deep metaphysical realist grounding; that conventional truth works is the only explanation it requires. As Nāgārjuna puts it, 'Everything accords with that with which emptiness accords' (*MMK* XXIV.14).

Thus when Nāgārjuna goes on to tell us he has no thesis, he expects us to resolve the apparent paradoxicality of this remark by reverting to the rough terrain of conventional truth; he does not intend for us to conclude that the ultimate truth is inexpressible. If the Mādhyamika is any sort of mystic at all, he is not the sort who traffics in ineffability theses.

Now some caution is required here, since the present point approaches the heart of the dispute between the so-called Svātantrikas and Prāsaṅgikas (the terms were coined by Tibetan scholars). Bhāvaviveka is classified as a Svātantrika because he maintains that Madhyamaka *reductios* involve propositions the truth of which is acknowledged by both the Mādhyamika and his dialectical adversary. A so-called Prāsaṅgika like Candrakīrti denies this: he maintains that only the opponent need accept the premises of a Madhyamaka refutation, for the Mādhyamika accepts no statement as true. What is really at issue here is the question of the Mādhyamika's standpoint. If analysis reveals that no statement is ultimately true, if nothing is 'analytically findable', where does that leave us who perform the analysis? Candrakīrti would seem to suggest that we stand outside all conceptual construction (*prapañca*). Analysis reveals the domain of ultimate truth to be empty, and this in turn undermines conventional truth as well. The resources of conventional truth to which we help ourselves when performing the Madhyamaka *reductios* are no more real than the mirror image we use when styling our hair. The Prāsaṅgika propounds a kind of linguistic idealism: all reals are mere conceptual constructions, posits that we accept only because of the demands of our discourse. True insight (*prajñā*) is attained when, frustrated at every turn in our efforts at grasping ultimate truth, we abandon our discursively fictionalizing practices and acquire the non-discursive awareness of suchness (*tathātā.*).

So Matilal is, after all, correct to attribute the IME thesis to Nāgārjuna—if we interpret him after the manner of Candrakīrti. I want to suggest, though, that there are irresolvable tensions in such a reading, that the Svātantrika account of Madhyamaka philosophical practice yields a more plausible picture of the *reductios* and our response to them. One clear indication of the tension in the Prāsaṅgika view is that Candrakīrti himself criticizes the Yogācārin and the Pudgalavādin for maintaining the strict ineffability of, respectively, the pure particular, and the person.[19] In each case, Candrakīrti wonders how we are to make sense of the thesis that there are entities of some determinate sort that nonetheless allow of no possible description. Now Candrakīrti might feel himself immune from this sort of criticism, since he (quite consistently) advances no positive claims concerning the ultimate nature of reality. But as Matilal himself helps us see, there are other kinds of metaphysics in the immediate neighbourhood, such as Advaita Vedānta.[20] The Advaitin likewise maintains that the ultimate nature of reality is strictly ineffable, that the world of conventional practice is ultimately illusory, and that true insight is to be attained through appreciating the contradictions inherent in all philosophical theorizing. How, then, are we to distinguish between the final positions of Madhyamaka and Advaita Vedānta? It appears that we cannot. Yet Madhyamaka is not Advaita.

The argument for the last claim is simple. At *MMK* v.1–3, Nāgārjuna argues that no coherent account may be given of the relation between an entity and its defining characteristics. It is possible to view a wide array of Indian metaphysical theories as making up a spectrum of responses to this argument.[21] On the left we have the radical nominalism of Yogācāra Sautrāntika, which solves the problem by denying anything other than the pure particular. In the middle of the spectrum we have Nyāya, with its posit of the inherence relation as a way of mediating between the particular and the universal. And at the right of the spectrum we have Advaita Vedānta, which solves the problem by denying all but the highest universal. It is also abundantly clear that none of these approaches succeeds at resolving the difficulty that Nāgārjuna points out. Indeed Candrakīrti himself, in his *Prasannapadā* commentary on *MMK* v.4–5, explains why the problem cannot be solved in any of these ways: we cannot make sense of the notion of an existent without employing both the concept of an essential nature and the concept of the bearer of that essential nature. Thus all such metaphysical accounts, including the Advaita account, must be rejected by the Mādhyamika. Yet if he subscribes to the IME thesis, then his own final position is indistinguishable from that of Advaita Vedānta. This should prove a profound embarrassment for Madhyamaka.

The Prāsaṅgika stance toward conventional truth also proves

dialectically crippling. Recall that Candrakīrti sees conventional truth as a mere net of conceptual constructions; its apparent utility for worldly practice itself rests on an illusion, for the 'eating' that is taken as proof of the 'pudding' is revealed by analysis to be, finally, nothing more than a conceptual construction. (Hence the label 'linguistic idealism'.) The only real utility conventional truth possesses lies in its capacity to generate the contradictions that impel us to transcend both it and (discursive) ultimate truth. But given this stance toward conventional truth, the Prāsaṅgika proves constitutionally incapable of seeking to understand those common-sense intuitions that motivate the metaphysical theories undergoing analysis. And the result is *reductios* that all too often are dialectically inept. Perhaps this is what really lies at the heart of the charge of sophist (*vaitaṇḍika*) so often made against the Mādhyamika. It is not as if one must, so to speak, earn one's stripes by holding some positive thesis before one may employ *reductio* arguments. It is rather that one is better equipped to formulate effective *reductios* if one possesses a sympathetic understanding of the views of the opponent, and one is more likely to arrive at such understanding if one holds some stock of beliefs in common with the opponent. Because he views himself as not committed to any thesis, ultimate or conventional, Candrakīrti is simply not very well situated to argue effectively against those who purport to describe the ultimate nature of reality.

Consider, by contrast, Bhāvaviveka's strategy in his *Prajñāpradīpa* commentary on *MMK* I.3. There, after rehearsing the arguments against origination of an existent from itself and from another, he explains why we might be tempted to suppose that an entity arises out of causes and conditions that are distinct existents:

> Other-existence of the causes and conditions is found due to intentness of the mind on the desire for what is productive of the arising of the existent—i.e. just by virtue of expectation. The being from another is added, it is a conventional concept, that is merely convention. Being from another is grasped conventionally.

While it is not ultimately true that entities arise either from themselves or from distinct causes, it is conventionally true that they arise from causes that are distinct existents. What this means is that given certain facts about human institutions and practices and certain facts about human psychology, the best account of an effect like a sprout sees it as arising out of causes and conditions (the seed, soil, moisture, warmth, etc.) that are viewed as distinct from it. Our temptation to inflate this account into a metaphysical theory concerning the ultimate nature of the relationship between cause and effect derives from our forgetting that it was derived on the basis of certain background assumptions about our psychology. Thus this account cannot represent the ultimate truth about causation. It does, though, explain the source of those intuitions that fuel the principal theories about the causal relation: that the effect pre-exists, in its unmanifest state, in its material cause; and that the effect is a new existent distinct from its material cause.

What I wish to draw attention to here is the way in which this account alters the dialectical situation from what we find in the case of a Prāsaṅgika like Candrakīrti. Where Candrakīrti merely seeks to show how the opponent's views lead him, through the exercise of philosophical rationality, to unacceptable consequences, Bhāvaviveka takes the further step of trying to bring out the grain of (conventional) truth in the opponent's metaphysical views. This then paves the way for the opponent's ultimate acceptance of the final result—that there is no ultimate fact of the matter concerning the relation between cause and effect. And this dialectical strategy is only made possible by the Svātantrika's attitude toward conventional truth. It is only because he views himself as sharing with his dialectical opponent a common realm of discourse—conventional truth—that Bhāvaviveka is able to reveal, and so undercut, the 'intuition pumps' that drive the search for ultimate truth.

Jñānagarbha gives the following as the Svātantrika definition of conventional truth:

> A mere thing which is not confused with anything that is imagined and arises dependently is known as correct conventional truth. Mere things are capable of effective action that corresponds to appearances. Mere things also depend for their arising on causes and conditions, so they are known as correct conventional truth. If something appears consistently to be caused in the cognition of everyone from a scholar to a child, it is correct conventional truth, because something exists that is consistent with what appears in cognition.[22]

When we translate from the material mode, in which this definition is couched, into the formal mode, we see that conventional truth is just warranted assertibility. A statement is conventionally true provided a normal cognizer would be disposed to assert it on the basis of the exercise of a reliable means of cognition. This is the real import of the key phrase 'corresponds to appearances' (*yathādarṣana*). What must be borne in mind here is that something can count as a reliable means of cognition only against the background of certain human institutions and practices. There is no such thing as epistemic warrant *überhaupt*; there is only epistemic warrant within a particular context determined by certain concrete interests and practices. This is the point that we can take away from Nāgārjuna's demonstration in *Vigrahavyāvartanī* that the means of knowledge cannot be ultimately established. There is nothing to fear from a correspondence theory of truth, for truth has gone epistemic. Or as Putnam puts it, 'The mind and the world jointly make up the mind and the world'.[23]

From this perspective, the Prāsaṅgikas' apparent avowal of the IME thesis looks like just so much more hankering after a substantive ultimate truth. And the linguistic idealism that underlies their flirtation with ineffability looks like just another metaphysical thesis—precisely the sort of thing we were meant to rid ourselves of. From this perspective the Prāsaṅgika charge that the Svātantrika holds back out of fear of emptiness[24] looks most peculiar. If anything it would seem to be the Prāsaṅgikas who are afraid of losing that privileged standpoint known as the ultimate truth. The Svātantrika line on the ultimate truth may be put as: the ultimate truth is that there is no ultimate truth. This is the point of their claim that the assertion, 'All existents are devoid of *svabhāva*' is only conventionally true.[25] The idea here is simply that we can rid ourselves of those reificationist tendencies that fuel our own suffering and our indifference towards the welfare of others only by abandoning the very idea that there is some final way—whether expressible or inexpressible—that the world is. The point of the Madhyamaka dialectic, of the *reductios* on various metaphysical theories, is not that we should throw away the ladder after we have climbed up it. It is rather that we should stop hankering after a ladder to climb.

So Matilal is right to characterize Nāgārjuna as a mystic who subscribes to the IME thesis only if we read Nāgārjuna after the manner of Candrakīrti, and not after the manner of Bhāvaviveka. It will be clear which reading I prefer. The Svātantrika line strikes me as both more coherent and more fruitful than the Prāsaṅgika line. Thus considerations of charity militate against our reading Nāgārjuna as mystic.

It was pointed out earlier, in our discussion of Nāgārjuna and scepticism, that Candrakīrti endorses, at the level of conventional truth, the Nyāya account of the means of knowledge. And the question was raised whether this should prove an embarrassment for the Naiyāyika. Now that we have seen what Candrakīrti's attitude toward conventional truth amounts to, the answer is clear. Nyāya is in no danger of being subsumed by Madhyamaka, at least not by Prāsaṅgika. Candrakīrti's endorsement amounts to no more than this, that the Nyāya account accurately portrays the logic of our preferred fairy-tale, with something like mystical intuition providing the necessary corrective. No one would be likely to mistake this view of the means of knowledge for the Nyāya position.

But there still remains the Svātantrika stance toward the means of knowledge. How does Nyāya stand with respect to it? Here the situation is complicated by the fact that Svātantrikas endorse the mainstream Buddhist view that there are only two means of knowledge—perception and inference—and not the four accepted by Nyāya. But when we set this dispute aside, we find convergences that should distress the Naiyāyika. In particular, Nyāya's externalist extensionalism begins to look rather like the view that something is a means of knowledge only against the background of certain human institutions and practices. There may yet be room for drawing a distinction between the Nyāya and Svātantrika views of the means of knowledge. For it may turn out on further investigation that the latter view amounts to a form of contextualism that is distinct from the extensionalism favoured by Nyāya. This should be the subject of further research. But that such an inquiry should be needed at all should come as something of a shock to the partisans of Nyāya.

III

It is clear that there are deep affinities between the projects of the sceptic and the mystic. Each is concerned to call into question at least some set of widely and deeply held convictions. There are also, of course, important differences between the two projects. It is not, as is sometimes claimed, that the sceptic lacks the soteriological aims of the mystic. Pyrrhonian scepticism has its ethical dimension; it too presents a vision of a life transformed—the tranquillity that is to be attained by going beyond belief. The difference lies rather in the fact that the sceptic offers us nothing to take the place of our vanished certainties; we are just somehow to make do on our diminished epistemic resources. The mystic, by contrast, pledges to more than make good our newly incurred epistemic deficit. Riches beyond compare—the world in its pure shining suchness—will be ours as soon as we abandon our misplaced confidence in discursive rationality. The sceptic counsels only epistemic humility as penance for our presumptuousness. The mystic instead offers the hope of final illumination at the end of our epistemic chastisement.

There is something of both the sceptic and the mystic in Nāgārjuna. Like the sceptic, Nāgārjuna advocates a kind of epistemic humility. Like the mystic, he holds out the possibility of a kind of knowledge that transforms our lives. Like both, he calls into question certain widely shared assumptions. But the target of Nāgārjuna's critical questioning is importantly different from that of both the sceptic and the mystic. The aim of his *reductios* is to call into question those essentialist assumptions that underlie metaphysical realism. Both the sceptic and the mystic, however, rely on essentialist assumptions. The sceptic assumes that 'human knowledge' is a natural kind with a real essence.[26] The mystic assumes that there is such a thing as 'the world as a whole' or 'pure thusness' to be grasped in mystical intuition. Nāgārjuna rejects essentialism in all its guises. Thus epistemic humility for Nāgārjuna is not nihilistic despair over the impossibility of ever attaining 'human knowledge'; it is just recognition of the radical contingency of any of our claims to knowledge. And transformative insight for Nāgārjuna is not the eternalist rapture of final insight into the innermost essence of the world; it is just the realization that nothing bears its own essential nature. Thus it is that *Catuḥstava*, a work traditionally attributed to Nāgārjuna, describes the Buddha as 'having transcended the duality of existence and non-existence without having transcended anything at all'.[27]

There is, then, something right as well as something wrong in Matilal's choices of scepticism and mysticism as ways of explicating Nāgārjuna. My focus here has been on what I think is wrong. What must never be lost sight of, though, is that the Nāgārjuna we encounter in Matilal's writings is an eminently able philosopher who raises profound and troubling challenges for the Naiyāyika. Matilal's Nāgārjuna might not be a Mādhyamika; but neither is he a straw man constructed just to make Nyāya appear victorious. Matilal's Nāgārjuna is testimony to Matilal's own commitment to the search for truth.

NOTES

1. Thus he prefaces his discussion of Nāgārjuna as sceptic with the following cautionary note: 'By calling Nāgārjuna a sceptic, or rather by using his arguments to delineate the position of my sceptical opponent of the *pramāṇa* theorists, I have only proposed a probable extension of the application of the term "scepticism"' (Bimal Krishna Matilal, *Perception: An Essay in Classical Indian Theories of Knowledge*, Oxford: Oxford University Press, 1986, p. 50). Likewise, 'We may doubt whether Nāgārjuna can strictly be called a mystic', Bimal Krishna Matilal, *The Logical Culmination of Indian Mysticism*, Oxford: Clarendon Press, 1977, p. 17.
2. Matilal, *Perception*, pp. 46–68.
3. He also (ibid., pp. 53–4) accuses Nāgārjuna of committing the fallacy of composition, of making the illicit passage from Nyāya fallibilism (the universal possibility of doubt) to the possibility of universal doubt. My own view is that this attribution is incorrect, but my focus will be on the attribution of the KK thesis to Nāgārjuna.
4. Indeed for Barry Stroud (*The Significance of Philosophical Scepticism*, Oxford: Clarendon Press, 1984) the problem of our knowledge of the external world is *the* sceptical problem.
5. J.N. Mohanty's *Gaṅgeśa's Theory of Truth* (Santiniketan, India: Visva Bharati, 1966) contains an extremely useful discussion of this debate.
6. Of course, for the Mīmāṃsākas this position also played a role in the support of their claim that the *Vedas* are intrinsically authoritative.
7. Thomas McEvilley ('Pyrrhonism and Mādhyamika', *Philosophy East and*

West 32, 1982) presents a strong case in favour of such an identification. As will emerge shortly, I have misgivings about the adequacy of his reading from the Madhyamaka side. But from the side of Pyrrhonism, one might question his claim that Sextus does have 'one positive ontological doctrine', namely 'the universality of relation'. This sounds rather too close to dogmatism for Sextus's liking.

8. See *Outlines of Pyrrhonism* I.164ff. For a characteristic use of the trilemma, see *Against the Logicians* I.340–3.
9. Michael Williams, 'Scepticism without Theory', *Review of Metaphysics* 41, 1988, p. 581.
10. See Christian Lindtner *Nagarjuniana: Studies in the Writings and Philosophy of Nagarjuna*, Copenhagen: Institut før Indisk Filologi, pp. 87–93. Lindtner's attribution of this work to Nāgārjuna may be questioned. In its favour is that in the work's treatment of Nyāya epistemology use is made of the argument of the three times that is alluded to at *Nyāya-Sūtra* II.1, 8–15 but is not found in *Vigrahavyāvartanī*. It is, in any event, clear that the work is the product of an orthodox Mādhyamika.
11. I explore the consequences of Nāgārjuna's trilemma in greater detail in my 'The Madhyamika Critique of Epistemology II', *Journal of Indian Philosophy* 8, 1980; and 'Nāgārjuna as Anti-Realist', *Journal of Indian Philosophy* 16, 1988.
12. Candrakīrti, *Prasannapadā*, ed. by Pandeya, p. 23.
13. Bimal Krishna Matilal, *The Word and the World: India's Contribution to the Study of Language*, Delhi: Oxford University Press, 1990.
14. For the explicit linking of the 'I have no thesis' position with a Tarski-style treatment of semantic paradoxes, see Matilal, ibid., p. 149. A similar, though somewhat more compressed, treatment of Nāgārjuna's claim not to have a thesis is to be found in Matilal, *Logical Illumination of Indian Mysticism*, pp. 7–15. Paul T. Sagal, 'Nāgārjuna's Paradox', *American Philosophical Quarterly* 29, 1992, also explores the possibility that Nāgārjuna's position generates a semantic paradox.
15. Hilary Putnam, *Reason, Truth and History*, Cambridge: Cambridge University Press, 1981.
16. See David Anderson, 'What is "Realistic" about Putnam's Internal Realism?', for a discussion of this reception of the argument, and for an illuminating account of the place of the argument in Putnam's larger project.
17. See, e.g. *Perception*, pp. 66–7; and *The Word and the World*, pp. 153–4.
18. Here is another respect in which Madhyamaka differs from Pyrrhonism. The Pyrrhonian would resist drawing the universal conclusion: the final suspension of belief that brings about tranquillity is not logically grounded, it is just psychologically induced.
19. For Candrakīrti's critique of the Yogācāra thesis that the *svalakṣana* is ineffable, see *Prasannapadā*, Pandeya, ed., pp. 21–2. For his critique of the Pudgalavādin thesis that the self is ineffable, see *Madhyamakāvatāra* 6.146–9 (C.W. Huntington, *The Emptiness of Emptiness: An*

Introduction to Early Indian Madhyamaka, Honolulu: University of Hawai Press, 1989, p. 175).

20. Matilal explicitly links Madhyamaka and Advaita at, e.g. *Logical Illumination of Indian Mysticism*, pp. 15–17, *Perception*, p. 51, and *The Word and the World*, p. 154.
21. This is not a historical claim: I am not claiming that these views in fact resulted from the consideration of Nāgārjuna's argument.
22. Malcolm David Eckel, *Jñānagarbha's Commentary on the Distinction Between the Two Truths*, Albany: SUNY Press, 1987, p. 75.
23. Putnam, *Reason, Truth and History*, p. xi.
24. This is, for instance, the impression conveyed throughout Peter Della Santina, *Madhyamaka Schools in India*, Delhi: Motilal Banarsidass, 1986.
25. See, e.g. Eckel, *Jñānagarbha's Commentary*, pp. 87–90.
26. For a discussion of the claim that scepticism requires a form of essentialism about knowledge, see Williams, *Unnatural Doubts*, Oxford: Blackwell, 1991, pp. 101–11.
27. Lindtner, *Nāgārjuniana*, pp. 152–3.

REFERENCES

ANDERSON, David, 'What is "Realistic" about Putnam's Internal Realism?', *Philosophical Topics*, forthcoming.

BHATTACHARYA, Kamaleswar, 'The Dialectical Method of Nāgārjuna (Translation of the *Vigrahavyāvartanī* from the original Sanskrit with Introduction and Notes)', *Journal of Indian Philosophy*, vol. 1, 1971, pp. 217–61.

Bhāvaviveka, *Prajñāpradīpa*, ed. by Pandeya.

Candrakīrti, *Prasannapadā*, ed. by Pandeya.

DELLA SANTINA, Peter, *Madhyamaka Schools in India* (Motilal Banarsidass, Delhi, 1986).

ECKEL, Malcolm David, *Jñānagarbha's Commentary on the Distinction Between the Two Truths* (SUNY Press, Albany, 1987).

HUNTINGTON, C.W., *The Emptiness of Emptiness: An Introduction to Early Indian Madhyamaka* (University of Hawai Press, Honolulu, 1989).

LINDTNER, Christian, *Nagarjuniana: Studies in the Writings and Philosophy of Nāgārjuna* (Institut før Indisk Filologi, Copenhagen, 1982).

LOPEZ, Donald S., *A Study of Svātantrika* (Snow Lion Press, Ithaca, 1987).

MATILAL, Bimal Krishna, *The Logical Illumination of Indian Mysticism* (Clarendon Press, Oxford, 1977).

——, *Perception: An Essay in Classical Indian Theories of Knowledge* (Oxford University Press, Oxford, 1986).

MATILAL, Bimal Krishna, *The Word and the World: India's Contribution to the Study of Language* (Oxford University Press, Delhi, 1990).

McEVILLEY, Thomas, 'Pyrrhonism and Mādhyamika', *Philosophy East and West* 32, 1982, pp. 3–35.

MOHANTY, J.N., *Gaṅgeśa's Theory of Truth* (Visva Bharati, Santiniketan, India, 1966).

PANDEYA, Raghunath, ed., *The Madhyamakaśāstram of Nāgārjuna*, with the Commentaries Akutobhayā by Nāgārjuna, Madhyamakavṛtti by Buddhapālita, Prajñāpradīpavṛtti by Bhāvaviveka, and Prasannapadā by Candrakīrti (Motilal Banarsidass, Delhi, 1988).

PUTNAM, Hilary, *Reason, Truth and History* (Cambridge University Press, Cambridge, 1981).

SAGAL, Paul T., 'Nāgārjuna's Paradox', *American Philosophical Quarterly* 29, 1992, pp. 79–85.

Sextus Empiricus, *Outlines of Pyrrhonism*, R.G. Bury, trans., *Sextus Empiricus*, vol. 1 (Harvard University Press, Cambridge, 1933).

——, *Against the Logicians*, R.G. Bury, trans., *Sextus Empiricus*, vol. 2 (Harvard University Press, Cambridge, 1935).

SIDERITS, Mark, 'The Madhyamaka Critique of Epistemology I', *Journal of Indian Philosophy* 8, 1980, pp. 307–35.

——, 'The Madhyamaka Critique of Epistemology II', *Journal of Indian Philosophy* 9, 1981, pp. 121–60.

——, 'Nāgārjuna as Anti-Realist', *Journal of Indian Philosophy* 16, 1988, pp. 311–25.

——, 'Thinking on Empty: Madhyamaka Anti-Realism and Canons of Rationality', *Rationality in Question: On Eastern and Western Views of Rationality*, eds, Shlomo Biderman and Ben-Ami Scharfstein (Brill, Leiden, 1989), pp. 231–49.

STROUD, Barry, *The Significance of Philosophical Scepticism* (Clarendon Press, Oxford, 1984).

WILLIAMS, Michael, 'Scepticism Without Theory', *Review of Metaphysics* 41, 1988, pp. 547–88.

——, *Unnatural Doubts* (Blackwell, Oxford, 1991).

Chapter 5

Relativism and Beyond: A Tribute to Bimal Matilal

Michael Krausz

Anyone fortunate enough to have had significant contact with Bimal Matilal was enriched by this learned and gentle soul. His absence is deeply felt by his friends and the academic community at large. Just before he passed away, Bimal Matilal and I were embarked on a joint project on relativism. It seems fitting, therefore, that my contribution to this volume in his memory should draw upon some of the material that we were working on together.

I RELATIVISM

I begin by recalling some of Matilal's comments about the relation between relativism and philosophical scepticism, especially as it concerns Indian sources. As regards scepticism, he remarked that if one reads *Outlines of Pyrrhonism* (compiled by Sextus Empiricus, *c.* AD 200) and *Madhyamaka-śāstra* (of Nāgārjuna, *c.* AD 150) one cannot fail to be struck by the similarity of the arguments, methods, and motivations presented in these two classics from two entirely different traditions. Nevertheless, Matilal held that this is a coincidence—a remarkable coincidence to be sure. The ultimate aims of the two authors were different. One was expounding the nature of scepticism while the other was fostering the case for Buddhism.

Further, Matilal remarked that R.G. Bury, translator of Sextus Empiricus, noted that, 'the new insight into strange habits and customs, which was given by the opening up of the East' (Bury, xxxi) was somewhat responsible for the rise of scepticism. The ethnographic argument of Sextus was to persuade the sceptics to suspend 'judgements as to the natural existence of anything good or bad or (in general) fit or unfit to be done' (Sextus, III.235). The overriding question in this context was whether anything is naturally good or naturally evil. Athenian sceptics, such as Sextus Empiricus, were influenced by accounts of different alien cultures, their apparently deviant and strange manners, apparently outrageous and bizarre habits and customs, and seemingly unintelligible beliefs. The bewildered philosophers found in such reports grounds for scepticism, not only about values but also about truth and knowledge about the way the world is.

But scepticism is not entailed by the observation of a bewildering variety of tastes and interests or by a diversity of knowledge-claims and value-claims. A halfway house may be a sort of relativism, according to which truth-claims and value-claims are to be relativized to the culture within which they are made. And a further position is also compatible with the same empirical evidence of cultural diversity, even if it is perhaps unwittingly arrogant. That is absolutism, which is often dogmatic in that it characteristically takes one particular (usually the dominant) culture's knowledge-claims and value-claims to be the closest to being absolutely right, while the alternative culture's claims are taken as irrational or immature or generally inferior. But we should notice that the dogmatism that is characteristically associated with absolutism may well be detached from absolutism *per se* if one couples one's absolutism with fallibilism, as does Karl Popper. That is, one may well couple the belief that there are absolute truths or values with the concession that at any particular time one cannot presume to have captured those truths or values.

The relativist offers something of a compromise between the sceptical and the absolutist positions. The relativist holds—contrary to the sceptic—that there are truths and values, and that one can know them. But—contrary to the absolutist—our knowledge of them is relative to the cultures or frameworks in which they appear. The point of the relativist's position is that there is no culture-neutral or framework-neutral standard according to which one can decide something to be true or false, or good or bad.

Interestingly, the sceptic may be in accord with the absolutist in that he too may require that knowledge involves absolute standards. In turn, while the sceptic denies the existence of such standards and thus denies the possibility of knowledge, the absolutist affirms the existence of such standards and endorses the search for them. In turn, the relativist contents himself with the more modest programme of seeking knowledge according to local standards only.

Protagoras, among the Greeks, is regarded as the first and foremost relativist, and there is a strong tradition of modern commentators who have analysed the Protagorean position and have shown the paradoxicality or self-refutability of the formulation ascribed to him.[1] The kind of case adduced by Protagoras, however, is problematic as regards the position often ascribed to him. For example, that the same water feels cold to one person and not to another person at the same time are observer-dependent descriptions. This description might lead one to conclude that water is both cold and not cold at the same time, and this has led to the view that relativism entails contradictory descriptions of the same thing. Yet it is unclear whether a property of the water as such is here being described in a contradictory way, or whether two phenomenal descriptions are offered with no particular commitment to the claim that a property of the water is being described at all—except perhaps for its dispositional property that it may cause two persons to experience it differently. If Protagoras indeed had these sorts of cases in mind when formulating his relativism (the historical sources are scant) he might not have committed the kind of logical contradiction with which he has so often been charged. Certainly, the Protagorean motto, 'Man is the measure', suggests the contradiction mentioned, but it need not. Of course, Socrates and more recent commentators have rightly shown that the contradictory construal is incoherent or false. But, again, that motto is also consistent with the claim that intentional objects may appear to be contradictory in form but are not, just in case the referents of their attributions are not identical. It is for this reason that we should be careful not to foist an indefensible formulation of relativism on to Protagoras. We should then consider a more fine-grained variety of relativism. Indeed, relativism has been used in no single way in philosophical literature. Consequently, not all versions of relativism would be unseated by a given refutation of a particular version of it.

We may broadly characterize relativism as holding that such (detachable) values as truth, rightness, meaningfulness, appropriateness, goodness, aptness, and so on, are relative to the context in which they are affirmed. Characteristically, such contexts are formulated in terms of contextual frameworks. The range of contextual frameworks may extend from highly localized person-specific or occasion-specific states to that of conceptual schemes, practices, cultures, societies, traditions, points of view, Weltanschauungen, forms of life, Noemata, worlds, linguistic frameworks, modes of discourse, systems of thought, disciplinary matrixes, constellations of absolute presuppositions, perspectives, or the like. Negatively, relativism denies the viability of grounding pertinent claims in *a*historical, *a*cultural, or contextless or absolutist terms. Relativism amounts to a denial of a God's Eye view of the universe, or a denial that anyone will ever have that view. It amounts to a denial of a view from nowhere. The insistence that knowledge should be absolute is to set standards for knowledge unattainably high.

Anti-relativists often think that relativism entails that there can be no common standards to adjudicate competing claims. They fear that relativism allows that anything goes. They fear that relativism precludes the possibility of critical evaluation and opens the way to intellectual nihilism. Such fears may well be justified for an extreme relativism (where contextual frameworks range over specific persons or occasions) but not for a moderate relativism (where such frameworks range over such wider social and temporal states as those mentioned). A more moderate relativism allows that some competing claims may fall under local adjudicating standards. Still, the anti-relativist may respond that wherever the pertinent contextual framework gives out—no matter how widely construed—at such a limit the possibility of adjudicating standards is disallowed. And that is where the threat of nihilism or irrationalism arises again. The relativist may well respond in turn that that reflects the fact that there just *is* a limit to the range of adjudicating standards. Yet local knowledge obtains within a designated context.

Relativism may be formulated for certain sorts of claims and not others. In predicating a value to a particular kind of statement (e.g. truth for cognitive claims, goodness for moral claims, or beauty for aesthetic claims) piecemeal relativism holds that the value (true, good, beautiful) be relative to a pertinent context. Global relativism holds that all statements are valued relative to some context. According to piecemeal relativism one may be anti-relativist with respect to bivalent truth but not with respect to goodness or beauty, or further with respect to reasonableness, rightness, appropriateness, aptness, and so on. Yet further, one might be an aesthetic relativist while denying cognitive and moral relativism. The point is that one might affirm a relativism with respect to some of these values without being committed to relativism with respect to others. No general position of relativism with respect to all such values is necessitated by the affirmation of relativism with respect to one or some of them. Consequently, one needs to ask, in a piecemeal way, with respect to what sort of relativism is one speaking of relativism. In short, there is a range of possible combinations of relativisms, and its alternatives depend upon one's presumed variables and their domains of application.

Relativism generally claims that there is no single scheme by virtue of which the way the (cognitive, moral, or aesthetic) world is is rightly to be described or explained. The relativist opposes the absolutist who holds, say, that positivistic science is the only privileged conceptual scheme to capture the way the world is. The idea of a description of 'the way the world is' is itself contentious as between a relativist and an absolutist. Indeed, the relativist questions whether there is a determinate way in which the world is independent of knowledge about it. All that can be known is restricted to the procedures of knowing within a particular historical or cultural context, and the relativist makes no claim about the way the world is beyond the results of those procedures.

The classical 'self-refuting' argument against relativism runs roughly along the following lines. If relativism is true then the thesis of relativism itself must be relatively true. It would be contradictory to affirm that relativism is true in an absolute sense. But while one could affirm that relativism is true in a relative sense, the counter-argument goes, to say that relativism is only relatively true has no general force. In order for the thesis to have general force it should include itself and should be presumed to be absolutely true. But that, again, would be contradictory.[2]

In response, first, one might observe that there is no reason to rule out of court any non-general thesis of relativism. That is, the claim that the thesis of relativism is a thesis embraced locally does not of itself show that it has no content or is not locally defensible. Local knowledge is knowledge nonetheless.[3] Rather along lines suggested by Nelson Goodman, the aim of justifying local claims, including the thesis of relativism itself, need not be the establishment of a general or a universal or an absolutist claim but may well be in the name of unpacking local understanding.[4]

As well, the self-refuting argument is centrally concerned with the truth of relativism. Yet there is no reason not to reformulate it in other terms. There is nothing in the relativist view that requires that it be characterized in terms of truth rather than, say, reasonableness, appropriateness, aptness, or the like. These latter values are not bivalent as truth is usually understood to be.

Contradiction obtains where 'truth' is held univocal, as when P is said to be true and not-P is said to be true in the same sense of truth. Yet for the relativist, 'truth' systematically equivocates between designated contexts. Under these conditions contradiction does not obtain. Indeed, such equivocation is necessitated by relativism. The thesis of relativism itself should therefore be understood as true, reasonable, appropriate, and so on, depending upon the designated relativized sense of pertinent values.

Consider another strategy to avoid the criticism of self-referential contradiction. That is, reconstrue the would-be competing claims by pluralizing their referents. Put otherwise, P and not-P would be reconstrued as P and not-P'. The referent of P and not-P would not be identical. Such a manoeuvre would avoid the contradiction. For example, the proposition that the shortest distance between two points is a straight line and that it is not appears to be contradictory. But it ceases to appear so when its referents are unpacked in the following way. The shortest distance between two points is a straight line in Euclidean geometry, and the shortest distance between two points is an arc in Riemanian geometry. That which is a 'straight line' varies depending upon the geometry. It seems that under these conditions the respective claims, P and not-P, no longer compete. T.S. Kuhn has deployed such a pluralizing manoeuvre with respect to the 'worlds' of different paradigms in the history of science. In this way, apparent contradictions are deflatable into cases of benign pluralism.

We should distinguish absolutism from universalism, which has often mistakenly been taken to oppose relativism. A relativist need not oppose the anthropological claim that all peoples do exhibit certain common traits or norms. For example, there is consensus amongst anthropologists that there are no cultures in which there is no prohibition against the practice of incest, or the practice of eating the flesh of one's own child, or which licenses aggression against its own members. Other features universal to all cultures include functioning according to some moral order, some symbolic system of communication, some self-referring indexical matrix, and so on. These are universals in a descriptive sense. The relativist resists the thought only that these universal are derivable from some yet further set of injunctions that go beyond the confines of the *de facto* norms and practices of the cultures concerned. Were anthropologists to discover a yet unknown culture that morally sanctioned incest, for example, the relativist would not then claim that the culture *should* not allow such a practice on the strength of some ahistorical or acultural absolutist principles.

The confusion of universalism with absolutism has been encouraged by a confusion between relativity (the existence of 'bizarre habits' as we said) with relativism. What has been called relativity is the philosophically innocuous empirical observation that different practices are exemplified in different cultures. The term relativity might best be replaced by the term particularity. For example, Ogibwe (Canadian) Indians regard it as rude to look directly into the eyes of those with whom they speak.[5] Such claims of particular differences should be contrasted with universality—the equally innocuous empirical claim that all cultures share certain traits, such as those mentioned. In turn, we should distinguish these empirical distinctions between particularity and universality from the philosophically interesting conceptual distinctions between relativism and absolutism. Absolutism holds that there always are cross-contextual standards of evaluation, according to which one might judge that the pertinent practice in one context is better than that in another context. It further holds that such standards are themselves not mandated in context-dependent ways. The particular/universal distinction is innocuous from an evaluative point of view The relative/absolute distinction is not.

Some philosophers have spoken of universal human needs by virtue of which a universal morality may be formulated. Yet the relativist—who is not bound to resist such a programme—would caution that even the formulation of such needs is bound by contextual constraints. For example, while it is usually taken that all peoples value human life as such, and on that basis one might be tempted to formulate a universalist morality, such a valuing of human life is not in fact mandated by one's humanness as such. For those cultures whose members understand the very idea of human life in different terms—for example, Hindus and Christians, in different terms, understand human life to continue after bodily death—it is unclear that there would be a consensus about human needs. For those who value a life after bodily death, in different terms, the range of needs thus identified could well be at variance with each other as well as with the range of needs specified by one who disbelieves that there is human life after bodily death at all. The point is that human needs are not identified in an acultural way. On the contrary, they are dependent upon cultural constructions.

That human needs are not intrinsic but are a function of their place in some cultural matrix can be seen in the case of begging in India. A neophyte in India might see the (near) naked beggar as appealing to the visitor's intrinsic humanness. Although the suffering of many beggars is real and grievous, the very idea that one has come to the 'bottom' of one's humanness as exhibited in his or her nakedness is itself intelligible still within a cultural frame. This frame can be seen in certain practices pursued by beggars themselves who periodically decorticate a limb of an infant to heighten the pity they hope to evoke from more affluent passers-by. The very idea of nakedness construed as a 'bottom' of humanness is itself a cultural construction, itself capable of manipulation. In this sense, there is no truly naked person. There is always a context—indeed even a market context—in which even one's very body is a commodity. Correspondingly, even beggars are choosers.

All this does not preclude the relativist from formulating a morality, perhaps even a universal morality—one in which the inflicting of gratuitous suffering is barred and universal human rights are affirmed—so long as it is understood that even these salient notions are themselves culturally contexted.

II BEYOND

Efforts to secure would-be absolutes are vulnerable to the Buddhist claim of the impermanence of all things. Even the 'absolutes' are impermanent. But the matter does not rest there. The very distinctions between permanent and impermanent and between absolute and relative are human products. Just as they may have been formulated and used in particular contexts for certain purposes, they cease to be useful at other times. Paradoxically, because all things are contexted, the idea of permanence cannot be permanent. But it does not follow that in the end all things are impermanent either, for impermanence too is contexted and it too finally drops out of any fixed constellation of concepts.

Such 'deconstruction' obtains too for the distinction between self and other. Such a Buddhist idea has a significant effect upon one's theorizing about ethnocentrism, which is characteristically formulated precisely in terms of the distinction between self and other. If there is no self—or if one 'overcomes' the self—the question of ethnocentrism drops out. While the relativist embraces ethnocentrism as ineliminable, and while the absolutist seeks to deny it from the start, the Buddhist 'deconstructs' it. For the Buddhist, in the end, the question of ethnocentrism falls by the wayside.[6]

Matilal held that there is an interesting parallel between self-referential arguments against relativism and those against Buddhism, scepticism, and negativism. And he offered an interesting response to them. Matilal noted that a self-referential argument was levelled against the Madhyamika Buddhist doctrine of 'emptiness', *śūnyatā*, by its Naiyāyika critic. It was noted by Nāgārjuna in the opening section of his *Vigrahavyāvartanī* that his Nyāya opponent brought against his *śūnyatā* doctrine the following charge: if every being or everything is *śūnya* (i.e. empty of its own-nature or own-being), then this doctrine itself is empty in the same way. The doctrine refutes itself, for there is at least one non-empty thing. The paradox is similar to the one mentioned by Jean Buridian in his *Sophistimata:* the statement, 'No statement is negative', is itself self-refuting in a similar way. Matilal believed that this sort of argument exposes the paradoxicality of certain construals of relativism, scepticism, and the doctrines of emptiness and negativism.

The relativist might reply, again, that relativism is true in a *relative* sense, and that his critics unfairly foist upon him their own absolutistic conception of truth in order to push him over the precipice of self-contradiction. Matilal suggested that this line of defence is similar to the defence that Nāgārjuna gave of his philosophy of *śūnyatā* or 'emptiness'. Specifically, if the doctrine of emptiness is itself 'empty' then it does no harm to the situation it paradoxically addresses. Rather, the doctrine becomes *non-assertable*; utterance of it becomes only 'empty' sound. It may be that everything is empty but one cannot assert that. Otherwise, one would have to attribute non-emptiness to its assertability-conditions, which one cannot do. Or, one can go on playing with 'empty' sounds according to the existing rules of the game.

This argument, for which Matilal had sympathy, rests upon making a distinction between making an assertion (a speech-act) and truth. The critic says that the proposition 'everything is empty' is, if true, false; then it is false. But the reply is that 'p is, if true, false' is actually expandable as 'p is, if asserted to be true, false'. So construed, it does not follow that p is false. From simply 'p, if true, is false' the conclusion 'p is false' does not strictly follow. According to Matilal's reconstructed argument, it is conceivable that p may be true even when it has never been nor will ever be asserted by anyone. This is in line with the intuition of the metaphysical realist who holds that what is true need not be asserted by somebody. Matilal suggested that assertability is a dispositional property of truth, just as destructibility is a dispositional property of atom bombs. The non-asserted p will not falsify p, much as the presence of atom bombs has not yet destroyed the earth. Matilal recalled that Jean Buridian made a similar point about the statement, 'no statement is negative'. Suppose God had destroyed all negative statements in the world, and had done it in such a way that nobody could make or create a negative statement. Then it would be a fact that no statement is negative, although we could not assert that 'no statement is negative'. Otherwise, we would court a contradiction, an inconsistency, an incoherence, or something of the kind. Matilal suggested that a relativist may adopt a similar stance, according to which 'all knowledge-claims are relative' (call it p), but he cannot take p to be absolute. 'P' has to be relative also, perhaps relative to the ways he has arrived at it, or relative to a hither to-unknown or yet-to-be-uncovered framework.

Matilal suggested that it is quite possible for someone to (mistakenly) predicate 'true simpliciter' to a claim which in fact is true relative to a framework, one which the person may not know. He demonstrated the point when considering the statement, 'My finger touches the button'. This may be true, but only in relation to common-sense realism where perceptually one cannot discover any gap between one's finger and the button. However, physical science holds that there will always be some gap, however small, between any finger and the button. The finger can only be placed in closer and closer proximity to the button. Now suppose that the subject is ignorant of physical science. He may affirm his truth-claim in some absolute sense. He may hold it to be true simpliciter. Yet the subject can be shown (as school students are shown in physics classes) that his statement is true relative to common-sense realism. Hence the initial truth-claim can be said to be relative to some yet-to-be uncovered framework. In this case, unless one learns a bit of physics, one would not be able to distinguish the two frameworks of common-sense and physical science. Consequently, one cannot argue that the truth-claim here is simply true instead of relatively true.

In his argument Matilal deployed the realist's distinction between truth and the assertion of truth. He justified doing so on the grounds that the attack on relativism, scepticism, Buddhism, and negativism typically comes from those who construe absolute truth in a realist way. For him, it is only fair to use the distinction between assertability and truth which absolutists typically endorse.

Matilal left us with the vexing question, then, of what positive intentional attitude—if any—should we take up with respect to the theses of relativism, scepticism, Buddhism, and negativism. Among his other substantial achievements, Bimal Matilal has cleared the ground for us to confront this pivotal issue, for which we may be grateful.[7]

NOTES

1. Maurice Mandelbaum, 'Subjective, Objective, and Conceptual Relativisms', *The Monist* 62:4, 1979, pp. 403–23. In his *The Concept of Mind* (London: Hutchinson & Co., 1963, p. 195), Gilbert Ryle said about relativism that it is 'self-commentary, self-ridicule and self-admonition'.
2. Mandelbaum (ibid.) extends the argument against a relativist conception of truth to one against a relativist conception of evidence.
3. See Clifford Geertz, *Local Knowledge: Further Essays in Interpretive Anthropology*, New York: Basic Books, 1983.
4. See Nelson Goodman, *Ways of Worldmaking*, Indianapolis: Hackett Publishing Co., 1978.
5. See Rupert Ross, *Dancing with a Ghost: Exploring Indian Reality*, Markham, Ontario: Octopus Publishing Group, 1992.
6. In the Buddhist tradition, in the limit, there is a zero–zero relation between a self and a person (Anatma). In contrast, in the Hindu tradition, there is a one–many relation between a self (Atma) and persons. In certain North-west American Indian tribes there is a many–one relation between selves and a person. (See Rom Harré, *Personal Being*, Oxford: Basil Blackwell, 1983; and *Social Being*, Oxford: Basil Blackwell, 1979. See also Roberto Assogioli, *Psychosynthesis*, New York: Pshychosynthesis Research Foundation, 1971.) In the so-called Judeo-Christian tradition, there is a one–one relation between a self and a person.
7. For elaboration of themes discussed here, see Rom Harré and Michael Krausz, *Varieties of Relativism*, Oxford: Basil Blackwell, 1979.

Chapter 6

Whose Experience Validates What for Dharmakīrti?

Richard P. Hayes

1 DHARMAKĪRTI'S THEORY OF KNOWLEDGE: AN OVERVIEW

It is well known that Dharmakīrti followed Diṅnāga in accepting that there are exactly two sources of new knowledge (*pramāṇa*), namely sensation (*pratyakṣa*) and inference (*anumāna*), and that the criteria by which these two means of acquiring knowledge are distinguished were somewhat different for the two philosophers. Diṅnāga claimed that the two sources of knowledge were distinct in that sensation deals only with sensible qualities, which are always particular (*svalakṣaṇa*), whereas inference deals only with intellectible properties, which are always general (*sāmānyalakṣaṇa*). Universals (*jāti*) and other general properties are never directly sensed, according to Diṅnāga. Rather, the intellect ignores subtle differences in sensible particulars and forms the notion that things that are not remarkably different are similar. This idea of similarity or generality is then attributed to what has been sensed so that one who is unwary may come to believe that he has actually sensed the similarity, rather than having imposed a mental construct upon sensation.

1.1 Two Sources of New Knowledge

Dharmakīrti's set of criteria for separating sensation from inference is more complex than Diṅnāga's. While accepting that sensation receives only sensible properties and inference deals only in intellectible properties, Dharmakīrti adds several further considerations. These important additions are spelled out in the first three verses of the chapter on sensation of his *Pramāṇavārttika*, where Dharmakīrti correlates the two types of knowable object to the two levels of truth recognized throughout Buddhist philosophy. Because the sensible particular has the capacity to realize an object (*arthakriyā-śakti*), it is real in the true sense (*paramārthasat*); because the intellectible universal lacks this capacity, it is only conventionally real (*saṃvṛtisat*).

1.2 *Arthakriyā-śakti*

The capacity to realize an object is a key criterion to distinguish between two types of reality, and yet, as Nagatomi,[1] Mikogami[2] and Katsura[3] have all pointed out, Dharmakīrti's notion of *arthakriyā-śakti* is said to be a generic feature of both kinds of knowledge, and not merely a feature that distinguishes sensation from inference. As a criterion of sensation, the term refers to the ability of a particular to serve as the cause of an effect. A particular sensible property causes a representation of itself to occur in the cognition of a sentient being whose sense faculties are stimulated by it. As a criterion of inferential knowledge, on the other hand, the term *arthakriyā-śakti* refers to the ability of an inference to guide a person successfully avoiding something undesirable or attaining something desirable. Thus, for example, if one were to have a desire to bring discontent (*duḥkha*) to an end, one might reflect on the advice of a person who knew how to achieve this goal.

1.3 The Buddha as a Source of Knowledge

A key aspect of Dharmakīrti's epistemological theory to bear in mind is that his overall purpose in writing his epistemological works seems to be to demonstrate that the teachings of the Buddha, and especially the four noble truths, are uniquely suited to guide people to the highest good, nirvāṇa. Other theories of the nature of discontent, its causes and the means of eliminating it, being either demonstrably false or not demonstrably true, are said to be attended by a greater risk of failure than are the basic teachings of the Buddha.

> The Compassionate One is well versed in strategies for getting rid of discontent, because it is a difficult task to explain the goal, which is not within the range of the senses, and the means of attaining it.[4]

Here Dharmakīrti clearly states his view that the goal of achieving nirvāṇa, which is the same as the cessation of the causes of discontent, is not available to the senses. The knowledge necessary to achieve nirvāṇa, therefore, is not considered to be a purely empirical matter. Rather, it requires the application of the intellectual faculty and is facilitated by the guidance of traditional Buddhist teachings (*āgama*).

> Examining things through both reasoning and the traditional teachings, one [who desires nirvana] inquires into the cause of discontent through the particularities of discontent, and inquires also into the impermanence that characterizes it.[5]

As Vasubandhu had done before him, Dharmakīrti spares no effort in showing that the Buddhist view of the human condition is uniquely capable of leading to nirvāṇa, since Buddhism is alone in recognizing that there is no enduring self (*ātman*), and that a false belief in such a self is the root delusion from which spring all unhealthy mental states, such as desire and aversion, from which in turn arise all counter-productive and harmful verbal and bodily actions.

> The realization that there is no self, which realization is incompatible with the cause, destroys it. The virtues and shortcomings of that cause become very clear to one who practices many methods repeatedly for a long time. And because of that lucidity of mind, the impression left by the cause is left behind.[6]

2 TENSIONS WITHIN DHARMAKĪRTI'S THEORY

In what has been discussed up to this point, there seems to be some inconsistency. On the one hand, it seems that it is the experience of the senses that grasps that which is true in the highest sense (*paramārtha*) of the word. On the other hand, it seems that the greatest good (*paramārtha*) is beyond the range of the senses and that one can be directed towards it only through sound reasoning. In the sections that follow, let me try to expand the problem inherent in Dharmakīrti's view of pure sensation (*pratyakṣa*), whereby it is portrayed on the one hand as the only means of acquiring knowledge of ultimate reality and is portrayed on the other hand as too weak to arrive at the knowledge necessary to enable one to achieve nirvāṇa, the greatest good.

2.1 The Limitations of Pure Sensation

Pure sensation, as described by Dharmakīrti, has two features that reduce its effectiveness as a means of acquiring knowledge of the four noble truths, which is supposed to be important in the attainment of nirvāṇa. The first of these features may be seen as intrinsic in that it is part of pure sensation by definition, while the second may be seen as an extrinsic feature that arises because of practical considerations.

An intrinsic feature of pure sensation, as expressed in *PV* 2.2, is that its subject matter is always a particular, which cannot be the subject matter of conceptual thinking and which is therefore inexpressible through language. But the content of the Buddha's awakening is not a particularity at all. Gautama became a Buddha by discovering the principle of dependent origination (*pratītya samutpāda*). An alternative account is that he became a Buddha by mastering the four levels of meditation (*dhyāna*) and acquiring three types of extraordinary knowledge. A third account is that he became a Buddha by understanding the four noble truths. It is the third type of account that is stressed most often by Dharmakīrti, but it may be worth examining each of the three patterns from the point of view of Buddhist epistemologists under the influence of Diṅnāga.

For philosophical purposes the most important version of dependent origination is the one that states the basic principle of causation in the words: 'This [effect] comes into being when that [cause] is present. This arises owing to the arising of that. This does not arise when that is absent. This ceases owing to the cessation of that.' The philosophical importance of this formula of dependent origination resides in part in the fact that it is plainly reflected in the definitions of evidence (*hetu*) offered by Diṅnāga and Dharmakīrti. Dharmakīrti says in the *Nyāyabindu*: 'an inferential sign has three characteristics: it must be known to be present with what is to be inferred, present only with what is like the subject and entirely absent in what is unlike the subject'.[7] According to Dharmakīrti, one thing X can serve as a sign of a second thing Y only if there is a natural relation (*svabhāva-pratibandha*) between X and Y, and X can be said to be naturally related to Y only if X is present when Y is present and absent when Y is absent. Speaking from a metaphysical point of view, Dharmakīrti says that there are only two situations in which this kind of natural relation is found: 1) when Y is a cause of X, and 2) when X and Y have exactly the same set of causes. The stock example of the first situation is that smoke can serve as a sign of fire only because fire is a cause of smoke, which means that smoke is present when fire is present and absent when fire is absent. The stock example of the second situation is that the fact that something is an oak can serve as a sign that it is a tree; because the set of causes that make the property of being an oak arise are exactly the set of causes that make the property of being a tree arise; in other words, it takes no more and no less to make a given particular thing a tree than it takes to make it an oak. The principle of causality stands behind Dharmakīrti's theory of inference to the same extent that it stands behind the Buddha's notion of how one attains freedom from distress. And so if the Buddha could say that to see dependent origination is tantamount to seeing the Buddha himself, Dharmakīrti would be entitled to say that to know the theory of inference is also tantamount to knowing the Buddha.

Now the question can be asked: what kind of knowledge is involved in seeing dependent origination? Is it sensation or inference? Given that the subject matter of a sensation can be only that which exists in the immediate present and that this type of cognition is said to be completely free of any admixture of recollections of the past or anticipations of the future, the knowledge described in the short formula of dependent origination cannot be sensation. For in order to know that X is present when Y is present and absent when Y is absent requires at least two moments, one of shared presence and a second of shared absence. Furthermore, one must retain the knowledge of one of these moments during the second moment, so that the second moment must involve some degree of thought on top of what is being immediately sensed. The cognition of even one instance of dependent origination, being the apprehension of a temporal process, is similar to the discernment of a melody, which can never be grasped if one is aware only of the note that is being played in the present instant. The full grasp of dependent origination is, however, much more than the apprehension of a single temporal process. It is really a generalization that is supposed to be true of all sentient beings at all times. The Buddha's first sermon is portrayed as his proclamation of a discovery that all desire anywhere eventually results in some degree of frustration of some kind. It is, in other words, a piece of knowledge that has all the characteristics that Dharmakīrti attributes to inference.

A second pattern of text that describes the Buddha's awakening relates that he entered into increasingly abstracted states of meditation. In the first state, it is said, the Buddha became aloof from the pleasures of the senses and entered into a state of elated intellectual reflection; the reflection stopped as he entered the second state, which was one of rapture and joy; at the third stage, rapture disappeared, leaving only joy; and at the fourth stage, joy disappeared and was replaced by equipoise, an emotionally balanced state free of both pleasure and pain. While in this fourth state, say the texts, the Buddha began to recall hundreds of thousands of his previous lives, including such details as his name, clan, diet and lifespan. Then he witnessed the dying and rebirth of all kinds of sentient beings, and he saw what kinds of conduct resulted in what kinds of birth. Finally, he 'directed the mind to knowledge of' the four noble truths, that is, the nature of distress, its cause, the fact that removing the cause would eliminate the effect, and the method of removing the cause.

Once again it can be asked whether it is sensation or inference that is involved in these three types of superior knowledge. The first type of knowledge is depicted as nothing more than recalling past events, and since this is grasping what has already been grasped (*gṛhītagrahaṇa*) it would not be regarded by Dharmakīrti as a case of *pramāṇa*, or acquiring new knowledge. The second superior knowledge, which consists in witnessing the deaths and rebirths of all kinds of sentient beings, is evidently the perception of a process that takes place over time. Moreover, observation of this process is said to have lead the Buddha to conclude that *in general,* beings who perform good deeds achieve pleasant rebirths, while beings who commit ugly deeds achieve unpleasant rebirths. As in the apprehension of dependent origination, this kind of knowledge would therefore have to be classed as a kind of inductive reasoning (*anumāna*). This leaves the third form of superior knowledge, which consists in grasping the four noble truths.

What kind of knowing is involved in grasping the four truths? As in all the cases discussed above, it is clear that the grasping of the four noble truths involves considerably more than the sort of pure sensation that Dharmakīrti says deals only with particulars and never with universals. Indeed, the four truths are typically presented as merely one of the many frameworks within which the general notion of causation may be discussed.

2.1.1 Bodhi: Arthakriyā-śakti and Paramārtha

If the Buddha's accounts of his awakening (*bodhi*) are to serve as a source of knowledge (*pramāṇa*) for others, then there is a sense in which those accounts are to be regarded as having *arthakriyā-śakti* and being *paramārtha-sat.* As we have seen, both of these terms are ambiguous, but it is possible to disambiguate them. The discussion in the preceding sections has shown that regardless which traditional report one follows, the content of the Buddha's awakening must be understood within Dharmakīrti's system as having the form of generalities rather than of particularities. Therefore, if one follows the definitions given by Diṅnāga and accepted with certain modifications by Dharmakīrti, the type of knowledge involved in the Buddha's becoming a Buddha was inference rather than sensation, that is, it was *anumāna* rather than *pratyakṣa.* Knowing this enables us to decide among the alternative possible meanings within each set of ambiguities. When applied to the four noble truths and so forth the term *arthakriyā* must be understood in the sense of realizing a goal rather than in the sense of causing a specific effect. And the term *paramārtha* cannot be understood in the sense of an ultimately real object in contrast to an object accepted as real by human consensus; rather, it must be understood in the sense of pertaining to the highest good, namely, nirvāṇa, in contrast with what is popularly (*laukika*) regarded as good in quotidian life. Indeed, Dharmakīrti is explicit in saying that nirvāṇa must ultimately be regarded as a fiction. Nirvāṇa is commonly understood as the cessation of rebirth, but rebirth itself is an idea that makes sense only if one imposes the notion of a unified self upon a group of discrete properties. But once this complex fiction of a self undergoing a series of lives, deaths and rebirths is given up, then so is the fiction that this elaborate process comes to an end.

> As long as one does not give up favouring oneself, one imagines oneself a victim of affliction and goes on suffering, and one does not live as a happy person. Even though there is no one who achieves liberation, it takes an effort to give up this false imagining.[8]

This suggests that nirvāṇa is not regarded as an ultimately real thing, since it is nothing more than the absence of the false belief in a self, and an absence is not a thing at all. That notwithstanding, nirvāṇa can still be regarded as the highest good, since nothing is better than being free of the delusion that serves as the root cause of all discontent.

2.2 The Problem of Verification

It has been shown that the content of the Buddha's awakening would be classed in Dharmakīrti's system as inferential knowledge that worked to the Buddha's own benefit (*svārtha*). But his teaching, which was based on the insights he gained through his awakening, was for the benefit of others (*parārtha*). And yet other people do not automatically benefit just by hearing this teaching; rather, for the teaching to be of any benefit to those who hear them, it must be confirmed or verified. What remains to be discussed is how this verification is to be accomplished.

As we have seen in some of the passages already cited from Dharmakīrti's work, the task of becoming free of discontent is one that takes an effort (*yatna*[9]) and the constant practise of methods (*upāyābhyāsa*[10]). This effort requires, among other things, thinking. And this thinking can be in itself a means of acquiring new knowledge.

A subjective cognition is not regarded as a source of knowledge, because it consists in grasping what has already been grasped. Thought is a source of knowledge, because it is the principal source of action upon things that one should avoid and things that one should welcome.[11]

It is at this point that we encounter a new problem. The problem now arising is that while repeated effort in thinking may lead to a correct understanding of things, it does not necessarily do so. In fact, if one begins with a false belief and repeats it constantly, the eventual result may be an almost unassailable delusion, one in which unreal things are experienced as vividly as if they were actually present to the senses.

Those who are mad with desire, pain or fear and those who are tormented by dreams of thieves and so forth see even things that are not present as if they were present before them.[12]

It is not the case that thinking is based passively upon what one has experienced, says Dharmakīrti, but rather, how one experiences things is affected by one's patterns of thinking and one's overall mentality:

For experience generates convictions of certainty according to the repetition of thoughts. For example, even though there is no difference in the seeing of visible properties, there are ideas of a corpse, an object of desire and something to be eaten.[13]

An ascetic, who has repeatedly practised the exercise of gazing at corpses until he can visualize them at will, will automatically perceive an attractive woman as a corpse; seeing her in this way protects him against lustful thoughts that might otherwise arise. A lecher, on the other hand, will see exactly the same visible properties that the ascetic saw, but he will perceive them as sexually exciting. And a dog, seeing exactly the same set of visible properties, will not be sexually aroused by them, for he is more likely to perceive them as a potential meal.

The point that Dharmakīrti intends to make through the example that the same woman makes different impressions on the ascetic, the lecher and the dog is evidently that one tends to form ideas about what one sees according to ideas that one already has in mind as a result of having immediate goals. The example, however, also invites a further question: can any of these ideas be considered more accurate or more in conformity with reality than the other two? If one were to apply only the criterion of whether the ideas have the capacity to achieve a goal, it would appear that none of these perceptions is inaccurate, since each has the potential of fulfilling the goal of the perceiver; the ascetic successfully fulfils his goal of protecting his chastity, the lecher his of being sexually excited, and the dog his of finding nutritious victuals.

In the various kinds of perception discussed by Dharmakīrti, we find two instances of perception in which someone interprets something that is not present to the senses as vividly as if it were actually present. One of these instances, which we have already discussed, is that of the yogin who visualizes an object through repeated practise. The other is that of a person who is so stricken by a fear of intruders that he misperceives a perfectly innocent person (or a harmless noise) as an aggressive intruder. Both of these experiences can be regarded as false cognitions or misperceptions, especially if the only criterion of accurate perception is that what one believes to be present to the senses actually is present to the senses. Nevertheless, Dharmakīrti regards the yogin's perception as a genuine source of knowledge (*pramāṇa*), while he regards the fearful person's alarming misperception of harmless sights and sounds as a bogus source of knowledge (*pramāṇābhāsa*). So now it must be asked: What differentiates the panic-stricken person's perception of a harmless person as an aggressor from an ascetic's perception of a living woman as a corpse? An answer to this may emerge by reviewing several different types of cognition that Dharmakīrti discusses.

As we have seen above, Dharmakīrti recognizes two radically different kinds of cognition: those that are purely sensory in that they involve no judgement, and those that are intellectual in that conceptual judgement plays a role. All sensations are caused by the functioning of physical senses. This is the case even when yogins 'see' things that are not really there, such as when they visualize living people as corpses and so forth; these acts of visualization are not regarded as the projection of mental images, but as a kind of sensation in which the organs of sense are somehow operating. Dharmakīrti takes care to distinguish these yogic visualizations from what we might call hallucinations. Hallucinations, unlike yogin visualization, are purely the product of the internal sense organ, located in the heart. Hallucinations involve a projection of an internal image into consciousness, along with a failure to be able to distinguish imagination from sensation. Therefore, a hallucination is at the root an intellectual error.

In addition to intellectual errors, there are, according to Dharmakīrti, also purely sensory errors, in which the judgement does not play a role at all. These might be called false sensations. False sensations, unlike hallucinations and dreams, do involve the senses. Moreover, Dharmakīrti insists that the errors that occur take place within the senses themselves, and not in the intellect. When one sees a rapidly twirling torch, one actually sees a circle of fire, even though there is in fact no circle to be seen. If the torch is twirling rapidly enough, one cannot help seeing the circle of fire, even if one knows intellectually that in fact there is not a continuous circle of fire. In this case, the intellect is required to correct the errors of the senses. Similarly, when one sees an enduring physical body or a continuing psychological self instead of a series of vanishing moments, this sensory illusion can be corrected only by the intellect, and this correction can occur only if the intellect is functioning within the constraints of sound reasoning. Presumably, what makes the yogin's superimposed vision of a corpse accurate for Dharmakīrti is the fact that the feelings of disgust and loathing that it produces are shown by reason, if not by the senses, to be just the sorts of feelings that it is suitable for a man to have towards a superficially attractive woman; the vision of the corpse, in other words, conforms to what reason shows an apparently attractive person's true nature to be. If this analysis is correct, it would seem to be in conflict with the claims that it is the experience of the senses that grasps the greatest good and that the greatest good is beyond the grasp of reason.

3 SOURCE OF THE TENSIONS

As we saw in section 2.1 above, Dharmakīrti's criteria for distinguishing pure sensation from judgemental conceptualizing is more multifaceted than Diṅnāga's. What remains to be seen in this final section is why Dharmakīrti felt it necessary to introduce these complexities, which, if the above analysis is correct, led him into apparent contradictions. While a full solution to this problem is beyond the scope of this paper, let me offer at least a sketch of the solution.

Diṅnāga's theory of cognition, as we saw above, posited a radical distinction between two kinds of cognition. Sensation provides knowledge of particular sensible properties, while reason provides knowledge only of intellectible properties that are derived from sensible properties by ignoring subtle differences among sensibilia. This theory suggests that intellectible properties such as genera are not only derivative but also to some extent distorted, in that they involve some loss of information; a bulky male wrestler and a trim female gymnast may be regarded as belonging to the same genus (*jāti*) only if all the sensible differences between them are factored out and discarded. This means that a general concept, for Diṅnāga, is always less rich in information than any given particular to which the concept might be applicable. Given that concepts are therefore always in some sense weaker than the particulars to which they apply, it is not easy to see how a piece of reasoning could ever stand as a corrective to a raw experience. Diṅnāga's radical division of cognitions into exactly two mutually exclusive classes would seem to favour an epistemological stance of radical empiricism, in which each moment of sensation validates itself and remains unassailable and ultimately incorrigible. Reason ultimately lacks the power to provide any new knowledge; at best, it can eliminate some interpretations of the sensible world that are logically contradictory to other interpretations. Moreover, reason lacks the force necessary to overturn the immediate intuitions of raw experience. If the perceiving mind *feels* like an enduring self witnessing a world of enduring substances that last for more than a moment, then there is no reason to doubt that feeling. The fact that an experience simply feels as if it contravenes Buddhist doctrine is insufficient reason to reject the experience; if anything, it would be a reason to doubt the doctrine.

The doctrine of radical empiricism may have its virtues, but it is clear that the virtue of being easily reconciled with classical Buddhist doctrine is not among them. Each of the three classical formulations of the Buddha's awakening, as was shown above in section 3.1, involves the use of the intellect to arrive at a correct interpretation of the world of experience. In other words, if one is determined to defend the view that the doctrines of Buddhism are something more than a diluted and distorted account of an experience that was, in the final analysis, unique to the Buddha and utterly private and therefore unavailable to anyone else, then one must try to show why reason has the power to correct some of the false views that arise from poorly interpreted experience. By trying to construct a system of epistemology that placed an emphasis on the unique value of Buddhist doctrine, while also trying to maintain the appearance that he was offering a commentary on the works of Diṅnāga, Dharmakīrti created a philosophical system that was at best convoluted and at worst self-contradictory.

The final answer to the question 'Whose experience validates what for Dharmakīrti?' appears to be this: the experience of the person whose interpretation of his experience is consistent with the basic doctrines of Buddhism validates exactly those doctrines. Thus, insofar as one's experiences confirm one's confidence in the Four Noble Truths, the doctrine of anātman, and the doctrines of karman and rebirth, then one is, by Dharmakīrti's standard, coming closer to the truth. While giving every appearance of trying to defend the doctrines of Buddhism by an appeal to experience and reason alone, independent of appeal to authority, Dharmakīrti ultimately makes a disappointing return to dogmatism.

NOTES

1. Masatoshi Nagatomi, 'Arthakriyā', *Adyar Library Bulletin* 31–2, 1967–8.
2. E. Mikogami, 'Some Remarks on the Concept of Arthakriyā', *Journal of Indian Philosophy* 7, 1979.
3. Shoryu Katsura, 'Dharmakīrti's Theory of Truth', *Journal of Indian Philosophy* 12, 1984.
4. *PV* 1.134.
5. *PV* 1.135.
6. *PV* 1.138–9.
7. *Nyāyabindu* 2.5.
8. *PV* 1.193cd–194.
9. *PV* 1.194.

10. *PV* 1.140.
11. *PV* 1.5.
12. *PV* 2.182.
13. *PVSV* under verse 58.

REFERENCES

KATSURA, Shoryu, 'Dharmakīrti's Theory of Truth', *Journal of Indian Philosophy* 12, 1984, pp. 215–35.

MIKOGAMI, E., 'Some Remarks on the Concept of Arthakriyā', *Journal of Indian Philosophy* 7, 1979, pp. 79–94.

NAGATOMI, Masatoshi, 'Arthakriyā', *Adyar Library Bulletin* 31–2, 1967–8, pp. 52–72.

Chapter 7

Seeing Daffodils, Seeing as Daffodils and Seeing Things Called 'Daffodils'

Arindam Chakrabarti

'Describe the aroma of Coffee,—Why can't it be done? Do we lack the words? . . .

P.I. Wittgenstein
§ *610*

When as a result of sense–object contact, someone has a perceptual awareness, do words or conceptual modes of presentation* have any necessary role to play in such awareness? This is a question which bothered Vātsyāyana, Diṅnāga, Udayana, Bhartṛhari, Frege and Husserl. More recently it has bothered Gareth Evans, Bimal Matilal and Michael Dummett.

The question concerns the exact content of a particular type of awareness which is called 'perception' (*pratyakṣa*). And it goes without saying that lots of such perceptions go without any *saying*. Yet, the only way to *talk* about their contents is to use names or descriptions of the items figuring as objects of the awareness—to use Matilal's Naiyāyika English—to specify linguistically what structures the experience 'bathes in' or what objects 'float in' the perception. To interpret the troublesome word 'unverbalizable' (*avyapadeśya*) in Gautama's definition of perception Vātsyāyana makes the following subtle point: Concept-words may be needed to articulate the content of each perception *from outside* when we ascribe the perception to someone. But that does not mean that the perceiver has *used* the concept or the word in identifying the object *from inside* the experience. The Sanskrit word '*iti*' (as) which he uses here to draw this distinction between seeing a colour and seeing it *as* that colour, is also the quotational device. Imagine an Indian villager who has not read Wordsworth being presented with a bunch of daffodils. He will see a bunch of daffodils but not *as* a bunch of daffodils. We can use the word to describe what he has seen without implying that he has any inkling of the word or the concept it articulates. Just because words and classificatory notions are needed for perception-*ascription* (*abhilāpa*) it does not follow that they are submerged in or required for *perception.*

Buddhists like Diṅnāga would only consider pre-predicative pure wordless sensations to be proper perceptual knowledge. Any predication, conceptualization or verbalization not only opens up the *possibility* of error, for them it necessarily distorts the pure individuality of the perceived particular. To identify is to classify, and for such arch-nominalists, to classify is to falsify.

So, strictly speaking, although inference and verbalized perception are admitted by the Buddhists as forms of knowledge on a pragmatic transactional level, knowledge in the absolute sense is confined to ineffable simple apprehension.

Bhartṛhari, the philosopher of grammar—in spite of his alleged Buddhist affiliations—went to the opposite extreme: No awareness is untouched by language. Implicitly or explicitly even babies in arms or animals cannot have experience which is not at least potentially perforated with words. To summarize Matilal's recent reformulation, two major arguments that the pan-linguist uses to prove his contention go like this:

*A*1. No intentionality without linguistic structure.
*A*2. No consciousness without intentionality.

Therefore, no consciousness without linguistic structure.

Furthermore,

*B*1. Every awareness illuminates something outside itself.
*B*2. Illumination consists in articulation of structure of the object.
*B*3. No articulation of structure is possible without a speech-like grid.

Therefore, awareness consists in use of speech (implicitly or explicitly).[1]

Notice that *B*1 will be right away rejected by Subjective Idealists like Diṅnāga. Also the Buddhist account of intentionality, I suspect, will have to be reductive such that—to use Frege's example—within the context of the idea of a heavy shell the distinction between *parts* and *properties* will be smudged off. Heaviness is taken (by the realist) to be a property of the shell. But once the identities, heaviness = *mental image* of heaviness (or, difference from non-heavy things), shell = *idea* of shell—are established the one mental entity becomes *part* of another and the exemplification relation is just eliminated. If 'aboutness' is construed in such intimately internalistic terms, the need for linguistic or predicative structure will melt away from the core of experience-episodes. As the object-ward arrow turns back reflexively to the inherent form of the awareness itself, awareness is liberated from all obligations to both language and the world!

Standard Nyāya epistemology stands somewhat in between Bhartṛhari and Diṅnāga on this issue. Unlike Buddhists, the Naiyāyika believes not only that perception most often *does* have predicative structure whether it is actually verbalized or not—but that proper knowledgehood attaches only to such predicative/judgemental perception. The argument for this is straightforward:

> Nothing but a *true* awareness can be knowledge. An awareness is *true* just in case the predicated property actually belongs to the item which figures as the subject. So, truth presupposes predicative structure. Hence, perception can be *knowledge* only if it has a predicative structure.

To put the point a bit differently: Only that which runs the risk of error can count as knowledge. This basic assumption of Nyāya which Potter calls 'fallibilism'—goes against the grain of the epistemologically over-ambitious Buddhist. Diṅnāga not only requires perception to be non-erroneous (*abhrānta*) but would ideally regard only that piece of awareness as *knowledge* which *could not* be erroneous. Insofar as unmistakenness or 'non-promiscuousness (*avyabhicāritva*) is part of even the Nyāya definition of perception, we cannot *claim* to have 'seen' something unless we have seen it correctly. But for the actual *truth* of the experience the external world has to co-operate. Given such externalism, perception for Nyāya, needs to be unmista*ken* but not unmista*kable*. And as Socrates saw in the *Theaetetus*, if perception has a simple particular content, we either see it or do not see it, there is no possibility of seeing it *falsely*. It takes *two* to make room for error and that is why truth also requires the duality of the subject and the predicate or of the object and the concept.

Before it was given a taxonomic twist (viz. that the definitional *Nyāyasūtra* 1.1.4 also *classifies* perception into two types—qualificatory and non-qualificatory) the occurrence of the word 'unverbalizable' (*avyapadeśya*) in that aphorism was interpreted by Vātsyāyana as applicable to *all* perceptions.

Matilal underscores Vātsyāyana's ingenuity of interpretation in the following words:

> In the sense-perception of a child (who has not yet learned the words to designate things) words do not play any significant role. When a person learns the name of a thing and perceives that thing, he says that it is called such-and-such. But as far as his awareness of that object is concerned it does not differ very much from the case of a child's perception. This shows that designation by name is not an essential factor in our perceptual process or cognitive act . . . the distinction, made for the first time, between conception and its phonological realization, may be attributed to Vātsyāyana. . . . A child may be said to have concepts before he acquires the corresponding words.[2]

However the matter is not so easily settled. Like a typical creative commentator after paying due tribute to Vātsyāyana, Matilal proceeds to develop points left unexplained by him. In his later work, *Perception*,[3] he formulates the problem succinctly in terms of the following principles:

*P*1: For all *x* there is an *f* such that, if *x* is presented to awareness then *x* is presented under the qualifier *f*.

*P*2: For all *x*, and for all *f*, if *x* is presented under the qualifier *f* then *prior to such presentation*, *f* must first be presented separately.

New Nyāya seems committed to both the principles. Notice that it is only a prior awareness of *f* which is presupposed and not a prior awareness of *x* on its own before 'x as f' can be experienced. If I hear a certain utterance as qualified by a southern accent I must have a prior awareness of what it is for an accent to be southern, but I surely do not need to have a *prior* awareness of that particular pronunciation-token. Still, this sounds like a strange doctrine to many Western ears because, somewhat like in Buddhist thought, *particulars* have been often taken to be *intuited* or *given* prior to propositional awareness in Western thought, whereas immediate bare acquaintance with universal features (like southernness, flowerness or waterness) has rarely been proposed. Also, should we find *P*1 to be inconsistent with the received Nyāya wisdom that the form taken by pre-judgemental perception of a cup is best verbally expressed as: experience of cup *and* cupness—a *conjunctive* as against predicative piece of cognition. Isn't the cup—the bare object identified once without any mode of presentation at all? Remember that even thisness can serve as a mode of presentation. Nyāya has no qualms about indexical or even uniquely exemplifiable qualifiers, whereas perceiving an altogether unqualified particular is hardly ever admitted as possible. In being presented as an element of *cup-and-cupness* the cup does not *wear* a cloak but it is *with* its cloak.[4]

Now if we take these two principles equally seriously we seem to get involved in a vicious regress. If perceiving any old x causally requires perceiving it under a mode f and if the mode has to be perceived *prior* to this predicative perception then, substituting x with f, even the mode needs to be perceived under a further mode g—and so on.

Bhartṛhari's solution—which Matilal calls 'Nativism' was that f need not always be presented in experience because the basic word-concepts are beginninglessly and innately present in human consciousness. Even if I have not learnt the name 'gold', the simple universal which is to be its meaning is entrenched in my consciousness such that even on the initial presentation of a sample piece of that metal I can apply the concept to it. Such a Cartesian solution was not to a Naiyāyika's taste. So the Naiyāyika lays down a further caveat to P_2:

E: When I know an ultimate simple universal (a natural kind or an unbreakable titular property), I *may* know it without any mode of presentation.

Doesn't this exception-clause look like just an *ad hoc* measure to stop the regress? Our understanding of speech is necessarily qualificative. Indeed, according to received Nyāya wisdom, it is *doubly* qualificative. Someone's statement: 'Bush is stubborn' can be understood only if the understander represents (committedly or otherwise) Bush under stubbornness, while stubbornness itself is brought under some mode of presentation like: *being a character-trait.* Thus, if simple universals are fit to be designated by words—which they obviously are—then they have to be presented under a further mode of presentation. The exception-clause seems out of place!

At this point we seem to face a real dilemma: either we admit that the universal barely presented is ineffable or innate to consciousness so that each effable experience contains an unsayable element in it or we admit that even the qualifying universals can and may figure as meanings of 'nessy' words (e.g. 'goldness', 'cupness', 'stubbornness') and thus are presentable under further cloaks—thereby keeping open the possibility of a regress.

For very deep metaphysical reasons, Dummett has always been interested in defending a variety of unverbalizable knowledge which is somewhere in between propositional *knowledge that* and *practical knowledge how.*[5] He thinks that our knowledge of a language is this kind of implicit knowledge and that is part of the reason why a truth-conditional theory cannot do full justice to what is known when a language is known. Now, Frege as a realist is committed to *all* our knowledge having a thought content which even if not actually put into a sentence, *can* be so expressed because *in the order of explanation* thoughts are nothing but meanings of sentences. Although we can never dig out any notion of a Fregean thought from Nyāya theory of epistemic contents, there too one finds the same *realistic* tendency to assume that everything that is knowable is sayable. The Buddhist anti-realist, on the other hand, as we have already noted, typically considers the paradigm cases of knowledge to be beyond words. (Of course, I am not suggesting that the Buddhist's ineffable sensation has any similarity with Dummett's implicit knowledge which is partly an *ability*—a concept Buddhists would find hard to stomach—except that they are both unverbalizable.) Very recently[6] Dummett looks back at Husserl to excavate the source of what he calls Analytic Philosophy's Cardinal Priority thesis: viz. *language is prior to thought—in the order of analysis.*

He quotes a passage from Husserl which sounds like an outright denial of the possibility of non-qualitative perception. Husserl almost seems to repeat the Naiyāyika's Principle *P*1—that whatever is experienced must be experienced as falling under a *type*. Dummett, then, locates in Frege the Nyāya-like principle that an object (reference) can be given to us only under this or that mode of presentation (=sense). But he thinks this principle eventually clashes with Frege's explicit doctrine that grasping a thought denuded of *all* its verbal clothing *is* a logical possibility. Matilal notices a similar tension in Udayana. On the one hand Udayana, he thinks, was so much influenced by Bhartṛhari that he had to admit that any experience worthy of being called knowledge (*pramā*) has to be word-impregnated because otherwise it would not be up for public epistemic appraisal. On the other hand, there was the principle of prior acquaintance with the qualifier which—to stop the regress—had to be admitted as non-predicative and prelinguistic.

After unearthing this tension Matilal comes very close to suggesting that we forget about *Nirvikalpaka*—or prior bare acquaintance with universals—when we are doing epistemology. Even so-called inchoate wordless experience of an infant, he remarks, 'May lack full-blown concepts but then it is only unconceptualized, not unconceptualizable.'[7]

We may now wonder on what *basis* does Matilal settle the issue somewhat in favour of Bhartṛhari and Frege, and against the knowledgehood of prelinguistic perception? Doesn't he himself remind us that even Bhartṛhari had to soften his original pan-linguistic claim a little bit when he gives the example of the man running (barefoot) on grass having tactile sensation of the individual blades of grass without any remote possibility of verbalizing such sensory awareness? Just commenting that such *in*communicable experience could not be analysable in epistemology *because* they are unverbalized does not convince us that they are not perceptions *at all*. True, we cannot *tell* what it is that we are aware of when we have such inchoate cognitions, as animals and infants cannot. But unnameability of the object does not entail non-existence of any object. So we cannot argue that inexpressible awareness is no awareness because it *lacks* object-directedness. Yet that is precisely what Matilal concludes. The so-called bare sensory grasp of an infant, he decides, does not amount to awareness. Perhaps, Matilal was trying to do justice to the explanatory priority of fully conceptualized predicative perception because, as Dummett comments:

> Most unverbalized thought on the part of adult human beings is related to fully verbalized thought as a sketch to a finished picture; it can therefore be explained only in terms of that of which it is a sketch.[8]

Perhaps, further, Matilal is not totally *denying* that languageless experience can and does happen. He could be justifying Udayana's downgrading it as non-knowledge by pointing out that since it cannot be put into words we better keep silent about it. Bhartṛhari, of course, would have retorted that even the unsayable gets a place in his wordy world by being called 'unsayable'. After all '*avācya*' (unverbalizable) *is* a word, isn't it?

I have heard it occasionally being suggested that Professor Matilal merely interpreted Indian philosophy but did not change it. I think the above recounting of his guarded but radical reasoning to the effect that Nyāya epistemology looks neater if we prune off non-qualificative awareness altogether—shows that he *did* change philosophy in the same sense as a Raghunātha Śiromaṇi or a Russell did. And of course to that extent he ran the risk of refutation.

NOTES

* Since writing this paper, I have revised my views quite substantially on this issue. I now think that it is possible and natural to possess and apply conceptual modes of presentation without using or even knowing the words for them. So this 'or' in the expression 'words or conceptual mode of presentation' cannot be so glibly glossed over. My new understanding, which enables me to see Matilal's point in a more sympathetic light, is argued for in a forthcoming paper, 'Sense-perception, Concept-possession and Knowledge of a Language'.

1. Bimal K. Matilal, *The Word and the World*, Delhi: OUP, 1990.
2. *Epistemology, Logic and Grammar in Indian Philosophical Analysis*, Mouton, 1971, pp. 25–6.
3. *Perception: An Essay on Classical Indian Theories of Knowledge*, Oxford: Clarendon Press, 1986, pp. 344–54.

4. See Śaṃkara Miśra's, 'Bhedaratnam'—Discussion of nirvikalpaka Pratyakṣa.
5. *The Logical Basis of Metaphysics*, Harvard University Press, 1991, pp. 93–7.
6. See his paper, 'Thought and Perception', in Bell and Cooper, eds, *The Analytic Tradition*, Blackwell, 1990.
7. *Perception.*
8. See his paper, 'The Relative Priority of Thought and Language', in *Frege and Other Philosophers*, p. 324.

Chapter 8

Negative Facts and Knowledge of Negative Facts*

Brendan S. Gillon

INTRODUCTION

Negative facts have perplexed Western philosophers ever since the time of Plato.[1] But the philosophers of Europe and America have not been the only philosophers to have been perplexed by them; classical Indian philosophers too have pondered their nature. My interest here is to explore how the reflections of these classical Indian philosophers, transposed into the contemporary philosophical idiom, might enrich current metaphysical thinking about negative facts; and what I shall conclude is that at least one of these philosophers has a view of negative facts and knowledge of them, which, when so transposed, is very plausible indeed.

I shall begin by asking the fundamental ontological question of whether or not negative facts exist and then sketch various replies which European and American philosophers have given to it. Since these replies have not led to any decisive answer to the question, I shall then ask two other questions: the more specific ontological question of whether or not absences—surely paradigmatic examples of negative facts—exist; and the related epistemological question of what is known when the absence of something is said to be known. Answers to these questions comprise an important part of classical Indian philosophy; and I shall outline their answers to them, concluding that the most plausible answers to these questions are those of Jayanta Bhaṭṭa, who maintained that absences do indeed exist and that they are known not only by inference but also by perception.

I NEGATIVE FACTS

Do negative facts exist or not? In one sense, of course, negative facts do exist. After all, if each of the following sentences should be true:

(1.1) There is not a hippopotamus in this room.
(1.2) Socrates is not alive.

(2.1) There is a dog in this room.
(2.2) Putnam is alive.

then, clearly, each of them expresses some fact.

This, however, is not what is meant by those who maintain that there are negative facts. What the proponents of the existence of negative facts mean is: first, that all facts can be partitioned (i.e. exhaustively divided) into two sets, positive and negative, neither of which, on its own, comprises the totality of facts making up the world; and, second, that facts of both kinds would obtain, even if there were no minds to know them.

Scepticism about the existence of negative facts is based, I believe, on two considerations. The first consideration is that 'negative facts are nowhere to be met with in experience'.[2] It is not clear, however, that this is so. Certainly common sense militates against this conclusion; for, common sense tells us that, if one simply looks into a refrigerator, and if there is broccoli in it, then one sees that there is broccoli in it, and if there is no broccoli in it, then one sees that there is no broccoli in it.

The sceptics' second consideration is that the division of facts into positive and negative ones is a spurious one, being an artefact of natural language. Looking again at the example sentences above, one might conclude that there is neither anything negative pertaining to the facts expressed in the pair of sentences in (1) nor anything positive pertaining to the facts expressed by the sentences in (2)—except, perhaps, that the first two facts are expressed by sentences with the adverb 'not' and the second two are expressed by sentences without it. Some sentences, remarked upon by Frege,[3] make the point. As he observed, the very same thing can be expressed both by sentences with negative particles and by ones without them.

(3.1) Christ is immortal.
(3.2) Christ lives for ever.

(4.1) Christ is not immortal.
(4.2) Christ is mortal.
(4.3) Christ does not live for ever.

The conclusion, however, is not convincing, unless it can be shown, first, that every sentence expressed with a negative particle can be expressed by an equivalent one without any negative particle, and second, that the ontic commitments of the paraphrase are fewer than those of what is paraphrased. As I shall show below, it is difficult to meet these two desiderata jointly.

On the basis of the implicit assumption that every sentence with exactly one negative particle of the form N(S) has a version without the negative particle of the form S, Demos suggests that any sentence of the form N(S) can be paraphrased by a sentence of the form 'that which is inconsistent with S is true'.[4] For example, a sentence such as 'this rose is not yellow' can be paraphrased as 'that which is inconsistent with "this rose is yellow" is true' (or, more idiomatically, as 'that which is inconsistent with this rose's being yellow is true').

But, in the relation of inconsistency, what is inconsistent with what? Sentences as such are not inconsistent with one another, except insofar as they are expressions of things which are themselves inconsistent with one another. Thus, to answer the question just raised, one must first determine what sentences are expressions of. As Husserl makes clear in his *Logical Investigations*,[5] sentences are expressions, on the one hand, of propositions, which are the meaning correlates of sentences, and on the other hand, of states of affairs, which are the object correlates of sentences.[6] Inconsistency, then, is a relation either between propositions or between states of affairs. And so, to eschew acceptance of negative facts through a paraphrase of the kind suggested by Demos is to accept the existence of either propositions or states of affairs.

Russell, in his *Philosophy of Logical Atomism*,[7] rejects the existence of both. Propositions, he says, are not included in 'an inventory of the world', and so 'are not what you might call "real" '. And while states of affairs which obtain—that is, facts—are included in an inventory of the world, and hence are real; states of affairs which do not obtain are not in the inventory, and hence are not real. To these two explicit premises must be added this implicit one: real relations must have real relata—from which it follows that, if even one of the relata of a relation is not real, the relation itself cannot be real. Therefore, inconsistency cannot be a real relation between propositions, since propositions themselves are not real. Neither can inconsistency be a real relation between states of affairs, since at least one of a pair of states of affairs related by inconsistency cannot obtain, that is, cannot be a fact, and hence cannot be real. Thus, Russell concludes, 'it is simpler to accept negative facts as facts'.[8]

It might be thought that Russell's concerns could be met by supposing that inconsistency between a pair of states of affairs is supervenient upon inconsistency between actual properties ascribed to the very same actual individual substance or between actual relations ascribed to the very same actual individual substances taken in some fixed order. Thus, for example, the inconsistency between a state of affairs of a particular rose being red and a state of affairs of the very same rose being yellow, might be thought of as supervenient upon the inconsistency of the property of being red and the property of being yellow. Thus, to say that a particular rose is not yellow is to say that it has a property inconsistent with that of being yellow.

Alternatively, having discarded states of affairs as real and accepting only actual properties, relations, and substances as real, one might dispense with the Demos's paraphrase altogether and invoke one which mentions the relation of non-identity instead. Thus, to say that a particular rose is not yellow is to say that every property which this rose has is distinct from the property of being yellow.

These last two alternatives meet with one serious problem: *impersonalia*, that is, states of affairs expressed by uniclausal sentences with impersonal subjects (e.g. 'it is raining' and 'it is sunny'). As Reinach[9] has pointed out, such states of affairs can be inconsistent with one another, for example,

(5.1) It is sunny.
(5.2) It is overcast.

and their inconsistency cannot be recast as an inconsistency among constituents of factual states of affairs.

In short, either one accepts the existence of negative facts or one accepts inconsistency between either propositions or states of affairs. But, even if one accepts the latter alternative, one still must accept the existence of at least one negative fact, namely the fact that there is the negative relation of inconsistency.

It seems that the move to eliminate negative facts, or more generally, to eliminate negative states of affairs, founders on the need to invoke a relation which is itself negative. It might be thought that perhaps this predicament results from a misassignment of the burden of proof: the burden of proof lies, not with the opponents of negative facts, but with their proponents. In other words, it is not that one must show how negative facts can be eliminated, rather it is that one must show that there are reasons to posit their existence in the first place.

Russell, in his *Philosophy of Logical Atomism*, offers reasons to posit the existence of negative facts. Recall that Russell's logical atomism has, as a central tenet, that a logically ideal language, in displaying the logical form of the sentences of natural language, would thereby lead one to a correct ontology. According to Russell,[10] the atomic sentences of a logically ideal language which are true are rendered true by positive facts, while its false atomic sentences are rendered false by negative facts. In other words, the correspondence theory of truth needed for the logically ideal language requires two relations, correspondence, whereby facts render atomic sentences true, and what one might call counter-correspondence, whereby facts render atomic sentences false. Counter-correspondence is to be distinguished from a simple failure of correspondence. In the latter case, there simply is no fact to render a sentence true; in the former case, there is a fact to render the sentence false.[11]

Oaklander and Miracchi[12] have maintained that Russell did not have to posit the relation of counter-correspondence, and hence that he had no need to posit its factual relata of negative facts. They suggest that, instead, atomic sentences are rendered false, not by counter-correspondence, but by a mere failure of correspondence.

The suggestion is a compelling one. Recall that the classical bivalent semantics for sentential logic is based on a truth-value assignment of exactly one of the truth-values of true and false to each of the atomic sentences of the language: that is to say, it is based on a total bivalent function from the set of atomic sentences to the truth-value set {T, F}. From this perspective, Oaklander and Miracchi's suggestion amounts to this: the total bivalent function from the set of atomic sentences to the truth-value set {T, F} can be replaced by a partial univalent function from the set of atomic sentences to the truth-value set {T}.

If the only task the relations of correspondence and counter-correspondence are required to do were to ground the truth-value assignment of atomic sentences, then the suggestion by Oaklander and Miracchi would be irreproachable. However, I do not believe that their only task is to ground the assignment of truth-values to atomic sentences. Surely it is necessary, in addition, to distinguish atomic sentences by what they express. Now, the differences among what true atomic sentences express can be grounded in the relation of correspondence: two true atomic sentences are expressively different by dint of their being made true by two different facts. But in what can be grounded the differences among what false atomic sentences express? Russell can ground the differences among what false atomic sentences express in counter-correspondence: two false atomic sentences are expressively different by dint of their being made false by two different (negative) facts. Oaklander and Miracchi's suggestion to give up the relation of counter-correspondence and the negative facts which go with it, leaves Russell with all false atomic sentences expressing the same thing, namely nothing.

Of course, the posit of negative facts and the relation of counter-correspondence is not the only way to address the problem of grounding the expressive difference among atomic sentences. An alternative is to posit a second relation of correspondence, call it expressive correspondence, whereby the expressive difference among atomic sentences is grounded in the differences among the states of affairs they express. On Russell's view, then, there are two verification relations, correspondence and counter-correspondence, and two kinds of facts, positive (positive states of affairs which obtain) and negative (negative states of affairs which obtain). On the alternative view, there are two kinds of states of affairs, namely those which obtain and those which do not, and one verification relation, namely correspondence.

This alternative is unacceptable to Russell, since, as we saw before, Russell, the logical atomist, was unwilling to countenance the existence of states of affairs which are not facts. Yet Russell's fellow logical atomist, the early Wittgenstein, was perfectly willing to do so;[13] and, as a result, he can, and does, reject the existence of negative facts.[14] Thus, even with the assumptions peculiar to logical atomism, one is not compelled to accept the existence of negative facts; though, as Hochberg[15] points out, the acceptance of the existence of some counterpart for negative facts, such as the acceptance of the existence of states of affairs, is inevitable.

The reasoning just given for the acceptance of the existence of negative facts depends on assumptions peculiar to the doctrine of logical atomism; yet, even accepting these assumptions, one is still not compelled to embrace the existence of negative facts. The question arises: are there reasons which are not dependent on assumptions peculiar to a philosophical doctrine which nonetheless compel the acceptance of negative facts? I believe so; but they lie in an approach to the question of the existence of negative facts, different from the kind considered thus far.

II ABSENCES AND KNOWLEDGE OF ABSENCES

The approach is to consider paradigmatic instances of negative facts and see how such instances are to be accounted for within a larger metaphysical theory. A natural candidate for paradigmatic instances of negative facts is absences; and an obvious metaphysical setting within which to consider them is the epistemological setting of their knowability. Absences and their knowability has been well explored by classical Indian philosophers.[16] In what follows, I shall review the various metaphysical positions taken with respect to this pair of questions and defend the conclusion of Jayanta Bhaṭṭa, namely that not only do absences exist but they can be known by mere perception.

Most classical Indian philosophers, like most contemporary philosophers, include perception and inference among the standard means of knowledge. Thus, for example, if I should enter a colleague's office and see him sitting at his desk, I know that he is sitting at his desk because of my visual perception of him there; and if, from my office window, I should see smoke rising from a building in the distance, I know that the building is on fire because of an inference from my observation of the smoke arising from it.

But now suppose that I enter a colleague's office and my colleague is not there. How do I know that he is not there? One obvious answer, tendered by the classical Indian philosopher Kumārila Bhaṭṭa (AD 620–680), is that I fail to perceive him in his office.[17] While this answer has the ring of common sense, there are reasons to baulk at accepting it. To begin with, perception and inference are, in one sense, faculties or cognitive capabilities, whereas non-apprehension (*an-upalabdhi*), the act of failing to perceive something, is not. There is, however, a more important reason to abandon the view that non-apprehension of something is a means of knowledge of its absence—something may be present yet not be perceived. Suppose that when I enter my colleague's office, it is night-time, the lights are out, and as a result his office is completely dark. Under these conditions, I would fail to see my colleague, even though he might very well be present.

There are essentially four possible metaphysical positions to account for what it is that I know when, after entering my colleague's office, I come to know that he is not present there. Each of the positions has been adopted and defended by certain classical Indian philosophers. On the one hand, some take the absence of my colleague from his office as a brute, negative fact. Of them, some hold knowledge of this fact to be perceptual; while others hold it to be inferential. On the other hand, some hold that the absence of my colleague from his office has no real ontic status at all, and believe that what there really is in the situation is just the sum of all the things present in the office. These latter philosophers hold that knowledge of my colleague's absence is just knowledge of what is present, though some believe the knowledge results from perception, while others from inference.

The first classical Indian philosopher known to have understood and addressed the problem which confronts the view of Kumārila Bhaṭṭa's is his older contemporary, Dharmakīrti (AD 600–660). According to Dharmakīrti, I know something, when I correctly report, having entered my colleague's office under the right conditions, that my colleague is absent from it; but what I know is not a negative fact and it is not known by perception. Rather, what I know is a simple (positive) fact and it is known by inference.

Recall that, for a philosopher like Demos,[18] the relevant inference is based on the relation of inconsistency between the simple (positive) fact I observe upon entering my colleague's office, on the one hand, and the state of affairs of my colleague's presence, on the other hand. To be sure, Dharmakīrti holds that the relation of inconsistency plays a role in many inferences which warrant one's claim to know some allegedly negative facts. For example, Dharmakīrti holds that from the presence of heat one can infer an absence of cold, since heat and cold are inconsistent (with respect to the same locus). But it is not his view that inference based on the relation of inconsistency plays a role in the case at hand. Rather, Dharmakīrti maintains, one reasons as follows: the causal conditions of perception known to me are such that, if my colleague were present in his office, I would see him; I do not see him; therefore, he is not present. The inference is counterfactual.[19]

One obvious objection to Dharmakīrti's position is one raised by Taylor[20] against Demos[21] and Russell.[22] Taylor argues that there is no more justification for the assertion that inference is involved in the case where one looks at a circle with no dot in it and observes that there is an absence of a dot from the circle, than there is for the assertion that inference is involved in the case where one looks at a circle with exactly one dot in it and observes that it has a dot. Taylor's point is that the two epistemic situations are phenomenologically the same in all relevant respects, and that since inference is not required to account for knowledge in the second case, it is not required in the first case either.

Taylor's point certainly seems to hold in the case at hand. Suppose I am standing at the threshold of my colleague's office door and looking into his office, my view unobstructed and the room suffused with light. If my colleague is in the office and I report that he is, it seems that, if asked how I know, I can answer only that I see that he is there; and if my colleague is not in his office and I report that he is not there, it seems that, if asked how I know, I can answer only that I see that he is not there.

There are, however, other cases where the epistemic situation seems less phenomenologically similar than what is suggested by either Taylor's case or the case I just gave. Suppose that I drop a screw and suspect that it has fallen into a hole which I cannot see into but which I can put my hand into and completely search. Suppose that, after doing so, I report that the screw is not in the hole. If asked how I know, I would answer: if it had been there, I would have touched it; but since I searched for it by hand and did not touch it, it must not be there.

Indeed, given the right setting for the case where I report the absence of my colleague from his office, one can readily imagine situations in which inference does seem relevant to the determination of an absence, in the way it is relevant to the determination of the absence of the lost screw from the hole. Suppose that I and a companion have some reason to believe that my colleague might be in his office but hiding. If I were to go in and look about his office carefully and then report that he is not in the office, I might very well justify my report with counterfactual inference: I looked thoroughly; and if he had been there, I would have seen him.

It might be countered that the phenomenological dissimilarity in the epistemic situations of the knowledge of absences and knowledge of presences, evinced in these last two cases, results, not from the cognitive process at the time when the absence came to be known, but from the cognitive process at the time when pressed for a justification. If that is the case, then a deeper issue is now raised which undercuts Taylor's original argument: namely the issue of the accessibility of cognitive processes to introspection—an issue which space does not permit me to pursue here, as important as it is.

Taylor's objection aside, the question remains: does Dharmakīrti's counterfactual inference obviate the need to invoke the existence of absences? It seems not to. Suppose, for example, I know that my colleague is absent from his office. Let me symbolize this as $\neg$ p. Now, according to Dharmakīrti, I know this, because

I have inferred it from two premises which I already knew: first, the counterfactual conditional that if my colleague were present in his office, then I would know that he is present there; and second, I do not know that he is present in his office. Let me symbolize them as p > Kp and ¬ Kp respectively (where 'K' denotes the epistemic operator corresponding to the English 'a knows that' and '>' denotes the connective for counterfactual conditionals). But this gives rise to a dilemma: how do I know that I do not know that my colleague is present in his office? In other words, how do I know that ¬ Kp? On the one hand, if Dharmakīrti holds that I perceive that I do not know that my colleague is present in his office, that is to say, that I perceive ¬ Kp, then Dharmakīrti accepts thereby not only the existence of negative facts, in this case, negative mental facts, but also their perceptibility, both of which he wants to deny. On the other hand, if Dharmakīrti holds that I infer that I do not know that my colleague is present in his office, that is to say, that I infer ¬ Kp, then there must be an inference to ground that claim, just as there is an inference to ground the initial claim that my colleague is not present in his office, that is, ¬ p. An infinite regress of inferences becomes inescapable, despite Dharmakīrti's protestations to the contrary.[23]

A different position regarding negative facts and knowledge of them is the one adopted by the followers of Kumārila Bhaṭṭa. Though they maintained, unlike Dharmakīrti, that there are negative facts; nonetheless, they seem to have adopted his view that knowledge in such cases is derived only from counterfactual inferences.[24] Their view, of course, suffers from the very same dilemma just raised with respect to Dharmakīrti's view, namely the dilemma of having either to concede the perception of some negative facts or to accept, for any negative fact said to be known, an infinite regress of inferences of negative facts.

A younger contemporary of Kumārila Bhaṭṭa, Prabhākara (AD 650–720), who shared many of his philosophical assumptions, adopted the view that what I know, when upon entering my colleague's office I come to know that he is absent from it, is some visually perceived, positive fact, that is, some presence. But what would be the plausible candidate for such a presence? According to Prabhākara, it might be the presence of his chair. What else could there be to the absence of my colleague from his chair than the presence of his chair? After all, as Buchdahl[25] has observed, one needs to draw only one picture to draw a picture either of my colleague's chair or of his chair without him in it. Yet, Prabhākara's metaphysical opponents were sceptical: they pointed out that, if my colleague's absence were identical with the presence of his chair, then anytime I perceive his chair, which includes the times that my colleague is in his chair, I should know that he is absent. This, according to the opponents, is patently absurd.

Prabhākara apparently sought to extricate himself from this objection by denying that the presence of my colleague in his chair is the same as the presence of his chair without him in it: in the former case, there are two presences, that of my colleague and that of his chair; but in the latter, there is only one, namely that of his chair. Opponents of Prabhākara suspected that this distinction cannot be made without directly or indirectly relying on an assumption of the existence of absences: for, as they pointed out, to distinguish the presence of my colleague in his chair from the presence of only his chair requires the use of such expressions as 'only' ('*eva*') or merely ('*mātra*'), which are implicitly negative expressions, since to say that there is only one presence is to say that there is at least one presence and there is no more than one presence.[26]

Prabhākara faced another objection. How is he to account for the fact that, when I see simply my colleague's chair, what I come to know is his absence and not one of the infinity of other absences which are also then present? One reply is to maintain that there is a difference between what I see and what I remark upon, and that this difference obtains equally in cases of the perception of what are unquestionably presences as well as in the cases of perception of what are allegedly absences. For the very same reason that when I look at my colleague's desk and remark that there is a pad of paper on it but fail to remark to myself that there is a pencil on it, I remark on his absence from his chair but I do not remark on the absence of the departmental chairman from it.[27]

An alternative to Prabhākara's position is one advocated by Jayanta Bhaṭṭa (fl. AD 890). He agrees with Prabhākara that perception is the relevant means for knowing that my colleague is absent from his office, but disagrees with him that all that is relevant to my knowledge of his absence is my perception of some presence. Rather, Jayanta Bhaṭṭa holds that absences are brute facts unto themselves and that my colleague's absence in the case described above is known through perception. Jayanta Bhaṭṭa, like others of the same school, thought of perception as a causal process, the linchpin of which is contact between the object of perception and the appropriate sense organ. Thus, for example, in the tactile perception of the pencil in my hand, the key causal link is the contact of the organ, the skin, with the pencil. This view of perception, however, raised an immediate problem: if contact between an object of perception and the appropriate sense organ is the key link in the causal process of perception, how can there be perception of absences, for surely it is absurd to think of any sense organ as coming into contact with an absence.

Jayanta Bhaṭṭa's response to this objection is to revise the definition of sense perception: he restricted the above definition to the perception of positive facts and provided a special clause to account for perception of negative facts. According to Jayanta Bhaṭṭa, this move was not *ad hoc;* and at least within his overall metaphysical framework, it had independent motivation, for he believed that, though universals do not come into contact with the sense organs, they nonetheless can be perceived. In general, something is perceived just in case it is a physical thing suited to a sense organ and in contact with it, or it is a universal inhering in a physical thing and both the universal and the physical thing are suited to the sense organ, or it is an absence related to a place and both the absence and the place are suited to the sense organ.

To a contemporary philosopher, Jayanta Bhaṭṭa's response to this objection might seem both *ad hoc* and implausible. But let us not so readily discard his view of absences and their knowability. The implausibility of his response derives, not from his acceptance of negative facts and of their perceptibility, but from other assumptions, common to pre-modern thinkers, which modern metaphysics is coming to abandon. These assumptions are: that perception requires contact; that the relata of the causation relation are things; and that the mental relata of the relation of perceptual causation are the thing-like things, images. It seems to me that the most interesting aspects of Jayanta Bhaṭṭa's metaphysical views are retained, even when these assumptions are revised into a form more congenial to current views of these metaphysical issues. Let us see how this can be done.

To begin with, in spite of its prevalence in the pre-modern world, the view that contact between the object of perception and the appropriate sense organ is the linchpin in the causal process resulting in perceptual knowledge, is simply untenable, except in some very attenuated sense of contact. Descartes is perhaps the last serious thinker to hold a contact theory of visual perception; and even in his theory, the contact between the eye and the object of vision is mediated by a long chain of things, each in contact with some other.[28] If one considers what is really essential in Jayanta Bhaṭṭa's epistemology, it is his view that perceptual knowledge results from a causal process; while contact between the object of perception and the appropriate sense organ plays an utterly ancillary role.

Another aspect of Jayanta Bhaṭṭa's metaphysics requiring some modification is his view of causation. To see what modification is required, we turn to the pioneering work by Zeno Vendler[29] on the expression of causation in natural language.[30]

We are all familiar with metonymy, the figure of speech whereby one word, typically a noun, literally denoting one thing, is used to denote something else, by dint of some well-known connection between the first thing and the second. Thus, for example, it is correct and idiomatic to say:

(6) The crown granted an amnesty to the rebels,

where the word 'crown', which literally denotes ornamental headgear worn by a monarch, is used to denote a monarch. Nor is metonymy a mere literary device, as the following sentences are intended to show.

(7.1) Have you read any (works by) Shakespeare?
(7.2) The White House (i.e. the President of the United States) has just made a statement.

Even entities which are not physical objects can be denoted metonymically. Thus, for example, what is forbidden by the first sentence below is not track shoes, but rather the activity of wearing track shoes in the locker room. Its literal paraphrase is given by the second sentence.

(8.1) Track shoes are forbidden in the locker room.
(8.2) The wearing of track shoes in the locker room is forbidden.

Metonymy also appears commonly and naturally in ordinary parlance pertaining to causation, as the following sentence and its literal paraphrase illustrate.

(9.1) This nail caused our flat tire.
(9.2) The puncturing of our tire by this nail caused it to go flat.

Metonymy, then, enables one to speak of a nail as a cause and a flat tire as an effect. Moreover, philosophical parlance easily abides such usage, as illustrated by the following remarks from Kant's *Critique of Pure Reason*:

> A room is warm, while the outer air is cool. I look around for the cause, and find a heated stove. Now the stove, as cause, is simultaneous with its effect, the heat of the room.[31]

In light of the role of metonymy in talk about causes and effects, it would be a mistake to permit such humdrum usage to inveigle one into believing that causation is a relation between physical objects. What, then, is the causation relation a relation between?

Since Vendler's insightful work, it is common to distinguish, at least *prima facie*, among physical objects, states of affairs, and events.[32] Events have both spatial and temporal location, since they take place or occur at a time and a place. Physical objects, however, have only spatial location, since they cannot be said to occur or take place. Facts have neither spatial nor temporal location.[33] What Vendler[34] shows is that, when metonymy is eliminated, causes and effects turn out to be states of affairs. And among the states of affairs which can serve as causes or effects are negative ones. The cause of an automobile accident, for example, might be a driver's not seeing a red light; and the driver's not seeing a red light, in addition, might be the effect of his being distracted by a passenger in his car. Moreover, absences are negative states of affairs, and they too can be both causes and effects. A negative state of affairs might be a cause: an absence of light in a room might cause someone to stumble. A negative state of affairs might be an effect: a short-circuit caused there to be an absence of electrical power. And a cause and its effect might both be negative states of affairs: an absence of oxygen in a room might cause someone to go unconscious.

The final assumption to be revised pertains to the specific relata of perceptual causal processes. The widespread pre-modern model is that the relata are physical things, on the one hand, and images or mental pictures, on the other.[35] But this model simply won't do for propositional knowledge.[36] If a proposition is a proper characterization of what is the result of an act of propositional knowledge, then what is the object of that act? The usual, if implicit, answer is a state of affairs or a fact.[37] The current consensus, sometimes implicit, sometimes explicit, is that facts are the cause of an epistemic state, itself a fact or state of affairs, whose content is expressed by a proposition.

Once some of Jayanta Bhaṭṭa's metaphysical assumptions are reformulated to overcome their *prima facie* implausibility, a rather plausible account emerges of what it is that one knows when, in the right circumstances, one comes to know, say, of a colleague's absence from his office after having entered it, and how one knows what one knows in those circumstances: what one knows is the negative fact of a colleague's absence from his office and one knows it by means of perception. The plausibility of this account stands in contrast to the other three metaphysical alternatives, explored above, advocated by Prabhākara, Dharmakīrti, and Kumārila Bhaṭṭa. As was shown, the view of Prabhākara, namely that so-called negative facts are just presences of some kind, fails to say what, exactly, such presences are, without covertly relying on some form of negative fact; while the views of Dharmakīrti and the followers of Kumārila Bhaṭṭa, namely that only inference is a source of knowledge for whatever is known when an absence is said to be known, founder on the fact that the appropriate inferences must be grounded, inevitably, in a perception of something which is itself said to be an absence—which entails, contrary to their views, that at least some cases of knowledge of absences are cases of perception.

NOTES

* The first version of this paper was presented on 29 March 1990 to the American Philosophical Association, Pacific Division. A second version was presented on 26 March 1993 to the University of Ottawa's Department of Philosophy. I am grateful to the audiences on each occasion for their questions and comments. I would also like to thank Paul Forester, Richard Hayes, Graeme Hunter, Stephen Menn, Karl Potter and Mark Siderits for their suggestions and discussions of earlier versions of this paper.

1. *Sophist.*
2. Raphael Demos, 'A Discussion of a Certain Type of Negative Proposition', *Mind* 5: 26, 1917, p. 189.
3. Gottlob Frege, 'Negation', *Beiträge zur Philosophie des deutschen Idealismus*, 5: 1, 1919, pp. 143–57, English trans., Geach and Blackwell, 1952, p. 125.
4. Though Demos most frequently resorts to the word 'opposite' in his paraphrases, he makes it clear that both 'contrary' and 'inconsistent' are suitable equivalents. Demos, 'A Discussion of a Certain Type of Negative Proposition', p. 191.
5. Edmund Husserl, *Logische Untersuchungen*, Halle, Germany: M. Niemeyer, 1900, English trans., Findlay, 1970, Investigation I, ch. 1, esp. secs 11–13.
6. Bertrand Russell, 'The Philosophy of Logical Atomism', *Monist*, 5: 28–9, 1918, Reprint, ed. by Pears, 1985, Lecture III, p. 77.
7. Barry Smith, 'An Essay in Formal Ontology', *Gratzer Philosophische Studien* 5: 6, 1978, art. 2, discusses the connection of Frege's distinction between sense and reference, well known to Anglo-American philosophers, with this distinction of Husserl's.
8. Later, Bertrand Russell, *An Enquiry into Meaning and Truth*, New York: Allen and Unwin, 1940, pp. 73–4, 81–2, rejects, without argument, the existence of negative facts.
9. Adolf Reinach, 'Zur Theorie des negativen Urteils', in A. Pfänder, ed., *Münchener Philosophische Abhandlungen. Festschrift for Theodor Lipps*, 1911, sec. 12.
10. Russell, *The Philosophy of Logical Atomism*, Lecture III, p. 78.
11. See Russell, ibid., Lecture I, pp. 46–7; Lecture II, p. 51; and Lecture III, p. 71.
12. L. Nathan Oaklander and Silvano Miracchi, 'Russell, Negative Facts and Ontology', *Philosophy of Science*, 5: 47, September 1980, sec. III.
13. Ludwig Wittgenstein, *Logish-Philosophische Abhandlung: Annalen der Naturophilosophie*, 1921, trans., Pears and McGuiness (*Tractatus Logico-Philosophicus*), 1961, 2.04.
14. Ibid., 2.05 and 2.06.
15. Herbert Hochberg, 'Negation and Generality', *Noûs*, 5: 3, 1969, pp. 325–8.

16. A good overview of the various positions and arguments is Jayanta Bhaṭṭa's *Nyāya-Mañjarī*, end of the first *āhnika*. K.S. Varadacharya, ed., *Nyāya-Mañjarī of Jayantabhaṭṭa with Tippaṇi Nyāyasaurabha by the Editor*, 2 vols, Mysore: Oriental Research Institute, University of Mysore, 1969, v. 1, pp. 130–65; J.V. Bhattacharya, trans., *Nyāya-Mañjarī: The Compendium of Indian Speculative Logic*, New Delhi: Motilal Banarsidass, 1978, pp. 102–32.
17. See his *śloka-vārttika*, ch. 9 (Abhāva-vāda: On Negation), verse 11.
18. Demos, 'A Discussion of a Certain Type of Negative Proposition'.
19. See Dharmakīrti's *Nyāyabindu*, ch. 2, sūtras 12ff. (Dalshukh Malvania, ed., *Paṇḍita Durveka Miśra's Dharmottarapradipa, Being a Sub-commentary on Dharmottara's Nyāyabinduṭīkā, a Commentary on Dharmakīrti's Nyāyabindu*, Patna: Kashiprasad Jayaswal Research Institute, 1955 [Tibetan Sanskrit Works Series: 2], 2nd edition, 1971, pp. 101ff); and his *Pramāṇa-vārttika*, ch. 3 (*Svārthānumāna*), verses 3–6 and his own commentary thereto Raniero Gnoli, ed., *The Pramānavārttikam of Dharmakīrti: The First Chapter with Autocommentary, Text and Critical Notes*, Rome: Istituto Italiano per il Medio ed Estremo Oriente, 1960 (Serie Orientale Roma: 23), pp. 4–6. This point is treated in detail by Brendan S. Gillon, 'Dharmakīrti and His Theory of Inference', in Matilal and Evans, eds, *Buddhist Logic and Epistemology*, 1986.
20. Richard Taylor, 'Negative Things', *The Journal of Philosophy*, 5: 49, 1952, pp. 443–5.
21. Demos, 'A Discussion of a Certain Type of Negative Proposition'.
22. Russell, *An Enquiry into Meaning and Truth*.
23. See his *Pramāṇavārttika*, ch. 3 (*Svārthānumāna*), v. 3 and the commentary thereto, Gnoli, *The Pramānavārttikam of Dharmakīrti*, pp. 4–5.
24. These philosophers, unlike Dharmakīrti, but like most other classical Indian philosophers, considered counterfactual reasoning (*tarka*) neither as a form of inference (*anumāna*) nor, in general, as a standard means of knowledge (*pramāṇa*). The case of counterfactual reasoning yielding knowledge of negative facts was set aside by them as a means of knowledge unto itself, called 'non-apprehension'. For further details, see Dhirendra Mohan Datta, *Six Ways of Knowing: A Critical Study of the Vedānta Theory of Knowledge*, Calcutta: Calcutta University Press, 1932, 2nd revised edition, 1960, Book III, ch. 4.
25. Gerd Buchdahl, 'The Problem of Negation', *Philosophy and Phenomenological Research*, 5: 22, 1961, pp. 176–7.
26. For linguistic confirmation of this view of 'only', see Laurence R. Horn, *A Natural History of Negation*, Chicago: The University of Chicago Press, 1989, p. 249.
27. This point has also been made by Buchdahl, 'The Problem of Negation', p. 175.
28. See, for example, his *La Dioptrique*.
29. Zeno Vendler, *Linguistics in Philosophy*, Ithaca, New York: Cornell University Press, 1967, chs 5 and 6; and Zeno Vendler, 'Causal Relations', *The Journal of Philosophy*, 5: 54, 1967.

30. Vendler's discussion is confined to English, but most of what he says holds with little or no modification of other languages.
31. Kant, *Critique of Pure Reason*, B 247–8, trans., Norman Kemp Smith.
32. It is customary in the specialist literature to use the term 'event' to include such entities as processes and states, due to a lack in English of an appropriate inclusive term.
33. See Vendler, *Linguistics in Philosophy*, ch. 5.12, for further discussion.
34. Ibid., ch. 6; and 'Causal Relations'.
35. Once again, see, for example, Hume's *An Enquiry Concerning Human Understanding*, secs IV, V and VII *passim*.
36. Russell (*The Philosophy of Logical Atomism*, Lecture IV, sec. 1, pp. 82ff), in his criticism of James and Dewey, exhibits an awareness of this problem.
37. See, for example, Alvin Goldman, 'A Causal Theory of Knowing', *The Journal of Philosophy*, 5: 64, no. 12, 22 June 1967, Reprint, ed. by Roth and Galis, 1970, p. 82.

REFERENCES

BHATTACHARYA, J.V., tr., *Nyāya-Mañjarī: The Compendium of Indian Speculative Logic* (Motilal Banarsidass, New Delhi, 1978).

BUCHDAHL, Gerd, 'The Problem of Negation', *Philosophy and Phenomenological Research*: v. 22, 1961, pp. 167–78.

CONRAD-MARTIUS, H., ed., *Gesammelte Schriften* (of Adolf Reinach) (M. Niemeyer, Halle, Germany, 1921).

DATTA, Dhirendra Mohan, *Six Ways of Knowing: A Critical Study of the Vedānta Theory of Knowledge* (Calcutta University Press, Calcutta, 1932), 2nd revised edition, 1960.

DEMOS, Raphael, 'A Discussion of a Certain Type of Negative Proposition', *Mind*: v. 26, 1917, pp. 188–96.

Dharmakīrti, *Pramāṇa-vārttika*, edn: Gnoli, ed., 1960. Partial English translation: Hayes and Gillon (tr.), 1991.

Dharmakīrti, *Nyāya-bindu*, edn: Malvania, ed., 1955. English translation: Stcherbatsky, 1930, vol. II.

FINDLAY, J.N., tr., *Logical Investigations*, 2 vols (Humanities Press, New York, 1970 [Internation Library of Philosophy and Scientific Method]). English translation of Husserl, 1990, 2nd revised edition.

FREGE, Gottlob, 'Negation', *Beiträge zur Philosophie des deutschen Idealismus*: v. 1, 1919, pp. 143–57. English translation: Geach and Black, trs, 1952, pp. 117–35.

GALE, Richard, *Negation and Non-being* (Basil Blackwell, Oxford, 1976

[American Philosophical Quarterly Monograph Series: Monograph No. 10]).

GEACH, Peter and Max BLACK, trs, *Translations from the Philosophical Writings of Gottlob Frege* (Basil Blackwell, Oxford, 1952), 2nd edition, 1960.

GILLON, Brendan S., 'Dharmakīrti and His Theory of Inference', in Matilal and Evans, eds, 1986, pp. 77–87.

GNOLI, Raniero, ed., *The Pramānavārttikam of Dharmakīrti: The First Chapter with the Autocommentary, Text and Critical Notes* (Istituto Italiano per il Medio ed Estremo Oriente, Rome, Italy, 1960 [Serie Orientale Roma: 23]).

GOLDMAN, Alvin, 'A Causal Theory of Knowing', *The Journal of Philosophy*: v. 64, no. 12 (22 June 1967), pp. 357–72. Reprint: Roth and Galis, eds, 1970, pp. 67–88.

HAYES, Richard P. and Brendan S. GILLON, 'Introduction to Dharmakīrti's Theory of Inference as Presented in *Pramāṇavārttika Svopajñavṛtti* 1–10', *Journal of Indian Philosophy*: v. 19, 1991, pp. 1–73.

HOCHBERG, Herbert, 'Negation and Generality', *Noûs*: v. 3, 1969, pp. 325–43.

HORN, Laurence R., *A Natural History of Negation* (The University of Chicago Press, Chicago, 1989).

HUSSERL, Edmund, *Logische Untersuchungen* (M. Niemeyer, Halle, 1900), 2nd revised edition 1913. English translation: J.N. Findlay, tr., 1970.

JAYANTA Bhaṭṭa, *Nyāya-Mañjarī* <End of Āhnika I>, edn: Varadacharya, ed., 1969. English translation: Bhattacharya, tr., 1978.

JHĀ, Gangānātha, tr., *Śloka-vārttika: Translated from the Original Sanskrit with Extracts from the Commentaries of Sucarita Miśra (The Kāśikā) and Pārthasārathi Miśra (Nyāyaratnākara), Bibliotheca Indica*, 1900, nos 965, 986, 1017, 1055, 1091, 1157 and 1183.

——, tr., *Prabhākara School of Mīmāṁsā* (Allahabad University, Allahabad, 1911), Allahabad University Series: v. 1.

KUMĀRILA Bhaṭṭa, *Śloka-vārttika*, edn: Śāstri, ed., 1978. English translation: Jhā, tr., 1900.

MALVANIA, Dalshukh, ed., *Paṇḍita Durveka Miśra's Dharmottarapradīpa, Being a Sub-commentary on Dharmottara's Nyāyabinduṭīkā, a commentary on Dharmakīrti's Nyāyabindu* (Kashiprasad Jayaswal Research Institute, Patna, 1955), Tibetan Sanskrit Works Series: 2, 2nd edition, 1971.

MATILAL, Bimal K. and R.D. EVANS, eds, *Buddhist Logic and Epistemology* (D. Reidel Publishing Co., Dordrecht, 1986).

OAKLANDER, L. Nathan and Silvano MIRACCHI, 'Russell, Negative Facts, and Ontology', *Philosophy of Science:* v. 47, no. 3 (September 1980), pp. 434–55.

PEARS, David, ed., *The Philosophy of Logical Atomism* (Open Court, La Salle, 1985 [Open Court Classics]).

PEARS, David and B.F. MCGUINNESS, trs, *Tractatus Logico-Philosophicus* (Routledge and Kegan Paul, London, 1961), 2nd impression, 1963.

PFÄNDER, A., ed., *Münchener Philosophische Abhandlungen. Festschrift für Theodor Lipps* (Barth, Leipzig, 1911).

PRABHĀKARA, *Bṛhatī*, edn: Śāstri, 1934. English translation: Jhā, tr., 1911.

REINACH, Adolf, 'Zur Theorie des negativen Urteils', in A. Pfänder, ed., 1911, pp. 196–254. Reprinted in H. Conrad-Martius, ed., 1921, pp. 56–102. English translation: Smith, Barry, tr., 1982.

ROTH, Michael D. and Leon GALIS, eds, *Knowing: Essays in the Analysis of Knowledge* (University Press of America, Lanham, 1970), 2nd edition, 1984.

RUSSELL, Bertrand, 'The Philosophy of Logical Atomism', *Monist:* v. 28, 1918, pp. 495–527; v. 29, pp. 32–63, 190–222, 345–80. Reprint: Pears, ed., 1985.

——, *An Enquiry into Meaning and Truth* (Allen and Unwin, New York, 1940).

ŚĀSTRI, Svāmī Dvārikādāsa, ed., *Ślokavārttika of Śri Kumārila Bhaṭṭa: With the Commentary Nyāyaratnākara of Śrī Pārthasārathi Miśra* (Tara Publications, Varanasi, 1978), Prāchyabhārati Series: v. 10.

ŚĀSTRI, S.K. Rāmanātha, ed., *Bṛhatī: edited with Śālikanātha Miśra's* Ṛjuvimālapañcikā and Bhāṣyapariśiṣṭa, 3 parts (University of Madras, Madras, 1934), Madras University Sanskrit Series.

STCHERBATSKY, Th., *Buddhist Logic*, 2 vols (The Academy of Sciences of the USSR, Leningrad, 1930; rpt, Dover Publications, New York, 1962 [Bibliotheca Buddhica: 26]).

SMITH, Barry, 'An Essay in Formal Ontology', *Gratzer Philosophische Studien:* v. 6, 1978, pp. 39–62.

——, tr., 'On the Theory of Negative Judgement', in Barry Smith, ed., 1982, pp. 315–77. English translation: Adolf Reinach, tr., 1911.

——, ed., *Parts and Moments, Studies in Logic and Formal Ontology* (Philosophia Verlag [Analytica], Munich, 1982).

TAYLOR, Richard, 'Negative Things', *The Journal of Philosophy:* v. 49, 1952, pp. 433–49.

VARADACHARYA, K.S., ed., *Nyāya-Mañjarī of Jayantabhaṭṭa with Tippaṇi Nyāyasaurabha by the Editor*, 2 vols (Oriental Research Institute, University of Mysore, Mysore, 1969).

VENDLER, Zeno, *Linguistics in Philosophy* (Cornell University Press, Ithaca, New York, 1967a).

——, 'Causal Relations', *The Journal of Philosophy*: v. 54, 1967b, pp. 704–13.

WITTGENSTEIN, Ludwig, *Logish-Philosophische Abhandlung. Annalen der Naturphilosophie*, 1921. Translation: Pears and McGuinness, trs, 1961.

Chapter 9

Happiness: A Nyāya-Vaiśeṣika Perspective

Wilhelm Halbfass

Happiness is one of the great and pervasive themes of Indian thought. Its meaning and ambiguities, its relevance for the human condition, its pursuit and its transcendence have been the subject of intense philosophical, psychological and soteriological inquiry and debate. In this process, words such as *sukha*, *ānanda*, *prīti* or *saṃtoṣa*, as well as their semantic cognates and their counterparts in the terminology of unhappiness and pain, have been used and discussed in a variety of ways.[1] The contributions of Nyāya and Vaiśeṣika may seem less central and significant than those of other schools. Yet they are by no means negligible insofar as the search for conceptual clarity and theoretical understanding is concerned. The Vaiśeṣika system in particular tries to do something that few other schools of thought have tried to do: It attempts to locate happiness in a table of categories, within a complete enumeration of what there is, to define its ontological status, and to determine its inherent nature and contents.

It is well known that the classical Vaiśeṣika system, as found in Praśastapāda's authoritative presentation, recognizes six fundamental categories (*padārtha*) or divisions of reality: substance (*dravya*), quality (*guṇa*), motion (*karman*), universal (*sāmānya*), particularity (*viśeṣa*), inherence (*samavāya*). In later presentations of the Vaiśeṣika system (or of the combined Nyāya and Vaiśeṣika systems), non-being (*abhāva*) is commonly added as a seventh category.[2] Within this list, happiness or pleasure (*sukha*) appears as one of the 24 qualities (*guṇa*). More specifically, it is, together with unhappiness or pain (*duḥkha*), desire (*icchā*), hate or aversion (*dveṣa*), etc., one of those attributes which inhere in the soul (*ātman*: one of the nine substances, or types of substance). In this context and framework, Praśastapāda attempts the elusive task of defining and classifying the state of happiness.

Sukha is essentially favourable or gratifying (*anugrahalakṣaṇa*, 'characterized by gratification'). It occurs in the presence of pleasant, desirable things, such as garlands (*sragādyabhipretaviṣaya-sānnidhye sati*). Provided that certain other causal factors, including good karma (i.e. *dharma*), are in place, it generates such symptoms as gratification, affection, brightness of the eyes, etc. (*anugrahābhiṣvaṅganayanādiprasādajanaka*). If the pertinent objects are not actually present, their recollection or anticipation may produce *sukha*. On the other hand, there is also a kind of happiness which is enjoyed by the wise (*vidvas*). It manifests itself even if there is no actual object of enjoyment, nor any recollection, desire or anticipation relating to it; it is based entirely on knowledge, inner peace, contentment and a special kind of good karma (*vidyāśamasaṃtoṣadharmaviśeṣanimitta*). *Duḥkha* is the opposite of *sukha*. It is essentially oppressive and distressing (*upaghāta-lakṣaṇa*); it is associated with undesirable objects and bad karma (*adharma*) and produces such feelings as anger (*amarṣa*), distress (*upaghāta*) and depression (*dainya*). Furthermore, *sukha* is inherently related to desire (*icchā*), while *duḥkha* is related to aversion (*dveṣa*). Desire is eagerness to obtain something either for oneself or for somebody else; as subdivisions or varieties of *icchā*, Praśastapāda recognizes not only sexual desire (*kāma*) or craving (*abhilāṣa*) for food etc., but also such selfless emotions as compassion (*kāruṇya*) and detachment (*vairāgya*).[3]

Several questions suggest themselves in connection with Praśastapāda's presentation. What is the meaning and function of *anugraha*, and in what sense is it the mark or characteristic (*lakṣaṇa*) of *sukha*? What is the precise meaning of the other terms used to describe and explain happiness? What is the positive content, if any, of *sukha* apart from the fact that it is the opposite of, and incompatible with, *duḥkha* or suffering? What is implied in its relationship with *icchā*, desire? What is the relationship and distinction between regular happiness or pleasure and the 'happiness of the wise', apart from Praśastapāda's own explicit statements? Beginning with the last of these questions, we may say that the 'happiness of the wise' is, unlike 'regular' *sukha*, not an actual affective response to an external stimulant, but a dispositional happiness based on the inner resources of a person. Its condition is not the gratification of desires, but their transformation and transcendence. It is the happiness of contentment, and associated with what other texts describe as the elimination of thirst or desire (*tṛṣṇākṣaya*).[4] Yet it remains a mere experiential attribute (*guṇa*) of the self (*ātman*) and needs to be distinguished from the kind of bliss which other systems identify as the very nature and essence of the self. But what is the content and identity of the experience of *sukha*, be it regular pleasure or the happiness of the wise? In what sense does the word *anugraha* define or describe the identity of *sukha*, in particular of regular pleasure and happiness? Praśastapāda first presents *anugraha* as the characteristic mark (*lakṣaṇa*) of *sukha*. But subsequently, he calls *sukha* the generative cause (*janaka*) of *anugraha* and of other mental and physiological phenomena. This seems to indicate that *anugraha* does not really describe the experiential content and inner nature of *sukha* itself, but rather one of its effects. In a literal sense, this would also seem to indicate that *sukha* as such is not a pleasant feeling or an agreeable state of awareness, but rather a mental condition which serves as a causal factor to produce such an awareness.

The commentator Śrīdhara explains that *anugraha* (which he calls an effect, *kārya*, of *sukha*) is a feeling or awareness relating to *sukha* (*anugrahaḥ sukhaviṣayaṃ saṃvedanam*).[5] The awareness of pleasure follows and accompanies pleasure, but is not simply identical with it. *Sukha* appears thus as an objective, reified mental factor which produces a number of effects, including the awareness of *sukha* itself. Śrīdhara says: 'By virtue of the occurrence of pleasure, an awareness of itself is produced; and this is something gratifying for the soul' (*sukhena-utpannena svānubhavo janyate, sa eva-ātmano 'nugrahaḥ*). Likewise, Vyomaśiva, Śrīdhara's predecessor among the Praśastapāda commentators, and Udayana, his successor, describe *anugraha* as 'cognition relating to pleasure' (*sukhaviṣayaṃ jñānam*) or simply 'cognition of pleasure' (*sukhajñānam*).[6]

Nonetheless, Śrīdhara does not hesitate to paraphrase the term *anugrahalakṣaṇa* in Praśastapāda's initial definition of *sukha* as *anugrahasvabhāva*, 'having *anugraha* as its essence'. Likewise, Vyomaśiva uses the expression *anugrahasvarūpa*.[7] In the context of classical Vaiśeṣika, this does not constitute any significant conflict or discrepancy. Here as well as in other instances, causal factors and internal data of awareness are not strictly separated.[8] While it may be true that *sukha* is the 'cause' of the feeling of gratification (*anugraha*), it also enters into the subjective realm of awareness and is experienced as its own agreeable content; at the same time, the sense of 'gratification' or 'being agreeable' may appear as the very essence of the *sukha* by which it is 'caused'. But what exactly is this agreeable content and essence associated with *sukha*? What is *sukha*, apart from the fact that we respond positively to it, that we desire and pursue it? What is it apart from the fact that it is opposed to *duḥkha*? Śrīdhara rejects the view that *sukha* is the mere absence of *duḥkha* (*duḥkhābhāva*). He insists that it has a positively identifiable essence of its own. This is, indeed, a fundamental postulate of the Vaiśeṣika school. It is not only taken for granted by Praśastapāda, but also stated in the *Vaiśeṣikasūtra* itself, which claims an irreducible identity and mutual otherness (*arthāntarabhāva*) for both *sukha* and *duḥkha*.[9] But are Praśastapāda's notions of *anugraha* and *upaghāta*, or related concepts, such as *anukūla* and *pratikūla*,[10] sufficient to ascertain and articulate this identity, and to avoid the danger of circularity in the definition of *sukha* and *duḥkha*? It is symptomatic that in his rejection of the view that *sukha* is just absence of *duḥkha* Śrīdhara does not simply rely on Praśastapāda's terminology or on the observation that we respond differently to *sukha* and *duḥkha*. Instead, he invokes the concept of *ānanda*, which is not normally used in Vaiśeṣika; it is most commonly associated with the Upaniṣadic tradition, where it represents the ideal of absolute, non-polar bliss.[11]

Śrīdhara's predecessor Vyomaśiva also rejects the thesis that happiness is nothing but the absence of pain. Likewise, it would be untenable to say that pain or suffering is nothing more than the absence of happiness or pleasure (*evaṃ duḥkhasya-api na sukhābhāvarūpatā-iti*). In the same context, Vyomaśiva criticizes and rejects the view that all worldly experiences, including those which seem to be happy and pleasant, are essentially painful and unpleasant for a wise, discriminating person (*na ca sarvaṃ vivekinaḥ svarūpato duḥkham*).[12] In principle, this view appears in various schools of Buddhism as well as Hinduism. Most specifically, it seems to be associated with the classical Yoga system, for instance with the following statement in Vyāsa's *Yogabhāṣya*: *duḥkham eva sarvaṃ vivekinaḥ*.[13]

Beginning with the *Nyāyasūtra*, the classical Nyāya texts agree with the Vaiśeṣika view that *sukha* can neither be reduced to the mere absence of *duḥkha*, nor to *duḥkha* itself. Yet the Nyāya treatment of the issue, and of the entire syndrome of pleasure and pain, happiness and unhappiness, is by no means simply identical with the Vaiśeṣika approach. The differences are as significant and instructive as the parallels; and they reflect the general fact that Nyāya and Vaiśeṣika, which came to form such a close alliance and combination in later times, were clearly distinct in their origins and early developments. Early Nyāya does not share the Vaiśeṣika commitment to a complete enumeration and classification of what there is. As a matter of fact, Vātsyāyana Pakṣilasvāmin's *Nyāyabhāṣya* suggests that such a project is unfeasible as well as soteriologically irrelevant. Instead, the Nyāya teachers propose a list of significant topics of epistemology, dialectics and metaphysics, for which they claim not only philosophical validity, but also soteriological applicability and relevance. The second item in this list, which begins with *pramāṇa*, 'means of knowledge', is *prameya*, 'object of knowledge'.[14] Twelve items are enumerated as soteriologically relevant 'objects of knowledge', beginning with self (*ātman*) and body (*śarīra*) and ending with pain (*duḥkha*) and final liberation (*apavarga*). It is conspicuous and significant that, unlike *duḥkha*, *sukha* does not appear in this list. However, this omission does not mean that the Nyāya texts disregard the topic of happiness. The *Nyāyasūtra* itself presents happiness or pleasure, together with desire, aversion, effort, pain and cognition, as a 'mark' of the soul: *icchādveṣaprayatnasukhduḥkhajñānāny ātmano liṅgam iti*.[15] Later on, the Sūtra states explicitly that the emphasis on pain does not imply that *sukha* as such does not occur at all.[16] The commentaries provide further explanations and clarifications concerning the mutual relationship and the systematic and soteriological implications of *sukha* and *duḥkha*.

The *Nyāyabhāṣya* states repeatedly that the explicit citation of *duḥkha*, together with the omission of *sukha*, in the list of objects of knowledge does not and cannot indicate a rejection or denial (*pratyākhyāna*) of *sukha*, which is an undeniable experiential reality, insofar as it is the content of an agreeable feeling (*anukūlavedanīya*).[17] The omission of *sukha* and the implicit suggestion that it amounts ultimately to *duḥkha* has to do with meditative practice (*samādhibhāvana*) and serves a soteriological purpose. It helps us realize the fact that all worldly pleasure is invariably accompanied by pain (*duḥkhānuṣaṅga*), and it teaches us that such pleasure should not be pursued (*sukhaṃ duḥkhānuṣaktam anādeyam*).[18] We have to realize that pain is inherent in the pursuit of worldly happiness, and that no happiness can be obtained without incurring unhappiness and pain (*sukhāṅgabhūtaṃ duḥkham, na duḥkham anāsādya śakyaṃ sukham avāptum*).[19] Such happiness is like milk (*payas*) contaminated by poison, or like food which is a mixture of honey and poison (*madhuviṣasaṃpṛktānna*).[20] Realizing this in thought and practice will lead us to detachment (*vairāgya*) from the world and prepare us for the goal of final liberation (*apavarga*) from the cycle of rebirth.[21] Compared to the Vaiśeṣika treatment of *sukha* and *duḥkha*, this presentation is obviously closer to classical Yoga. In a sense, it represents an intermediate position between Vaiśeṣika and Yoga.

Happiness must not be pursued, even though it may have a reality and identity of its own. Its pursuit itself implies attachment (*rāga*) and constitutes a form of bondage (*bandhana*). Vātsyāyana Pakṣilasvāmin does not hesitate to apply and emphasize this basic assumption in his discussion of final liberation (*apavarga: mokṣa*).[22] The goal of final liberation cannot be associated with the idea of eternal happiness (*nityasukha*); it cannot and should not be a 'desirable', 'attractive' goal in this sense. Otherwise, any attempt to reach such a goal would produce new attachment and perpetuate our worldly bondage. 'Unless we abandon our attachment to eternal happiness, final liberation cannot be attained, because attachment itself is known to be bondage' (*nityasukharāgasya-aprahāṇe mokṣādhigamābhāvaḥ, rāgasya bandhana-samājñānāt*).[23] Vātsyāyana discusses and rejects the Vedāntic view that eternal bliss is inherent in the very nature of the self, and that final liberation, which coincides with recognizing the true nature of the self, thus implies eternal happiness. For him, final liberation is, above all, complete and definitive absence of pain, as well as transcendence of both pleasure and pain.[24] Of course, the Vedānta teachers themselves were well aware of the problems inherent in the relationship between happiness or bliss on the one hand, and desire and attachment on the other. They proposed various approaches to deal with these problems, and to distinguish soteriological commitment from worldly forms of motivation.[25]

The opponents criticize the views on final liberation held in the Nyāya and Vaiśeṣika schools. They argue that such notions cannot explain or justify any kind of soteriological motivation. A popular verse states that 'being a jackal' in the Vṛndā forest would be preferable to 'liberation according to the Vaiśeṣika' (*vaiśeṣikī muktiḥ*).[26] Even within the two schools, the treatment of this topic is somewhat ambiguous and controversial. Bhāsarvajña[27] does not accept the view expressed in the *Nyāyabhāṣya.* Vātsyāyana himself does not categorically deny the possibility of happiness in the state of liberation. But he insists time and again that any kind of soteriological motivation by *sukha* would be counterproductive. Whether or not there is eternal happiness for the liberated self (*muktasya nityaṃ sukham bhavati athāpi na bhavati*)—what counts is that the seeker should have relinquished his attachment to eternal happiness (*nityasukharāga*).[28]

The classical Nyāya commentators Uddyotakara, Vācaspati and Udayana expand, elucidate and modify the statements found in the *Nyāyasūtra* and *Nyāyabhāṣya.* Uddyotakara associates himself closely with the Vaiśeṣika school and adopts many of its teachings for the Nyāya system. Udayana is both a Naiyāyika and a Vaiśeṣika; and he deals with the problems concerning *sukha* in a Nyāya as well as Vaiśeṣika context. With him, the debate reaches a new phase of reflection and conceptual analysis. We focus on his discussion of the topic of happiness, as found in his Vaiśesika commentary, *Kiraṇāvalī.*

Udayana first addresses the problems inherent in the attempt to define *sukha* in terms of *anugraha,* i.e. the fact that we experience *sukha* as agreeable and gratifying. It is obvious that *anugraha* itself is a problematic concept. Udayana also refers to the relationship between *sukha* (as a *guṇa,* i.e. quality of the self) itself, and the awareness of *sukha* (*sukhabuddhi*; here, we have to keep in mind that *buddhi* is another quality of the self). Finally, he examines the relationship between 'pleasantness', or that which constitutes pleasure (*sukhatva*), on the one hand, and 'desirability', or the fact of being desired (*iṣṭatva*), on the other. 'What about the suggestion that it (i.e. *sukhatva* as that which constitutes pleasure) is desirability (or desiredness)?—No, (this has to be rejected) because there is a discrepancy due to the (variety of) means for its accomplishment.—What if we say instead that *sukhatva* is exemplified by what is desired for its own sake (*svabhāvatas*; due to its own nature)?—No, (this has to be rejected) because of uncertainty (*anaikāntikatva*) due to absence of pain (which is also desired for its own sake).—What if we say that this (absence of pain), too, is desired for the sake of happiness?—No, (this has to be rejected) because the consequence would be that final liberation, which is absence of pain, is not a valid human goal (*puruṣārtha*).' In response to such problems and ambiguities, Udayana tries to provide a more precise and definitive determination of the meaning of *sukha*, or the *sukhatva* which constitutes *sukha* as such. 'Pleasantness', i.e. the nature of happiness or pleasure, is exemplified and made manifest by that which is desired for its own sake and at the same time (unlike absence of pain) does not require any reference to a positive counterpart (*pratiyoginirūpaṇānadhīnaṃ svabhāvata isyamānaṃ sukhatvābhivyañjakam*). Udayana also rejects the idea that happiness is mere absence of pain (*duḥkhābhāva*) or amounts to unhappiness and pain for the wise and discriminating ones (*vivekin*).[29] In his Nyāya commentary *Pariśuddhi* (a sub-commentary on Vācaspati's *Nyāyavārttikatātparyaṭīkā*), Udayana offers further reflections on *sukha*; he presents, for instance, a classification of different types of worldly pleasure.[30]

Udayana's discussion of the relation between pleasure (or happiness) and desirability (or the status of being desired; i.e. *iṣṭatva*) illustrates a crucial problem which accompanies, at least implicitly, a good deal of the debate concerning the nature and pursuit of happiness and pleasure.[31] To what extent does the status of being pursued and desired determine the meaning and function of *sukha*? To what extent does the content of *sukha* explain the desire to obtain it? What is the role of happiness and pleasure within the entire domain of human action and motivation? Can other human goals and purposes (*prayojana, artha*) and apparently independent modes of orientation be derived from or reduced to the pursuit of happiness and its complement, the avoidance of pain? An essential affinity and correlation between *icchā* and *sukha* on the one hand, and *dveṣa* and *duḥkha* on the other, is usually taken for granted by the philosophical schools. Yet *sukha*, in its varieties from sheer worldly pleasure to the detached and spiritual 'happiness of the wise', is not normally seen as the only purpose (*prayojana, artha*) of human existence, or as the only motivating force for human action. The Indian tradition in general recognizes a scheme of different values which are supposed to guide and motivate human action. This is the well-known scheme of the four 'goals of man' (*puruṣārtha*), i.e. religious merit (*dharma*), political and economic success (*artha*), pleasure, especially of the erotic kind (*kāma*), and final liberation (*mokṣa*). An older threefold group (*trivarga*) does not contain final liberation.[32] In this list, *kāma* seems to correspond most directly to *sukha* (and *mokṣa*, whatever its other associations may be, to absence of pain), while the other 'goals of man' seem to represent different and independent values and interests. But is this really so? Are they genuinely different from the pursuit of *sukha*? And if they are different, how can their special motivating power, their status as distinctive goals and purposes, be defined?

Among the Nyāya commentators, Uddyotakara and some of his successors address such questions explicitly. In his *Nyāyavārttika*, Uddyotakara denies that the 'goals of man' are really independent factors of motivation, and he does not hesitate to reduce the fourfold scheme to the pursuit of happiness and the avoidance of pain. He says: *kena punaḥ prayujyate? dharmārthakāmamokṣair iti kecit. vayaṃ tu paśyāmaḥ sukhaduḥkhāptihānibhyāṃ prayujyata iti.* All goals, if they are not happiness or absence of pain themselves, motivate the mind only insofar as they are means to this end: *sukhaduḥkhasādhanabhāvāt tu sarve 'rthāś cetanaṃ prayojayanti-iti.*[33] This sounds sufficiently clear. Yet it contains a significant problem of interpretation, for which it does not provide an explicit clarification. In what sense can the motivating power of all 'goals of man', or of all human interests, be reduced to, or translated into, the motivating power of *sukha* and *duḥkha*? Does *sukha* occur at different levels? Can it be filled with different meanings and contents, depending on the level and context of orientation? Do *artha* and *dharma* constitute happiness for those who are committed to them? Can we pursue happiness itself in the form of economic success or religious merit? Or is the commitment to and the pursuit of these goals only a less direct and immediate way of pursuing what is ultimately one and the same goal of pleasure and well-being, i.e. a readiness to seek happiness in the more distant future? Uddyotakara's expression *sukhaduḥkhasādhanabhāva*, 'being the means for (the attainment of) happiness and (the avoidance of) pain', together with the general background of Vaiśeṣika and Nyāya thought, suggests the second interpretation. Regardless of all other implications, the acquisition of *artha*, i.e. power and wealth, and the accumulation of *dharma*, religious merit, certainly imply the potential to enjoy pleasure and well-being at a later time. In the case of *dharma*, this later time would normally be located beyond the limits of our current existence, i.e. in heaven (*svarga*) or a future rebirth; *dharma* is the means to secure the reward of heavenly well-being, or the enjoyment of a pleasant rebirth.[34] Of course, this interpretation does not exhaust the rich potential of the *puruṣārtha* theory. But it is the prevailing approach not only in Nyāya and Vaiśeṣika, but also in other philosophical schools which are committed to the ideal of final liberation (*mokṣa*).

Earlier, we have referred to the *Nyāyabhāṣya*'s (and *Nyāyavārttika*'s) insistence that there should be no assumption of pleasure or bliss in final liberation. This would imply attachment, which is an ingredient of the world of *saṃsāra*. The addiction to enjoyment and the acquisition of its means is a symptom of a fundamental cognitive disease. Like other schools of Indian philosophy, Nyāya and Vaiśeṣika prescribe the 'medicine of knowledge and detachment'[35] against this cognitive disease. The attachment to happiness and its pursuit is something that has to be cured and overcome. Yet this has not prevented the Nyāya-Vaiśeṣika teachers from recognizing the irreducible reality and identity of *sukha*, and from making serious theoretical efforts to analyse and understand the idea and phenomenon of happiness.

This is not the place to compare the Nyāya and Vaiśeṣika analysis of pleasure and pain, happiness and unhappiness with other Indian approaches to this topic. Sāṃkhya and in particular

Yoga would suggest themselves. How do they correlate *sukha* and *duḥkha*? In what sense do they subordinate or even reduce *sukha* to *duḥkha*? How do they connect *sukha* with *rāga*, i.e. attachment and passion? How do they view its role in the processes of rebirth and transmigration, or in the turning of the 'wheel of *saṃsāra*' (*saṃsāracakra*)? Do they make any attempts to determine the ontological status of *sukha* and *duḥkha*? In what sense do they admit a transcendent bliss, which is free from attachment and supersedes the worldly, saṃsāric dichotomy of *sukha* and *duḥkha*?

The same set of questions could, of course, also be asked with reference to Buddhist thought which would be even more pertinent in this connection. In particular, the soteriological approach of early Nyāya echoes, but also modifies and transforms early Buddhist ideas, such as the notion of 'dependent origination'.[36] Some of the Nyāya statements concerning the relationship between the analysis of *sukha* and the practical realization of detachment (*vairāgya*) are clearly reminiscent of old and familiar Buddhist phrases.[37] However, by the time of Diṅnāga, Dharmakīrti and, on the Nyāya side, Uddyotakara, mutual critique and polemics had become far more conspicuous than any borrowing. Here as elsewhere, the concept of the self (*ātman*) is at the centre of disagreement and controversy. But this would be a topic for another, more elaborate paper.[38]

NOTES

1. On the role of *sukha, ānanda,* etc. in presystematic Indian thought, see G. Gispert-Sauch, *Bliss in the Upanishads,* New Delhi, Oriental Publishers and Distributors, 1977. Concerning the English terminology employed in this article, we use 'happiness', but also, if appropriate, 'pleasure' for *sukha.* This reflects the semantic range of the word *sukha,* but also a certain ambiguity in the English terms themselves.
2. On the Vaiśeṣika categories, cf. W. Halbfass, *On Being and What There Is: Classical Vaiśeṣika and the History of Indian Ontology,* Albany, State University of New York Press, 1992, pp. 69–87. The Vaiśeṣikasūtra lists only 17 qualities (*guṇa*).
3. See *The Bhāṣya of Praśastapāda, together with the Nyāyakandalī of Śrīdhara,* ed. by V.P. Dvivedin, Benares, E.J. Lazarus and Co., 1895, Reprint Delhi, Indian Books Centre, 1984, pp. 259–60.

4. See, for instance, *Yogabhāsya* II, 42 (quoting Mahābhārata XII, 171, 51; 268, 6: *yac ca kāmasukhaṃ loke yac ca divyaṃ mahat sukham/ tṛṣṇākṣayasukhasya-ete na-arhataḥ ṣoḍaśiṃ kalām*).
5. See Śrīdhara, *Nyāyakandalī* (as above, n. 3), p. 260.
6. See Vyomaśiva, *Vyomavatī* (in *The Praśastapādabhāshyam by Praśastadevācārya with Commentaries*, ed. by Gopinath Kaviraj and Dhundhiraj Shastri, Benares, Chowkhambha, 1924–30), p. 624; and *Udayana, Kiraṇāvalī*, ed. by J.S. Jetly, Baroda, Oriental Institute, 1971, p. 248 (this section is missing in the older editions of the text).
7. *Nyāyakandalī*, p. 259; cf. Vyomaśiva, *Vyomavatī*, p. 624.
8. Cf. Halbfass, *On Being and What There Is*, p. 100.
9. See *Vaiśeṣikasūtra* X, 2 (ed. Jambuvijaya; X, 1, 1 in the Upaskāra version): *iṣṭāniṣṭakāraṇaviśeṣād virodhāc ca mithaḥ sukhaduḥkhayor arthāntarabhāvaḥ*.
10. See *Yogabhāṣya* II, 14f; *Nyāyabhāṣya* I, 1, 9.
11. See *Nyāyakandalī*, p. 260. On *ānanda*, see also pp. 286f; Śrīdhara rejects the Advaita Vedānta notion of *ānanda*.
12. *Vyomavatī*, p. 624.
13. *Yogabhāṣya* II, 15; cf. also II, 14: *viṣayasukhakāle 'pi duḥkham asty eva pratikūlātmakaṃ yoginaḥ*. The Yoga analysis of *sukha* and *duḥkha* has parallels in Buddhist thought.
14. See *Nyāyasūtra* and *Nyāyabhāṣya* I, 1, 9; at the end of this section, the Bhāṣya mentions and dismisses the Vaiśeṣika project of enumeration and classification.
15. *Nyāyasūtra* I, 1, 10.
16. See *Nyāyasūtra* IV, 1, 54 (*vividhabādhanāyogād duḥkham eva janmotpattiḥ*) and 55 (*na, sukhasya-antarālaniṣpatteḥ*).
17. See *Nyāyabhāṣya* I, 1, 9: *duḥkham iti na-idam anukūlavedanīyasya sukhasya pratīteḥ pratyākhyānam*; IV, 1, 55: *na khalv ayaṃ duḥkhoddeśaḥ sukhasya pratyākhyānam*.
18. See *Nyāyabhāṣya* I, 1, 2; and below, n. 21.
19. *Nyāyabhāṣya* IV, 1, 57.
20. *Nyāyabhāṣya* IV, 1, 53; I, 1, 2.
21. *Nyāyabhāṣya* I, 1, 9: *duḥkham iti samādhibhāvanam upadiśyate. samāhito bhāvayati, bhāvayan nirvidyate, nirviṇṇasya vairāgyam, viraktasya-apavarga iti*.
22. See *Nyāyabhāṣya* I, 1, 2; I, 1, 22. Cf. Śrīdhara, *Nyāyakandalī*, p. 286.
23. *Nyāyabhāṣya* I, 1, 22.
24. See *Nyāyabhāṣya* I, 1, 22; and I, 1, 2: *kathaṃ buddhimān sarvaduḥkhocchedaṃ sarvaduḥkhāsaṃvidam āpavargaṃ na rocayet*: this resumes and supersedes the earlier question: *kathaṃ buddhimān sarvasukhocchedam acaitanyam amum apavargaṃ rocayet*.
25. See, for instance, *Brahmasiddhi of Maṇḍanamiśra* I, 2, ed. by S. Kuppuswami Sastri, Delhi, Sri Satguru Publications, 1984 (reprint), pp. 1–6.
26. A modified version of this verse, which refers to the Nyāya teacher Gautama instead of the Vaiśeṣika, appears in Bhāsarvajña's *Nyāya-*

bhūṣaṇa, ed. by Yogīndrānanda, Varanasi, Ṣaḍdarśan Prakāśan Pratiṣṭhān, 1968, p. 594. Śrī Harṣa, *Naiṣadhīyacarita* XVII, 75, associates the Nyāya-Vaiśeṣika concept of final liberation with the 'status of a stone' (*śilātva*), i.e., utter unconsciousness and inertia. Cf. Śrīdhara, Nyāyakandalī, p. 287: *pāṣāṇād aviśeṣaḥ.*

27. See Bhāsarvajña, *Nyāyabhūṣaṇa*, pp. 594–8. Bhāsarvjña presents a spirited defense of the idea of eternal happiness (*nityasukha*), complete with Upaniṣadic references (p. 595). He distinguishes *nityasukha* from all object-based (*vaiṣayika*, p. 595) happiness and emphasizes that the desire to attain it does not entail any kind of bondage.
28. See *Nyāyabhāṣya* I, 1, 22. This is not the place to discuss the role and meaning of liberation in most ancient Nyāya and Vaiśeṣika thought.
29. Udayana, *Kiraṇāvalī*, pp. 247ff.
30. See Udayana, *Pariśuddhi* I, 1, 4 (in *Nyāyadarśana of Gautama*, ed. by A. Thakur, Darbhanga, Mithila Institute, 1967, p. 258). Worldly happiness can be caused by habitual practice (*abhyāsa*), conceit (*abhimāna*), enjoyment of sense- objects or sensuous pleasures (*viṣayasaṃbhoga*), and wishful thinking (*manoratha*). Expanding on statements in Vācaspati's Nyāyavārttikatātparyaṭīkā, Udayana also discusses the relation between *sukha* and cognition (*jñāna*) and rejects the view that *sukha* is a special kind of cognition (*jñānaviśeṣa*); see *Pariśuddhi*, pp. 254ff.
31. The problem is obviously implied in Vaiśeṣikasūtra X, 2, which invokes the distinction between desired and undesired entities to explain the 'otherness' of *sukha* and *duḥkha*. Praśastapāda (see *The Bhāṣya of Praśastapāda*) refers to the awareness of something desirable (*iṣṭopalabdhi*) as a cause of *sukha*. *Sukha*, on the other hand, is supposed to explain the arising of *icchā*, desire. Udayana articulates the circularity (*iṣyamāṇatvād anukūlatvam, tena ca-iṣyamāṇatvam*) and insists that *sukha* is a special content or object of experience (*viṣayaviśeṣa*) which is naturally and inherently agreeable (*nisargānukūlasvabhāva*). Once again, and in accordance with the tradition, he presents 'being the content of an agreeable feeling' (*anukūlavedanīyatva*) as the basic characteristic of *sukha* (see *Pariśuddhi*, p. 256). Insofar, *sukha* itself has to be distinguished from enjoyment (*upabhoga*), which is the experience (*anubhava*) of *sukha* (p. 257).
32. On the *puruṣārtha* doctrine, see Ch. Malamoud, 'On the Rhetoric and Semantics of *puruṣārtha*', in *Way of Life. Essays in Honour of Louis Dumont*, ed. by T.N. Madan, New Delhi, Vikas, 1982, pp. 33–54; and 'Menschsein und Lebensziele. Beobachtungen zu den *puruṣārthas*', in *Hermeneutics of Encounter. Essays in Honour of G. Oberhammer*, ed. by F.X. D'sa and R. Mesquita. Vienna: De Nobili Research Library, 1994, pp. 33–54.
33. *Nyāyavārttika* I, 1, 1 (in *Nyāyadarśana of Gautama*, p. 14).
34. According to Jayanta, only the attainment of happiness and the avoidance of pain are direct and primary (*mukhya*) ends or purposes, whereas the means (*sādhana*) to achieve these ends are secondary (*gauṇa*) purposes;

see *Nyāyamañjarī*, ed. by S.N. Sukla and Dhundhiraj Sastri, Benares, Haridas Gupta, 1934–6, vol. 2, p. 126. In this sense, *artha* and *dharma* would have to be classified as secondary or indirect purposes.

35. The expression 'medicine of knowledge and detachment' (*jñāna-vairāgyabheṣaja*) is used by Śaṅkara, *Upadeśasāhasrī* XIX, 1. This idea is, of course, shared by many Hindu and Buddhist schools of thought.
36. *Nyāyasūtra* I, 1, 2 is obviously indebted to *pratītyasamutpāda*.
37. Compare, for instance, Vātsyāyana's *bhāvayan nirvidyate, nirviṇṇasya vairāgyaṃ, viraktasya-apavarga iti* (see above, n. 21) with the Buddhist phrase (found in Majjhimanikāya 22 and elsewhere): *evaṃ passaṃ . . . nibbindati, nibbindaṃ virajjati, virāgā vimuccati.* See also T. Vetter, *Der Buddha und seine Lehre in Dharmakirtīs Pramāṇavārttika*, Vienna, Arbeitskreis für Tibetische und Buddhistische Studien, 1984, p. 127, n. 1.
38. See, for instance, Dharmakīrti, *Pramāṇavārttika I (Pramāṇasiddhi)*, v. 222ff (ed. by Dwarikadas Shastri, Varanasi, Bauddha Bharati, 1968, pp. 77ff); v. 233 (p. 80) has the word *saviṣānnavat*, which appears also in Nyāyavārttika I, 1, 2. Cf. also T. Vetter, *Der Buddha*, pp. 120ff. Vetter's notes are not always convincing; p. 132, n. 1 obviously misinterprets the rhetorical question *kathaṃ buddhimān sarvasukhocchedam acaitanyam amum apavargaṃ rocayet* (see above, n. 24). Vācaspati's Ṭīkā on I, 1, 2 quotes Dharmakīrti, v. 202f.

Chapter 10

Causal Connections, Cognition and Regularity: Comparativist Remarks on David Hume and Śrī Harṣa

C. Ram-Prasad

A philosopher interested in Indian thought could do a lot worse than examining in detail that masterly work of critical analysis, the *Khaṇḍana-khaṇḍa-khādya,*[1] of the Advaitin Śrī Harṣa. In its pages are found some of the most sustained and penetrating studies of philosophical categories of middle classical Indian thought. 'Through his incisive critique of the Nyāya-Vaiśeṣika categories of *pramāṇas* in general, and of the definitions of Udayana in particular, he paved the way for the rise of the Navya-nyāya school . . . (his) trenchant criticism of Nyāya categories had a salutary effect on the Indian philosophic scene, and the philosophic sophistication of later authors of both Nyāya and Vedānta deepened as a result.'[2] I shall attempt here a reading of a short and suggestive passage on causality in the introductory section of the work. The interest in this passage lies in the fact that it addresses an aspect of causality which is not usually attended to in classical Indian discussions.

Bimal Matilal remarks in a brief passage that though there is some resemblance between the Nyāya position on causality and David Hume's, 'the doctrine of invariable sequence was not propounded [by Nyāya] in exactly a Humean spirit. For Hume, it is only the mind that spreads itself on external objects and conjoins them as cause and effect while nothing really exists between them to be so conjoined. . . . For Nyāya, invariable sequence is discovered by the mind but it exists between extramental realities. . . .'[3] While examining Śrī Harṣa's critical analysis of the Nyāya view of 'invariable sequence' (or concomitance), we shall see that there is a defensible interpretation of Hume (rather different from the conventional one Matilal follows), which would not only be different from the Nyāya view but, in fact, would be akin to Śrī Harṣa's own.

I QUESTIONS ON CAUSALITY DISTINGUISHED

To start with—and to indicate the unconventional nature of the issue with which I shall be dealing here—it will be of some use to distinguish between several sorts of questions about causality in Indian thought.

(i) What is meant by causality, causal factor or causal connection?
(ii) What kinds or sorts of causes are there?
(iii) Which elements—objects, entities or other ontological categories—count as what sorts/kinds of causes?
(iv) What is the nature of a cause, and the nature of the relationship between cause and its effect?
(v) How—on the basis of what—are the various definitions of the categories and concepts given above justified?

The point I want to make here is that the debates on causality in classical Indian thought concentrate almost exclusively on (ii), (iii) and (iv), whereas I want to focus on an interesting little passage in Śrī Harṣa where he seems to be addressing (i) and (v). This focus means that I shall have very little, if anything, to say about the classical questions on causality in Indian thought. On the contrary, my interest in (i) and, more importantly, (v), will deliberately be unconventional, primarily because I believe that it is worthwhile to look at a neglected issue which is undoubtedly present in Indian thought, and incidentally because I think there are some interesting comparativist ideas to be found here. The issue I deal with here appears out of necessity in isolation, but I am aware that it is part of a much more complex and elaborate philosophic architectonic constructed by Śrī Harṣa.

II THE CLASSICAL CONSENSUS ON MINIMAL CAUSAL REQUIREMENTS

It is startling but true that there is almost universal consensus in Indian thought about the answer to question (i); whatever the subsequent debate over (ii)–(iv), there is broad agreement over what, minimally, a causal connection should be. On second thoughts, perhaps it is not so surprising after all: it is because they agree what a causal factor should amount to, that the Indian philosophers can then put up various candidates and examine their natures.

In order to arrive at a consensual definition, let us look at these definitions:

1. Causality is a regulative/determinative antecedent of the effect (*kāryaniyataḥ pūrvabhāvaḥ*); it is a relation of obligatory dependence or regular connection.[4] 'Regulative' is used here as synonymous with 'determinative'. Cause C 'regulates' effect E if C's nature, time and place of occurrence extract conformity of nature, time and place of occurrence from E.
2. Causality is that which follows the rule, 'that being-so, this occurs' (*asmin sati idaṃ bhavati*).[5] Each state results in a concomitant occurrence, each change results in a concomitant transformation (*tad-bhāva-bhāvitva tad-vikāra-vikāritva*). The effect is regulatively dependent on its cause. '[W]e call dependence of the effect upon its cause the fact that it *always* follows upon the presence of that cause. . . .'[6] There is thus, to use Stcherbatsky's term, 'strict conformity' between fact and result.

   It is noticeable that the idea of causality as a regulative/determinative relationship, where one entity extracts conformity from another, is clearly common to both Naiyāyikas and Buddhists, opposed though they are to each other on other matters. The notion that whatever else is at issue, minimally, a causal factor is a determinative connection of some sort, is pointed out by Śrī Harṣa:
3. 'Equally for us, causality is a determinative prior connection.'[7]

In emphasizing the consensus, Śrī Harṣa seems to accept the Nyāya understanding of determinative priority.

A causal factor—strictly, an element in which causehood (*kāraṇatva*) inheres—must be considered, according to the Naiyāyika, as a determinative antecedent, whose determinative capacity consists in the possession of invariable concomitance (*anvayavyatireka*) with an effect.[8] Concomitance becomes thereafter, in Nyāya thought, determinative (*niyata*) antecedence. *Invariable concomitance* I shall define briefly thus:

If always when X occurs, Ψ occurs (or must have occurred), and never when Ψ does not occur does X occur, then X is invariably concomitant with Ψ.[9]

Given these thoughts, a causal factor is understood thus:

[C] C is a/the causal factor for E if E occurs when C occurs (or has occurred) and E does not occur when C does not occur; i.e. when E is invariably concomitant with C.

[C] is notable in that a version of it will remain unchallenged even by the Buddhists. Most of the debate in Indian thought is over the ontological relationship between C and E: i.e. whether there is some sense in which E is ontologically derived from or sustained by C, whether E must, in some sense, exist in C, whether C continues to exist after E has occurred, and so on; in short, the debates are over issues (ii)–(iv) mentioned above. But over issue (i) there is little debate: the Indian philosophers agree that there is a determinative invariance between two entities (objects, events, whatever) which represents a causal connection. This general agreement must be emphasized: for, whereas the debate over the existence of a causal connection may be challenged from both certain realist and certain idealist points of view in Western thought, even the Buddhist theory of interdependent (or 'combined dependent') origination (*pratītya-samutpāda*), which comes closest to challenging intuitive and pre-theoretic notions of causality, holds that[10] '[e]very origination obeys strict causal laws . . . it appears in accordance with strict causal laws'.[11]

It is in this context that Śrī Harṣa's epistemological challenge must be placed; for it is here that we come closest in classical Indian thought to what seems a challenge to the very notion of a causal connection. Though the question is directed, in the midst of a longer debate, at Nyāya realism, we can see it without too much trouble as a question directed at the consensus on (i). However, since a substantiation of that claim will involve tedious comparison with Buddhist views which he does not here mention, I shall stay faithful to the short passage involved, and concentrate exclusively on the Nyāya view. Nevertheless, the reader ought, I think, to keep the more general relevance of the criticism in mind throughout.

To put the matter at issue somewhat enigmatically: Śrī Harṣa asks question (v) above, and in doing so, seems to strike at the consensus on (i). If he is indeed questioning (i), he is committing himself to a position which seems extraordinarily difficult to defend; how difficult, we will see in a brief study of the non-causal regularity thesis which rejection of the consensus on (i) entails. It is here that some current discussion on the proper interpretation of Hume is relevant. However, I shall argue that in asking question (v), he is not, in fact, rejecting the consensus, but rather pointing out a critical epistemological constraint on its metaphysical claims. One comparativist consequence is that my favoured interpretation of Hume's view on causality will be seen to bear a close resemblance to the understanding of Śrī Harṣa presented here.

III THE CONSENSUS DEFINITION OF CAUSAL CONNECTION AS GIVEN BY NYĀYA

Causal efficiency is said to be found invariably with existenthood. Objects which are 'existent' in the realist way are causally efficient entities.[12] An entity is part of 'physical reality' if its occurrence is understood as spatiotemporally indexed. Then, the independent 'physical reality' is termed *sattā*, which I will translate here as 'existenthood'.[13] The overall strategy of the realist is to claim that a realist ontology (of objects inhered in by *sattā*) is invariably concomitant with causal efficiency, such that a world of 'real' objects is a world of 'real' causally efficient connections between those real/realist objects and their effects.

It is striking, but in keeping with the general suggestion about the realist–idealist consensus on causality in Indian thought, that the Buddhist position is exactly the same with regard to the correlation between entities and causality. It is said that whatsoever exists is itself a cause (*yā bhūtiḥ saiva kriyā*);[14] it is maintained that the ultimately existent is definitively causally efficacious (*artha-kriyā-āmarthaṃ paramārtha-sad ucyate*).[15] The disagreement, of course, has to do with what this existent is, whether pure point-instants or atomically constituted mid-size objects or something else. But importantly, that ontological category, howsoever construed, is thought of in terms of its possessing causal efficiency.

In the Nyāya case on the other hand, the relationship between existenthood and causal efficiency is sought to be established through determinative priority. If *y* is determined by *x*, then *y*'s occurrence is bound invariably with, because conditioned by, the occurrence of *x*. Thus, if *y* is to occur, *x* must be logically prior to it. But if it is so prior, its occurrence must be unconditioned by *y*, for it must be able to occur without the occurrence of *y* (this is what is taken to constitute priority). Then, Nyāya causality can be understood thus:

[NC] C is the causal factor for E when C is invariably priorly existent to E.

It is this understanding specifically which is challenged by Śrī Harṣa. A formulation of his understanding of the causal connection will eventually show itself to be his version of the best possible answer to question (v) asked in Section I.

IV A FIRST LOOK AT ŚRĪ HARṢA'S DEFINITION OF THE SPECIFICATION OF THE CAUSAL CONNECTION

We come now to the core of the issue. Śrī Harṣa gives his own understanding of causality; let us henceforth call it 'Śrī Harṣa's dictum':

> The specification [of the causal factor] is [given by] the cognition, 'this invariably exists prior to that'.[16]

This stands as an alternative to [NC] given above. On first reading, the claim seems to be that the relationship between two entities C and E, which we normally, intuitively, comprehend as a causal connection between C and E and therefore an ontologically 'real' determinative dependence of E on C, is not that at all; but rather, a mere idea we get from our experience of perceiving C and E occurring in a certain sequence. This is a dramatic and radical rejection of the consensus answer to (i); it denies that there is a determinative relationship between entities C and E, and holds instead that all there is to causality is the experience of C and E occurring in a certain way. C and E are here connected, not by the determinative dependence of E on the invariably prior existent C, but rather by the *impression* we have of E occurring after C as if it were determined by C. Nowhere is there any commitment to an independent connection between C and E. Were it not for our cognition, there would be no intrinsic concomitance and no ontological connection between C and E. Such an understanding of Śrī Harṣa's words would mean that the Advaitic view of causality was something like this:

[RC] C is the causal factor for E when C is cognized as invariably priorly existent to E.

This reduction of the notion of causality to the impressions of our cognitive apparatus bears a striking resemblance to the strategy found in the so-called regularity theory of causality in Western thought. An examination of this theory and its weaknesses will highlight the difficulties Śrī Harṣa would face, were he to be seen as adhering to a version of it.

V THE REGULARITY THEORY DESCRIBED

The regularity thesis can usefully be studied in the context of the 'conventional' interpretation of Hume, for it is the thrust of that interpretation that Hume was a regularity theorist. There are different questions here, which must be disambiguated so that I can say which concern me and which not:

(i) What is the prime objection to a general regularity theory?

(ii) Does what Hume say substantiate the charge that he was a regularity theorist?
(iii) What was Hume if he was not a regularity theorist?
(iv) What alternative is there to thinking of Hume's words as implying regularity?

I will have nothing to say about (ii), for it requires an exegetical assessment. I shall not give a detailed answer to (iii) as it would go well beyond the scope of this paper. With regard to (iv) I shall suggest an alternative interpretation of Hume's words on causality which seemingly support the regularity theory; this alternative interpretation, I suggest, looks like the alternative interpretation in the case of Śrī Harṣa. I shall look at (i) because Śrī Harṣa himself seems to anticipate just the sort of objection which a regularity theory would meet, and his response to it suggests that he could not have adhered to a regularity theory himself. This alternative interpretation of Śrī Harṣa not only is not a regularity theory, it is one which he claims cannot possibly be rejected by any of his opponents. The paper will end with a study of that claim.

Hume is taken to adhere to a regularity thesis that there is nothing but a constant conjunction between objects, which the mind interprets as being a relation between causes and effects.[17] Certainly, there are openly provocative passages where Hume rejects anything but a mental operation linking objects: 'Objects have no discoverable connection together; nor is it from any other principle but custom operating on the imagination, that we can draw any inference from the appearance of one to the existence of the other'.[18] This sort of dramatic presentation, where he does not write 'coolly', where he is 'carried away by his own rhetoric', is best accepted—and possibly discounted—as potentially in contradiction with his more 'cool, naturalistic account' of our beliefs; a piece of advice from David Pears[19] which I, only partly interested in Humean exegesis, am happy to follow.

There are, however, important passages which seem to lend themselves to the 'conventional' understanding of Hume as a regularity theorist, as when he argues that 'the nature of [causal] relations depends so much on human habits of thought based on observation'.[20] Importantly, there is his so-called second definition of 'cause':

An object precedent and contiguous to another, and so united with it in the imagination, that the idea of the one determines the mind to form the idea of the other, and the impression of the one to form a more lively idea of the other.[21]

This is in effect substantially the equivalent of Śrī Harṣa's dictum. Where Śrī Harṣa says there is a cognition, there Hume says is an idea or an impression; where Śrī Harṣa says that the cognition is of invariable priority, Hume says that the impression is of precedence and contiguity. It is on the basis of words like these that Hume came to considered a regularity theorist. By the same token, Śrī Harṣa's words seem to condemn him to a regularity theory. We may list the ideas of a regularity theory:

(1) Objects/events do not have anything in their constitution by virtue of which they are intrinsically related to other objects; the occurrence of the one is not invariably determined by the occurrence of the other.[22]
(2) But objects/events follow one another in some temporal sequence and spatial orientation and are so observed, often repetitively.
(3) This observation is conceptualized as a regulative sequence by observers and is labelled a causal one between the antecedent objects/events and the consequent ones.
(4) Therefore, what is called a causal connection is really only the mind's ordering of its experience of objects/events following in an invariant sequence, with the prior object/event taken as the cause of the consequent one.

It is obvious that these ideas fit in easily into a theory of causal connection which adheres to [RC].[23]

VI THE OBJECTION TO THE REGULARITY THEORY'S DENIAL OF CAUSAL CONNECTIONS

Given the argument that any theory which endorses [RC] must most naturally be committed to a regularity theory, it is clear that a general argument against the regularity theory will also be an argument against [RC]. If that is the case, then, interpreting Śrī Harṣa's dictum as equivalent to [RC] would amount to the discounting of his dictum once the regularity theory has been refuted. On the other hand, even if the regularity theory fails to persuade, and even if we grant that it sits naturally with [RC], if we can argue successfully that Śrī Harṣa's dictum ought to be interpreted in some other way, then the failure of the regularity theory will not lead to the discounting of Śrī Harṣa's dictum. The relevant parallel point would be that Hume too need not be seen as adhering to a regularity theory. To do all this, we must examine the argument against regularity.

The single most striking argument in Western thought against the regularity theory is: since it holds that there is nothing other than succession, it must also hold that there is no explanation for why things appear or are cognized in an ordered way; if any such explanation were to be given, that would only be to admit that there is in fact a causal basis for that order. In such a scenario, we may consider a world, call it R-world, where the constituent objects *ex hypothesi* change their natures in totally random ways and follow in a succession which observers cognize as regular. R-observers then think of the sequences they observe as causal sequences. Now suppose that the sequences they observe turn out to be identical to those observed in our actual world, or A-world. Then, R-world and A-world sequences would be identical sequences for the generation of the concept of 'a causal factor' for both R-observers and A-observers. The random sequence would be the required basis for the derivation of the concept of cause among observers, for, on the supposition, it is the same as the sequence in the A-world. A regularity theorist has to accept that there is no difference between the two worlds, since to distinguish between the random sequences in the R-world and the sequences in the A-world would require that there be something in the A-world which is not random; but this non-random factor would be the reason for the regularity, and allowing such a reason would be to give up pure regularity.[24]

Having given his definition of the characterization of the causal factor, Śrī Harṣa tersely presents an objection to it.

> It being in the form of an erroneous cognition, would this not lead to undesirable consequences?[25]

The argument here, which uses the classical Indian strategy of error-analysis, may be understood as follows. What is the difference between an erroneous cognition and a veridical one that X and Y are actually a cause C and an effect E? Take a scenario, R-scenario, in which X and Y are *ex hypothesi* randomly brought together in succession, such that X is an example of haphazard production (*adhītya-samutpāda* or *yadṛcchā-vāda*[26]); and the R-observer cognizes them in that succession. Suppose there is another scenario, A-scenario, in which A-observer cognizes X and Y in such succession. Now suppose that the R- and the A-successions are identical. Then, on the theory that the *cognition* of X and Y in succession is *all* that is required for the formation of the concept that there are causally connected C and E, a cognitively constructed apparent 'causal connection' must be supposed to exist commonly in both scenarios. If a causal factor is specified purely on the basis of a cognition that Y follows when X occurs, then in both scenarios, there being the R-cognition and A-cognition respectively of such regular succession, X and Y must be thought of as C and E. But, of course, this sounds distinctly odd, given that the hypothesis is that in the R-scenario, X and Y are haphazardly produced.

It is clear that the same absence of distinction which holds between the random and non-random sequences in the first case holds between the haphazard and regular sequences in the second. Whatever criticism holds for the one holds for the other; and it is damaging indeed. The criticism which has come up points in fact to only part of the problem. Obviously, says the anti-regularity theorist, if we wish to retain our idea of randomness or haphazard sequences, we must also admit that the regularity theory cannot distinguish between haphazard sequences and regular ones. But, the regularity theorist might object, that is precisely his point: whatever is *cognized* as a succession is what *constitutes* the concept of causal connection. Surely, the anti-regularity theorist is already assuming that his theory of causal connection is correct. To see the point of the regularity theorist's protest, consider the anti-regularity theorist's definition of the causal connection.

[NC] C is the causal factor for E when C *is* invariably priorly existent to E.

Then, the veridical cognition of a causal factor would be

[VCog.NC] The cognition that C is the causal factor for E is correct when the cognition is that [NC] holds and when [NC] holds.

Given this, the anti-regularity theorist can argue that a random or haphazard connection occurs when [NC] breaks down:

[~NC] C is randomly followed by E when [NC] fails to hold.

From this it follows that an erroneous cognition occurs under this condition:

[ECog.NC] The cognition that C is the causal factor for E is erroneous when the cognition is that [NC] holds and when [~NC] holds.

But the anti-regularity theorist cannot make the distinction between haphazard and causal sequences without assuming that there are causal sequences independent of cognition; he cannot, according to the regularity theorist, make the distinction between [NC] and [VCog.NC] and the further distinction between [VCog.NC] and [ECog.NC], because (the regularity theorist contends) [NC] does not hold.

But this puts the regularity theorist in an even worse position. His opponent can come back with a devastating objection: if the distinction between [VCog.NC] and [ECog.NC] is denied, then there is no ground for the distinction between erroneous and veridical cognition—and between random and causal sequences—at all. The distinction between the two simply disappears. The regularity theorist's claim is

[RC] C is the causal factor for E when C *is cognized* as invariably priorly existent to E.

But [RC] cannot—indeed, apparently, the regularity theorist seems to hold that it *should not*—generate [~RC] to give the ground conditions for haphazard sequences and error. Failure to do so means that there can be no further distinction between veridical and erroneous cognition either, for such a distinction depends upon the distinction between when a haphazard sequence occurs and when a causal sequence does.

In sum, if causal connections are just the cognitions of regular succession, then not only would a haphazard sequence, if cognized as regular, be counted as a causal connection; causal connections must themselves simply be haphazard sequences, since no prior distinction is allowed between haphazard and causal sequences. So, if causal connections are nothing but cognitions of regulative priority—or regularity—there can be no way of distinguishing between veridical and erroneous cognitions.

VII THE NON-REGULARITY THEORY OF HUME

If the regularity theory is so flawed as to be untenable, and Śrī Harṣa's dictum is interpreted as a regularity theory, why does he himself articulate the objection which demonstrates that untenability? The obvious answer is that the dictum does not represent a regularity theory, but rather, some other theory which is, however, still at odds with the Nyāya theory which has so far carried the burden of the argument against regularity.

I shall approach the reinterpretation of the dictum obliquely, through a comparativist comment on the alternative interpretation that Hume was not a regularity theorist. I think that the basic argument for why Hume ought not to be considered a regularity theorist is akin to the one I shall make for Śrī Harṣa; moreover, the alternative, non-regularity interpretation of the Humean account bears a close resemblance to Śrī Harṣa's. This alternative interpretation of Śrī Harṣa accounts for his reply to his own objection to the dictum. When we see that this reply could possibly be given only on the alternative interpretation and not the regularity interpretation of the dictum, a case would have been made for my understanding of Śrī Harṣa's view on causality and his probing and novel questioning of the epistemic basis of the consensus Indian view of causality as determinative priority.

The standard interpretation of Hume is that the concept of causality is the order the mind of the observer imposes on a sequence of objects/events which really are not causally connected.

This interpretation is supposed to come from, among other passages, the claim that an object 'precedent and contiguous' with another (and therefore, presumably, *nothing* but so sequentially occurrent) is 'so united in the mind' as to have an 'idea' of it 'determine' the 'idea of the other' (see Section V). He says that it is a 'false philosophy' to suppose that there is a causal connection between objects when 'we can never observe in them' such a connection.[27] But the question to ask, as Galen Strawson does,[28] is whether Hume is making an ontological (O) claim or an epistemological (E) one. He could be saying that we cannot assert that there is a causal connection, independently of our observation—or cognition—of regular succession because

(O) there *is* nothing like a causal connection and there *is* only regular succession

or he could be saying that it is because

(E) there is nothing we can *know* about causal connections apart from our cognition of regular succession.

As Strawson says, [E] is arguably true while [O] is wildly implausible. But it is not just that [O] is implausible; more importantly, holding it would be '*very seriously at odds with his strictly non-committal scepticism with regard to knowledge claims about the nature of reality—his strictly non-committal attitude to questions about what we can know to exist or know not to exist*, in reality'.[29] The point is that in order to be a regularity theorist, Hume must be willing to assert something about the world—namely the non-existence of causal connections—beyond our experience or cognition of it; but the very foundation of his cautious empiricism lies in the notion that all that can be said about the world is based on our experience of it. Hume cannot be taken as claiming 'it is known that there is no causality', as must be the case were he to be seen as adhering to a regularity theory, but claiming, 'it is not (and cannot be) known whether there is or there is no causality at all'.

Hume must be seen as claiming two things: (1) the 'uniting principle' between objects/events 'is not known to us in any other way than by experience';[30] and (2) it is the 'constant union' between objects 'alone, with which we are acquainted',[31] i.e. which we experience.[32] Given this understanding, let us look at the problematic Humean definition again.

> An object precedent and contiguous to another, and so united with it in the imagination, that the idea of the one determines the mind to form the idea of the other, and the impression of the one to form a more lively idea of the other.[33]

On the regularity interpretation, Hume asserts that (a) there is nothing but a regulative sequence between objects and that (b) the observation of one object determining the idea that the other follows is all there is to the concept that the two objects are causally related. But as David Pears suggests,[34] we could 'take the consecutive clause . . . to *specify the way in which the two objects need to be united* in the imagination, rather than to require that the observer *should actually make the causal inference*'. The conclusion I draw from this suggestion (which is not quite Pears') is that Hume must be taken to be (1) describing the way in which human observers actually arrive at our understanding in any particular instance of a causal connection, and *not* (2) asserting that if such observation did not occur, there would be no such thing as a causal connection between objects. The description is *all* that Hume thinks can be allowed on a strictly sceptical account of what our experience yields. The interpretation of Hume which emerges is not that he either denied causal connections or asserted that causal connections are purely cognitive constructs, but that he thought that, given that we derived our concepts from some experiential input, all that could be said about how we derived our concept of causality was that we had experience of regulative priority. This is not a regularity theory at all but a plea for cautious scepticism about ontological claims.[35]

VIII ŚRI HARṢA'S DICTUM REINTERPRETED

I have already argued that Hume's description of the causal connection is equivalent to Śrī Harṣa's dictum; both can be misinterpreted as being of the form of [AC]. My claim is that the interpretation of Hume given above, inasmuch as it rescues him from the regularity theory, points to a parallel treatment of Śrī Harṣa's dictum and its rescue from the [RC] interpretation. Given the sort of argument we saw in the case of Hume, it would be possible to argue that Śrī Harṣa's dictum need not be of the form [AC].

We must go back to question (v) about causality asked in Section I: how—on the basis of what—are the various definitions of the categories and concepts involved in causality justified? Most generally, how at all do we get the idea that causality *is* determinative priority?

Śrī Harṣa's dictum must be seen as his version of the best answer available to that question. The answer is that it is from our experience of the world that we get the idea of a determinative connection between objects. But this suggests that he is merely (1) *specifying what happens* when we try to establish a causal connection between C and E; he is (2) *not asserting what must be the case* if we are to claim a causal connection between C and E. On such an understanding, he is not making any assertions about the nature of—or more precisely, about the absence of any—causal connection; he is merely pointing out that we have at best the cognition of determinative priority to go on when attempting to specify a causal factor. On this understanding, the dictum should *not* be interpreted as

[RC] C is the causal factor for E when C is cognized as invariably existing prior to E.

This would be an assertion of what actually *is* the case in reality: i.e. (i) the causal factor *is* a cognition of priority, and (ii) the causal factor *is* nothing other than such cognition. Instead, the dictum must be understood, not as an assertion of what there is, but a description of how we make justifiable assertions in the first place. If that is so, the dictum must be interpreted thus:

[AC] C is determined/known as the causal factor for E when C is cognized as invariably existent prior to E.

This is the *specification of the procedure* which is followed when a metaphysician asserts that

[NC] C is the causal factor for E when C *is* invariably existent prior to E.

In other words, Śrī Harṣa is making the point that

[AC'] It is determined/known that [NC] because it is cognized that C is invariably existent prior to E.

An Advaitic Argument *via* the Reinterpreted Dictum

That he is focusing on this epistemic procedure is clear when he answers his own objection. The objection was that if the dictum held good, then there would be no distinction between correct and erroneous cognition, because there would be no distinction between haphazard and causal sequences. But, Śrī Harṣa argues, there need be so such worry. After all,

> just as you take a judgement that rests on three or four uncontradicted tests as satisfying the determination of the existenthood of the object, similarly, with the same judgement, we take such determination as being of the causal efficiency of the object.[36]

This is the answer he has to offer to the question articulated in (v). Philosophers arrive at their concepts of causal connections on the basis of uncontradicted cognitions of determinative sequences. That is how the legitimacy of claims about causal connections is determined. His opponents may all assert that there is a causal connection between C and E, and that C is a causal object; but by virtue of what are they able to assert that? This question should be considered as quite general: its target is the consensus on (i) given in Section I. The answer is this: whatever their specific ontologies, all the other schools assert that there are real causal connections which form part of the intrinsic nature of whatever are the elements of their respective ontologies. *They can justify their assertion that there are such connections only because, strictly speaking, they have experience of determinative priorities*; it is on the basis of such experience that Buddhists, Naiyāyikas, Mīmāṃsakas, all assert their specific views of causal factors. This point cannot be overemphasized: it is at the heart of the critique of causality. There is an ineliminable experiential/cognitive basis to the concept of causality. Without such cognition there could be no means of forming the concept. But if such cognition is ineliminable, the dictum must go through, since it holds that it is cognition of determinative connection which alone yields the concept of causality. The metaphysical point is that assertions about the existential status ('what there really is') of causality cannot justifiably be made, since justification depends upon the epistemic limits imposed by cognitive access, and dependence upon cognition makes it impossible to make assertions about what there 'really' is independently of the epistemic licence given by cognitive grasp. In effect, his opponents are epistemically no different from the Advaitin. *Everyone depends upon the cognition of determinative priority to determine the idea of causal connection.*

The question that comes to mind now is: why is this reply to the objection part of the 'cautious sceptic' interpretation given in [AC] and not simply a variation on the regularity theory given in [RC]? The answer is that a regularity theorist cannot give it consistently, for his theory fails to be epistemically modest. If Śrī Harṣa gives it himself, it is only because he is not a regularity theorist who holds the view that causal connections are purely cognitive constructs. To see this, let us consider his reply carefully.

Śrī Harṣa points out that all philosophers *test* their judgements. Testing consists in determining whether the knowledge-claim of the judgement under test is *contradicted* or not. What could possibly constitute such contradiction? Suppose the judgement under consideration is of the form

[J] C is invariably existent prior to E.

Then, if [J] were to be contradicted, there would be a cognition of the form

[~J] C' is not invariably existent prior to E: it is not the case that always when E occurs, C' occurs and if C' did not occur, E does not occur.

But [~J] would require there to be a distinction between C and C', such that there could be a possible scenario under which the occurrence of C' could be held to contradict the claim that C was the cause of E. But [~J] implies that there is a conception of the failure of determinative connection between C and E such that the cognition of C can be held to be a deviant (*vyabhicārin*) occurrence. So, if the idea of contradiction is to be secured, there must be a distinction between a determinative sequence C'-E and a deviant one C-E, so that the occurrence of the latter could be thought of as contradicting the claim inherent in the judgement consequent on the cognition of the former. But such a distinction would not be possible under [RC]. As we have seen in Section VI, under [RC], a cognition of a sequence would itself *constitute* a causal connection; it is just that cognition which makes a sequence fall under the concept of a causal connection. If that is so, then the sequence C'-E would, by virtue of being cognized, itself form *another* case of a causal sequence, rather than be a contradiction of the original cognition (which could then be determined as erroneous). [RC] simply makes every cognition of sequence a causal one; it allows no space for the concept of deviation and thus no space for contradiction. Instead, it allows every cognition to constitute a new causal connection. If Śrī Harṣa had indeed asserted that there were no causal connections in reality, then he would have had no space for the concept of testing through contradiction, and consequently, would not have put forward this picture of what he thinks all philosophers do.

However, there is no such problem if his dictum is interpreted as [AC']. Śrī Harṣa is perfectly willing to allow the distinction between correct and erroneous cognition; he is perfectly willing to say that there is a method for determining whether a cognition of causal connection is veridical or not. The very fact that he admits these things militates against the regularity interpretation of his dictum embodied in [RC].

As a matter of fact, when he is so interpreted, we come to the crux of the issue: if [AC] is correct, then [NC] is mistaken. If [RC] unjustifiably asserts that there is no causal connection in reality, equally, [NC] unjustifiably asserts that there is indeed causal connection in reality. To see this, let us go back to the claim that he does indeed admit that there is a need to distinguish between erroneous and veridical cognitions of causal connections. Could it be that if he is to be seen as having rejected [RC] and its denial of difference, he is committed to the difference embodied in [NC]? After all, we saw in Section VI that if we used [NC] and [~NC], we could distinguish between [VCog.NC] and [ECog.NC].

CONCLUSION: THE ADVAITIC CASE FOR EPISTEMIC MODESTY

This seems to imply that if he admits that there is a distinction between correct and erroneous cognition, Śrī Harṣa must be driven into an [NC] view, which would oblige him to abandon his dictum. But this is not so. His comment on the procedure of testing brings attention to bear on the conditionals in [VCog.NC] and [ECog.NC]: 'when [NC] holds' and 'when [~NC] holds' respectively. The question is: how can it be determined as to when these conditions hold? The answer is: when there have been tests that these conditions hold. Question: what do these tests consist in? Answer: in there being uncontradicted cognitions that they do so hold. But this is only to apply [AC] and [AC'] on such judgements of causal connections. This can be put in terms of the definitions we already have, so as to derive the Advaitic version of veridical and erroneous cognition.

Given

[AC'] It is determined/known that [NC] because it is cognized that C is invariably existent prior to E,

and

[~AC'] It is not determined/known that [NC] because it is not cognized that C is invariably existent prior to E,

we will have

[VCog.AC'] The cognition that C is the causal factor for E is correct when the cognition is that [NC] holds and when [AC'] holds.

and

[ECog.AC'] The cognition that C is the causal factor for E is erroneous when the cognition is that [NC] holds and when [~AC'] holds.

This is an epistemically modest conclusion. It imposes a critical epistemic constraint on those metaphysical claims about the 'real' nature of causality which go to form the consensus view. There is no justification for the assertion of [NC] other than through tests on the validity of the cognition that [NC]; and such tests do nothing else than satisfy [VCog.AC] (and not being subject to [ECog.AC']).

Thus Śrī Harṣa must not be seen as denying that there is a distinction between correct and erroneous cognitions of causal connection. His dictum must be understood as carefully avoiding making any assertions about the nature of causal connections which are not justified by cognitive determination through uncontradicted tests. His point is that even those who assert that there are real causal connections—and all the other major schools, in their own way, do so—are only depending on systematic (i.e. uncontradicted) cognitions to justify their claims. This cautious refusal to go beyond the bounds of cognition is motivated by a strong metaphysical theory about knowledge, but to defend his dictum on the basis of that theory would be to undertake another enterprise altogether.

NOTES

1. Chowkhambha Sanskrit Series, Benares, 1970; and Benares: Achyut Granthamala, 1969. Page references are given to the Chowkhambha edition first, followed by the other.
2. B.K. Matilal, 'Foreword' to P. Granoff, *Philosophy and Argument in Late Vedānta*, Dordrecht: Reidel, 1978.
3. *Logic, Language and Reality*, Delhi: Motilal Banarsidass, 1990, 2nd edition, p. 289.
4. See Udayana, *Nyāyakusumāñjali* (ed. by P. Upadhyaya and D. Sastri, Varanasi: Chowkhambha, 1957, pp. 41–60), for a discussion; the latter translated terms are from K.H. Potter, *Encyclopaedia of Indian Philosophies*, vol. II, Delhi: Motilal Banarsidass, 1977, p. 560; and Matilal, *Logic, Language and Reality*, p. 288.
5. Cp. Stcherbatsky, *Buddhist Logic*, vol. I, New York: Dover, 1962, p. 121: 'This being, that appears.'
6. Ibid., p. 121; the emphasis is mine to draw attention to the regulative aspect implied by 'always'.
7. *Pūrvasambandhaniyame hetutve tulya eva nau*; p. 36/24.
8. Udayana, *Ātmatattvaviveka*, ed. by D. Sastri, Varanasi: Chowkhambha, 1940, p. 19.
9. Note that under these conditions, it is not correct to assert that Ψ is invariably concomitant with X. In the classical example, smoke is

invariably concomitant with fire, but, of course, fire is not invariably concomitant with smoke.

10. Stcherbatsky, *Buddhist Logic*, p. 122.
11. Clearly, [C] could not be a sufficient account of all causal events, if the invariance is a merely empirical matter of finding no counter-instances; for instance, a donkey may be found at all times near a pottery, but could not be considered a causal factor for a pot. This consideration led the Naiyāyikas to draw up an expanding list of 'relevant' (*ananyathāsiddha*) causal factors. But it could be argued that, properly strengthened, [C] is a perfectly good definition. All that has to be done is to think of any candidate for C, not as one invariably found with the occurrence of E, but rather as one whose not being found would result in E not occurring. Thus, the removal of the donkey would not affect the production of the pot. An alternative suggestion is that causal factors need not be thought of as transitive: so that the father may be a causal factor for the potter and the potter for the pot, but the father may not be a factor for the pot.
12. Udayana: *Nyāyakusumāñjali*, ed. by Padmaprasadopadhyaya (Kashi Sanskrit Series 30), Varanasi, 1950, p. 413.
13. Cp. Matilal (*Perception*, Oxford: Clarendon Press, 1983), 'existent-ness', p. 379, for an explanation for this terminology, see my: 'Knowledge and the real world: Śrī Harṣa and the pramāṇas', *Journal of Indian Philosophy*, 1993.
14. Stcherbatsky, *Buddhist Logic*, p. 119.
15. Dharmottara, *Nyāyabinduṭīka*, in Dharmakīrti, *Nyāyabindu*, ed. by C. Sastri, Varanasi: Chowkhambha, 1954, p. 17.
16. '*Idam asmān niyata prāk-sat*' *iti buddhyāviśeṣāt*, p. 37/25.
17. R. Woolhouse, *The Empiricists* (Oxford: Oxford University Press, 1988), is a recent example of this 'conventional' interpretation; it seems to have been fairly popular earlier in the century.
18. David Hume, *The Treatise of Human Nature*, ed. by L.A. Selby-Bigge and P.H. Nidditch, Oxford: Oxford University Press, 1978, 2nd edition, p. 103.
19. David Pears, *Hume's System*, Oxford: Oxford University Press, 1990, pp. 96–7.
20. Ibid., p. 169.
21. Ibid., p. 170.
22. I have throughout kept to the Indian terminology of invariable concomitance, without identifying it with the Western concept of necessity. As has often been pointed out (e.g. by Karl Potter, 1977, p. 66), the formal notion of necessity, as for example, as true in all possible worlds, does not seem to have been held in Indian thought; at most, it is that there is no spatial or temporal instance in the actual world where a counter-instance is found. This has to do with the general epistemological and empirical orientation of Indian logical categories. On the other hand, invariable concomitance can certainly be assimilated into the stronger Western concept of necessity as one component: if something is necessary for something else, the former thing will certainly extract an invariable

concomitance from the latter. So, invariable concomitance becomes a minimal common factor for Western and Indian discussions.

23. Note that I am here only (1) providing a *description* of how the notion of cause is derived from what is ostensibly an acausal succession; I am *not* here (2) examining Hume's alleged claim about the *reason* for the formation of the notion of cause.
24. G. Strawson, *The Secret Connexion* (Oxford: Clarendon, 1989, pp. 24ff), gives the basic example from which I have deviated somewhat.
25. *Bhrāntaivaṃ-buddhi-gocare atiprasaṅgaḥ*; p. 37/25.
26. Stcherbatsky, *Buddhist Logic*, p. 122.
27. Hume, *Treatise of Human Nature*, pp. 168–9.
28. Strawson, *Secret Connexion*, pp. 10ff.
29. Ibid., p. 11; his emphasis.
30. Hume, *Treatise of Human Nature*, p. 169.
31. Ibid., p. 400.
32. This way, we do not have the awkward interpretation of Hume (which, however, I cannot here attempt to rebut) that he first went beyond experience and denied that there was any causality in the world, and then finding a need to explain the experience of regularity, brought it back as a purely mental imposition on an acausal world. (T. Beauchamp and A. Rosenberg, *Hume and the Problem of Causation*, Oxford: Clarendon, 1981, argue that this is what happens in Hume.)
33. Hume, *Treatise on Human Nature*, p. 170.
34. (1990), p. 118; emphasis Pears'.
35. Was Hume in any way a realist about causal connections? Was his epistemological caution in some way consonant with a belief in real causality? If belief in causality is clearly distinguished from knowledge thereof, such that a Humean could cheerfully admit the possibility, even the necessity, of epistemically unjustifiable beliefs in the form of natural dispositions, then, perhaps, Hume was a realist. But that is another issue.
36. *Yādṛśyā hi dhiyā tri-catura-kakṣābādhānavabodha-viśrāntayā vastu-sattva-niścayaste, tādṛśyaiva viṣayīkṛtasya mamāpi kāraṇatā niścayaḥ*; p. 37/25.

Chapter 11

Eṣa Dharmaḥ Sanātanaḥ: Shifting Moral Values and the Indian Epics

Robert P. Goldman

As the great popular handbooks on *dharma* in all its many varieties and contexts, the Sanskrit epics, the *Mahābhārata* and the *Rāmāyaṇa*, have fuelled, over the centuries, innumerable discussions and debates on the ethics of the characters and situations so colourfully represented in these monumental literary masterpieces.

The appearance of ethical dissonance—of the resort to conduct that can be viewed as unethical or in violation of *dharma* on the part of a character who is put forward in these texts as a paragon of righteousness—is taken as a matter of particular concern by their authors and audiences by the very nature and function of these documents. This, it would appear, is owing to what is perhaps the epics' most defining characteristic: their function as illustrated guides to *dharma* and *adharma* where all the varieties of social, political, and ethical behaviour—good and bad—and their consequences for their practitioners are conveyed not so much through lists of rules and prohibitions (although there is no shortage of these) as by gripping character portrayals and dramatic narratives.

Thus epic characters such as Lakṣmaṇa, Sāvitrī, Vibhīṣaṇa, Śakuni, and Durvāsas have become proverbial figures in the popular culture representing respectively loyalty, wifely devotion, fraternal treachery, deviousness, and irascibility. When the figures in question are not merely monovalent caricatures personifying a particular virtue or vice but are instead the complex, fully developed, and sometimes ambivalent principal heroes and heroines of the epics, it is only natural—given this emphasis on characters as exemplars—that perceived deviations from the theoretically rigid codes of conduct should be a source of considerable concern. When, as is generally the case, retellings of the epic stories are predicated upon an understanding that these characters are also historical manifestations of the highest divinities, it is only to be expected that the perception or representation of an action as ethically deviant should generate both discomfort and debate. This is, of course, most particularly the case with regard to Rāma and Kṛṣṇa, but also, to a somewhat lesser extent, in the case of the Pāṇḍavas.

Such debates are joined in many kinds of texts ranging from the epics themselves to the commentaries on the epics, literary and religious texts that derive from or ground themselves in the epic narratives,[1] as well as modern scholarly, journalistic, and popular writing.[2] Their outlines may also be discerned in medieval and modern renditions of the epic stories which may modify or even omit episodes perceived to present ethical problems. In some cases, an entire version may be predicated upon such an ethical concern.[3] Moreover, the framing of these debates, particularly in the case of those found in the epics and their sectarian exegeses, frequently serves to foreground important ethical and theological concerns with which various segments of Indian society have had to grapple in the more than two millennia since the composition and popular diffusion of the *Rāmāyaṇa* and *Mahābhārata.*

Writers, both scholarly and popular, who have dealt with the various ethical issues raised by the Indian epics, have often appeared to proceed from an unstated assumption that, because of the supposedly timeless and unchanging character of Hindu culture, the authors and audiences of the epics at whatever historical period, in whatever region of the country, and of whatever gender or social class essentially share the value systems and ethical norms that frame the actions of the characters represented in the ancient poems of Vyāsa and Vālmīki.[4]

In the following essay I will attempt to evaluate the validity of this assumption through a discussion of a few selected episodes in the *Mahābhārata* and the *Rāmāyaṇa* which have provoked some kind of ethical discussion at some point in their history. I will also, by way of contrast, discuss a few episodes that, one might think, would have stimulated such debate but which—at least until very recently—appear not to have done so. In the course of these discussions I shall offer some observations on the kinds of ethical norms put forward by the epic poets and the techniques of ethical revisionism that have been brought to bear on the poems and their principal characters by their varied and shifting audiences over the centuries.

To begin this discussion it will be useful to make two basic distinctions. The first of these is the distinction between episodes that are represented in the epics themselves as ethically problematic and those which are subjected to ethical scrutiny only at later periods. The second is that distinction which can be made between the treatment of ethically questionable behaviour represented in the *Rāmāyaṇa* as opposed to the *Mahābhārata.*

One additional caution that should be kept in mind throughout the following discussion is that the epics present very few ethical and moral absolutes. The *dharma* of the epics and the *dharmaśāstras* alike varies by its very nature according to differences of *varṇa*, *āśrama*, *puruṣārtha*, gender, time, place, and a host of other contextual variables. Moreover, commodious instruments for exceptions to even contextually conditioned codes of ethical behaviour, instruments such as the concept of *āpaddharma*, 'an action normally not sanctioned but permissible during times of great distress or calamity', make deviance from normative codes justifiable.[5]

Let me begin with one of the most central and timeless of Indian ethical values, Truth. Adherence to one's word, frequently regarded as synonymous with truthfulness, is surely one of the fundamental values of Indian culture throughout its history, not least of all in the age of the epics. Epithets such as *satyavāk* and *satyapratijña*, 'one whose words *or* vows are truthful', are among the poets' highest encomia for their heroes, and the strict keeping to one's given word, even when the consequences are catastrophic, is a central theme in many episodes.[6] Surely it is the single most critical element in the narrative development of the *Rāmāyaṇa* where the aged king Daśaratha's rigid adherence to his boons to Kaikeyī forces him to accede to Rāma's dispossession and exile and leads not only to the abduction of Sītā but to his own death. Similarly, Rāma's insistence on preserving his father's truthfulness at all costs explains his cheerful acceptance of the stunning reversal of his fortune even after Daśaratha's death and in the face of Bharata's entreaties.[7]

Nonetheless, despite the extreme gravity of this issue for the culture, reflected today in the motto of the Republic of India, *satyam eva jayate*, 'Truth alone prevails', the epic heroes are frequently permitted to escape the fulfilment of their given word. In some cases, when the fulfilment of a character's vow unexpectedly would entail the death of a beloved kinsman, the character is let off the hook through the use of a kind of sophistry in which 'death' is taken figuratively. Thus, in the central example of a moral dilemma chosen by Matilal, Arjuna vows to kill anyone who casts aspersion on his possession of his cherished bow Gāṇḍīva. When Yudhiṣṭhira does so, Arjuna, instead of killing his elder brother, follows the advice of Kṛṣṇa and speaks disrespectfully to Yudhiṣṭhira. According to Kṛṣṇa, such disrespectful treatment of an elder brother, at the hands of a junior is like 'death'.[8] Similarly Rāma, the very paragon of truthfulness, having vowed to put to death anyone who interrupts his colloquy with Death, merely banishes the intruding Lakṣmaṇa, arguing that for the latter banishment is the moral equivalent of death.[9]

Noteworthy here is the fact that in the *Rāmāyaṇa's Ayodhyākāṇḍa*, whose central theme is the inviolability of Daśaratha's promise to Kaikeyī, we learn that Rāma's disinheritance in favour of his younger brother is in fact the fulfilment of a promise given by the king to Kaikeyī's father at the time of his marriage, even earlier than his celebrated boons. There, during the course of his lengthy arguments designed to induce Bharata to accept his own exaltation to the kingdom in place of Rāma, he tells him, 'Long ago, dear brother, when our father was about to marry your mother, he made a brideprice pledge to your grandfather—the ultimate price, the kingship'.[10] This sudden, late revelation of the prehistory of the succession problem in Kosala presents—then immediately glosses over—a grave ethical problem. If Rāma is telling the truth (and given the text characterization of the hero, how can we imagine otherwise?), then it must follow that, in undertaking to install Rāma as *yuvarāja*, Daśaratha is knowingly violating a solemn contractual agreement with another king. If he was, with the apparent complicity of Rāma, able to do so with impunity and without any sense of transgression, why then is he later constrained to honour the pledges—really 'blank checks'—he makes to Kaikeyī?[11]

What is particularly striking about this seemingly undeniable ethical contradiction is the fact that it is not remarked upon by the poet or any of his characters. This is not, however, the case with his commentators, who are much exercised by it and exert considerable effort to justify the king's behaviour. Several of them cite passages from the *dharmaśāstra* that list specific circumstances—including love-making and the contracting of a love-marriage—under which it is permissible for a man to lie.[12] One commentator, the author of the widely known *Tilakaṭīkā*, Nāgeśa, makes an explicit distinction between the two promises. He says:

> In keeping with the *smṛti* passage that runs: 'An untruth is not to be condemned when it is told to women, in order to contract a love-marriage, to secure one's means of livelihood, when one's life is in danger, for the sake of cows and brahmins, or when [it is to avoid] violence,'[13] there is nothing wrong with this violation of his promise because of the fact that since Kausalyā was already in place [as his senior wife], his union with Kaikeyī was a love-marriage. Therefore, he [Vālmīki] utters [the verse beginning], 'And in the war between the gods and the demons (*devāsure ca*)'. In connection with that [promise] which had to do with the war between the gods and the demons, [with respect to the boons given to Kaikeyī after her help during this war], untruthfulness would be a grave wrong since this was a matter of rewarding help rendered. This is the real meaning.[14]

This incident and its treatment by the commentators is illustrative in two respects. In the first place, it gives an unusually clear example of an action taken by a significant epic hero, one of the great paragons of righteousness, which, although it raises no explicit ethical questions in the text of the epic itself, is clearly troubling to the scholiasts who exert themselves to exculpate Daśaratha. In such an example, we see a case in which it would appear that there is a significant shift in ethical expectations from the time of Vālmīki, perhaps the middle of the first millennium BC, to that of the Śrīvaiṣṇava commentators who are for the most part active in the sixteenth and seventeenth centuries AD.[15] The second point worth noting here is the commentators' recourse to a large and highly 'user-friendly' corpus of *smṛti* literature that can be mined, as it were, for rules (and exceptions) that can provide legal and ethical justification for an extraordinary variety of actions, even when they seem to be in violation of normative codes of behaviour. This is in many ways a defining characteristic of the *dharmaśāstras* where various—even contradictory—sets of rules are recorded making allowances for a broad spectrum of variables such as time, place, stage of life, degree of necessity, the social status of subject and object, etc.[16]

One of the most widely debated actions of Rāma is his killing of the monkey-king Vālin while the latter is engaged in single combat with his own brother and rival, Sugrīva. Criticism of what is often represented as a noteworthy departure from the *kṣatriya* code of battle is diverse and persistent in a variety of literary[17] and didactic texts and has been discussed by numerous scholars.[18] It is first raised, however, in the epic text itself.

In a well-known passage,[19] the stricken monkey reproaches Rāma for having shot him while he was engaged in battle with another. His reproach is harsh and includes a number of arguments. Vālin argues that Rāma has violated the rules of ethical behaviour in killing someone who has done him no harm, an offence compounded further by his having done so when his victim was off-guard. Vālin further argues that even if Rāma's action is characterized as hunting rather than combat, it is still wrongful since the skin, bones, flesh, etc. of monkeys are forbidden to people of high caste. He also implies that Rāma is a coward afraid to face him in a fair fight and a fool since he (Vālin) could have easily defeated Rāvaṇa and recovered Sītā for him.

But having heard Vālin's reproaches, Rāma dismisses them, refuting the monkey's arguments in a lengthy response.[20] His defence, like the accusation of Vālin, rests on a number of grounds. His basic argument is that the normal rules of chivalric combat did not apply in the case of Vālin since Rāma, acting as the agent of the rightful authority, Bharata, is executing an incestuous adulterer and not fighting an enemy. His supplemental arguments, however, are less convincing and some of them even appear to contradict his principal point. Thus, he argues that he had to kill Vālin since he had promised to do so as part of his agreement with Sugrīva and to fail to do so would be to commit the cardinal ethical sin of violating one's given word. This, however, can at best be a subordinate argument as it evades the question of the morality of the act itself. Promising to do a wrongful act would itself be wrong.

Moreover, Rāma attempts to dismiss Vālin's complaint on the grounds that the latter is a mere animal to be slain by a hunting king at his pleasure and therefore not entitled to the chivalrous treatment owing to high-born warriors. This line of argumentation, however, is in sharp contradiction to the theme of crime and punishment that constitutes Rāma's principal argument. Surely Vālin must either be held to the strict standards of sexual, social, and legal propriety enunciated by Rāma and thus subject to punishment for violating them, or he is to be regarded as a wild animal, utterly outside the range of human morality, and so to be killed at the whim of a hunter—but not both. Finally, Rāma attempts to stifle debate on the ethicality of his actions by invoking the divinity of kings and their immunity from censure.[21]

This episode raises a number of ethical questions, several of which are not addressed in the text or the commentaries.[22] Nonetheless, it seems that, at least for the majority of the poem's audiences, the questions raised by Vālin are seen as adequately answered by Rāma as signalled by the monkey's ultimate acceptance of Rāma's actions, his apology for his *lèse majesté*, and his acknowledgement of his own transgression.[23]

Still the fact remains that by the very inclusion of Vālin's denunciation and the commentators' need to justify Rāma's behaviour,[24] the text acknowledges the need to address the episode's ethical difficulties. Moreover, it is difficult to avoid the sense that Rāma's least compelling moral argument—the one based on his promise to Sugrīva and echoed in the arguments from adversity, necessity, self-preservation, and *realpolitik* adduced by commentators and scholar-apologists—is the one that most truly explains his actions in this case. Higher morality and juridical punishment notwithstanding, the real issue here is that the elimination of Vālin is part of a political deal between Rāma and Sugrīva. Circumstances have deprived both princes of their kingdoms and their wives and, since Sugrīva indicates his willingness to assist Rāma in the recovery of Sītā, the killing of Vālin is Rāma's part of a simple *quid pro quo*. In short it is politically expedient. This kind of expedient, as I will attempt to illustrate further with examples drawn from the *Mahābhārata*, is a normal part of the *kṣatriyadharma*. It is only in the rather special case of Rāma, who bears the burden of a moral perfection rarely laid upon other epic heroes, that there appears—on the part of the epic poet as well as commentators and apologetic scholars—to be a felt need to demonstrate that all his actions conform to a code of strict ethical and moral purity.[25]

This special effort on the part of the epic poet to justify Rāma's exercise of the *daṇḍa*, or violence sanctioned in the service of morality and the state, is also evident in the description of the young prince's scruples when he is instructed to kill the monstrous *rākṣasī* Tāṭakā in the *Bālakāṇḍa*.[26] On that occasion, despite the fact that he is repeatedly ordered and given elaborate justifications to do so by his recently acquired guru Viśvāmitra in order to rid the world of a noxious menace[27] and 'for the sake of cows and brahmins',[28] Rāma exhibits a marked resistance to killing a woman, an act that is both morally abhorrent and beneath the ethical scruples of a warrior.[29] Viśvāmitra, who is explicitly said to stand in *loco parentis vis à vis* Rāma,[30] is forced to repeat his command to the recalcitrant prince several times and must also buttress his exhortation by citing several examples from the mythological tradition in which heroic figures found it necessary to kill women.[31] In the end, despite his powerful ambivalence, Rāma complies with the *ṛṣi's* instructions and dispatches the ogress.[32]

Rāma's encounter with Tāṭakā is noteworthy in the present connection. For with its reiteration of the hero's moral scruples, it foreshadows several of Rāma's later hostile encounters with other transgressors belonging to inferior social and ontological orders. Most noteworthy among these are the episodes involving Śūrpaṇakhā, Vālin, and Śambūka.

The cases of Śūrpaṇakhā and Śambūka differ from that of Tāṭakā in that in the former instances Rāma deals out extremely harsh punishment to these lowly characters without hesitation or the articulation of ethical qualms on the part of anyone—poet, character, commentator, or scholar—at least until a quite modern period.[33] Indeed the low status of these two characters is closely connected with their punishments, although in the first case, a *rākṣasī* is punished for what appears to be the normative behaviour of her race, while in the second, a *śūdra* is executed for his practise of asceticism that is seen as the exclusive prerogative of the upper classes.[34]

It is noteworthy in the account of the mutilation of Śūrpaṇakhā that despite the cruel teasing she experiences at the hands of Rāma and Lakṣmaṇa, no ethical question about their actions is raised by the author or his commentators. Thus, when Rāma tires of the game and sees that it has worked the *rākṣasī* into a deadly rage against Sītā, he offhandedly remarks to Lakṣmaṇa that such non-Aryans 'cannot take a joke' and instructs him to cut off her nose and ears.[35]

In the case of Śambūka, social class, in this context a function of ritual status, is everything. The *śudra*'*s* transgression of the strict rules of *varṇāśramadharma* has had a dreadful repercussion in the otherwise flawless fabric of the Rāmarājya: the child of a brahmin has died. This is a grave thing, and it can only be reversed by the death of the offending *śūdra.*

A point that emerges with some clarity from these examples is that the ethical norms for interaction with someone seen as violating the behavioural norms articulated in the *smṛtis* depend as much upon the relative status of the violator in terms of social, species, and gender hierarchies, as upon the substance of the violation. Phrased somewhat differently, it could be said that, in the markedly hierarchical context of Aryan brahmanical civilization, the status relationship between any two individuals constitutes a critical element in their interaction. This point is made most forcefully in the two concluding arguments in justification of Rāma's killing of Vālin. In the end, because of the vast difference in status between the king and the monkey, there is no need for the former to justify any action he may choose to take against the latter however questionable such an action would be deemed if taken against a social equal. Indeed, as the passage makes clear, it is a serious breach of conduct for an inferior even to question an action of his or her superior.

The lack of ethical qualms with which the elites of the epic age treat people of the lower classes is shown even more clearly in two episodes of the *Mahābhārata.* The first of these is the encounter of the Pāṇḍavas with the tribal archer Ekalavya.[36] In this illuminating episode, the two cultural imperatives—the maintenance of hierarchically ranked *varṇa* functions and the preservation of the truth of the given word of a member of the elite classes—combine to force an action that must strike a more egalitarian sensibility as highly unethical.

Observing the young Arjuna's growing skill at archery and his ability to shoot in the dark, the great martial arts master Droṇācārya vows to make every effort to ensure that his protégé has no equal in skill as an archer.[37] He then instructs his pupil in the arts of combat. Learning of his extraordinary expertise, thousands of kings and princes come to try to master the *dhanurveda* at his feet. Among these would-be disciples is Ekalavya, the son of the ruler of the tribal Niṣādas, Hiraṇyadhanus.[38] Reflecting on the boy's lowly tribal status and out of regard for the sensibilities of his other high-born pupils, the brahmin refuses to accept him. Undaunted by his rejection, Ekalavya withdraws to the forest where he fashions an image of Droṇa out of earth. Worshipping this image as his master, he devotes himself to the practice of archery. As a result of his diligence and his faith in his chosen guru, he acquires extraordinary dexterity.

One day, when the Pāṇḍavas are hunting, one of their dogs comes upon the Niṣāda in the dense forest and begins barking at him. Ekalavya responds by firing seven arrows into the animal's mouth virtually simultaneously. When the Pāṇḍavas see the evidence of such skill at shooting, they are filled with amazement and much abashed. But when the identity of the archer is discovered, Arjuna is bitter and rebukes Droṇa, reminding him of his promise that none of his pupils would surpass him. Arjuna then takes him to task for having nonetheless produced a mighty disciple who exceeds him and everyone else in skill.[39] After a moment's thought, the *ācārya* leads the prince to the filthy, shabby tribal boy who, seeing his guru approach, makes all the traditional gestures of reverence. Droṇa now demands his teacher's fee, and Ekalavya humbly declares that there is nothing that he would not give his guru. Droṇa demands that the boy give him his right thumb. Hearing Droṇa's horrifying demand, Ekalavya, who is described as keeping his word and firmly devoted to truth, cheerfully and unhesitatingly cuts off his thumb and presents it to his teacher. As a result, Ekalavya's dexterity is diminished. In this way Droṇa's vow comes to be fulfilled, and Arjuna's chagrin is banished, for he is once again the foremost of archers.[40]

This episode appears to be read chiefly as yet another dramatic example both of the lengths to which people of high and low social status were supposed to go in order to preserve the truth of their vows and those to which an ideal disciple should be prepared to go to prove his devotion to his guru. It does not appear that the story is intended to reflect badly upon either Droṇa or Arjuna despite the former's cruelty and the latter's self-pitying envy. For nowhere in the text or elsewhere in the pre-modern literature have I come across any criticism of their actions in this episode.[41] What appears to be at stake here for the epic poet and his intended audience is the honour of the two high-caste principals in the story, the brahmin Droṇa and the *kṣatriya* Arjuna. The former's reputation for truthfulness and the latter's for skill at archery, appropriate to their respective *varṇas*, must not be tarnished. That the lowly Ekalavya surpasses them both in both of these virtues is at best an embarrassment, at worst, a threat to the ritual and social order. As for the Niṣāda prince himself, he is not characterized as a tragic figure. Indeed he does not matter as a character at all, serving merely as a foil against which the actions of his high-caste superiors are played out.

Still more disturbing to a post-enlightenment ethos with its egalitarianism and sense of social justice is the well-known episode in which Yudhiṣṭhira, having become aware of his rival Duryodhana's plot to have him, his brothers, and Kuntī burnt to death in a specially constructed firetrap, decides to destroy the treacherous minister Purocana—who is supposed to perpetrate the proposed arson—and effect their own escape.[42] In order to cover the Pāṇḍavas' flight and prevent any pursuit and further attempts at assassination, Yudhiṣṭhira decides to lure six innocent people into the inflammable house so that their charred corpses, mistaken for those of the heroes and their mother, will lead Duryodhana to believe that his plot has been successful.[43] Neither Yudhiṣṭhira's brothers, the paragons of *dharma*, nor their righteous mother objects to this callous plan. In fact, they readily participate in it. Under the pretext of offering a charitable feast for brahmins to which women are invited as well as men, Kuntī invites many people to the house. Among them is a hungry Niṣāda woman and her five sons. Their hosts ply them with wine so that, when the higher caste guests have taken Kuntī's leave and returned to their homes, the women and her sons remain in a drunken stupor. Then, as a strong wind arises which will fan the flames, Bhīma sets the house afire killing the Pāṇḍava's unwitting guests along with their secret enemy Purocana.[44] When the townspeople comb through the ashes the next morning, they find the charred bodies of the innocent woman and her sons and report to Duryodhana that Kuntī and the Pāṇḍavas (who have in fact made their escape) are dead.[45]

Nowhere in the text of the *Mahābhārata* or its most popular commentary is there a hint of any ethical or moral censure for this slaughter of innocents of which the only justification is the Pāṇḍavas' fear of pursuit.[46] Here again it is the low social and ritual status of the victims that enables the aristocratic heroes to kill them without either moral qualm or legal consequence. Much less shocking transgressions against individuals of high social and ritual status are explicitly shown in the epic to have moral and even eschatological consequences. Compare, for example, the insouciance and impunity of Yudhiṣṭhira in this episode with Indra's somewhat peculiar discourse to him on eschatology in the epic's concluding passage. There the king of the gods, explaining to Yudhiṣṭhira why the latter has had to undergo the physical and mental torture of seeing his wife and brothers undergoing the torments of hell, tells him that he was, in fact, expiating the sin of having deceived Droṇa by falsely giving him to understand that the latter's son Aśvatthāman had been killed during the Bhārata war.[47]

Clearly the weight placed upon Yudhiṣṭhira's half-truth which leads to the death of Droṇa, as contrasted with the epic's failure even to regard his cruel deception and murder of six innocent people as worthy of comment, can only be seen as a reflex of the vastly unequal social status of the respective victims. This kind of inequality, although perhaps exaggerated in the epics, is, in fact, perfectly consonant with the cultural tradition, explicitly codified in the *dharmaśāstras*, of indexing the gravity of a crime to the relative rank of the perpetrator and victim.[48]

It is true even in the contemporary world that harsher punishments are often meted out to offenders of low socio-economic status while the rich and powerful are frequently able to delay, mitigate, or even escape punishment for similar offences. By the same token, there are doubtless cases in which the seriousness with which a crime is regarded is in direct proportion to the status of the victim. Still, such things are generally regarded as violations of the egalitarian ideologies that have largely dominated social and political discourse both in the West and in the post-colonial societies of the so-called 'third world'. Very rarely do they reflect an explicit official legal code or social theory no matter how prevalent they may be in practice.

Few of the nations that have emerged as independent in the post-colonial period have a more elaborate set of egalitarian ideals and instruments at their foundation than does India. The Constitution of the Republic of India, authored by the revered and celebrated 'untouchable' barrister and freedom fighter Dr B.R. Ambedkar, is the foundation for a secular, egalitarian, and democratic state based upon universal adult suffrage which is officially committed to the elimination of the traditional inequalities of class, caste, community, etc. At the same time many segments of Indian society, as witnessed by the extraordinary popularity of the recent TV serializations of the *Rāmāyaṇa* and the *Mahābhārata*, the continuing attraction of such events as the Rāmlīlās, and the so-called Sangh Parivār's ability to mobilize Hindu sentiment around epic themes, still cherish these texts and their heroes and heroines deeply and are willing to see in them normative, emulable, and indeed exemplary models for the political, social, and religious life of the nation.

But since so many of the central actions of the epic heroes proceed from assumptions of profound and fundamental inequalities, it is only to be expected that they should present difficulties to a contemporary audience which regards these figures as earthly manifestations of an infallible godhead while at the same time remaining steeped in the egalitarian ethos that was constitutive of the Freedom Struggle and the Republic of India.

The most highly charged illustration of this is to be found in a comparison of the treatment of Rāma's banishment of Sītā as it is first recounted in the *Uttarakāṇḍa* of the *Vālmīki Rāmāyaṇa* with the treatment it ultimately received at the hands of Ramanand Sagar, the producer and director of *Rāmāyaṇ*, the popular television serialization of the epic.

Few episodes in the Rāma story are as deeply controversial, and there are many reasons for the depth of feeling it evokes. For here is an acknowledged injustice perpetrated by the hero against a character who is not merely of high status but is his very own beloved wife, the long-suffering heroine of the poem. Moreover, it is a wrong inflicted on her purely for reasons of political expediency and with Rāma's full awareness that she has done nothing to merit it. Indications of this controversy or at least of an ambivalence of feeling regarding it is—as in the case of the *Vālivadha*—to be found in the epic itself. By contrast with the case of Vālin, however, no one, least of all Sītā herself, is given the opportunity to directly challenge Rāma's actions. There are, however, some hints that dissent is, if not clearly articulated, at least conceivable.

In deciding to renounce Sītā, Rāma makes it quite clear that he is certain of her innocence.[49] Nonetheless, he resolves to repudiate her rather than face the scandal his taking her back has occasioned.[50] In instructing Lakṣmaṇa that it is he who is to undertake the 'dirty work' of deceiving the pregnant queen with the ruse of a pleasant outing only to cruelly abandon her in the wilderness, it would appear that Rāma anticipates some resistance from his brother. For he forbids him to question this decision, warning him that any attempt to interfere will incur his wrath.[51]

The passage makes it clear that Rāma is deeply saddened by the loss of Sītā,[52] but it gives no evidence of his being troubled by the ethical implications of what he fully understands to be an act of sheer political expediency. Indeed, given the apparent ethos of the presumed author and his intended audience, the cruel abandonment of Sītā is constructed in Vālmīki as a regrettable, but nonetheless necessary, means of preserving the legitimacy of Rāma as a moral ruler. For, in the moral and ethical universe of the epics, the legitimacy of a monarch and—most crucially—his heirs critically depends upon the people's perception that he exemplifies the moral code that he enforces upon others.[53] When universally acceptable testimony to Sītā's faithfulness—in the form of the infallible word of the *ṛṣi* Vālmīki himself—becomes available, Rāma is perfectly willing to take her back once again.[54]

It is not that Rāma does not see it as wrong to repudiate an innocent woman. He does, and he even begs Vālmīki's pardon for having done so.[55] It is just that in the epic world his action is regarded as less grave a flaw in a king than that of incurring the distrust and censure of the people.[56]

These hints of the moral and ethical difficulties inherent in the rejection of Sītā for what amounts to political expediency are greatly developed in some later versions of the *Rāmāyaṇa*. By the eighth century, for example, Rāma's sense of guilt and a general acknowledgement that he has acted improperly form the central themes of Bhavabhūti's great drama, *Uttararāmacarita*. Here, although the deed is the same, the response to it is more elaborately examined and the judgement of it—both on the part of Rāma and others—is considerably harsher. Most interestingly, this action, which is more or less uncritically constructed as the epitome of stern, dispassionate, and mature statecraft in Vālmīki is, in Bhavabhūti, written off as the result of youthful inexperience and a perhaps excessively rigid sense of political morality.

The stage for this is set immediately before the advent of the spy whose report of the people's discontent leads Rāma to his resolve to renounce Sītā. Upon hearing a messenger convey the exhortation of his guru Vasiṣṭha that he who is so young, and still new to the kingship should always strive to please the people, Rāma boasts that he would not hesitate to renounce love, compassion, happiness, and even Sītā herself in the pursuit of that goal.[57] Noteworthy in this context is the immediate and unequivocal approbation of this vow to sacrifice her on the part of Sītā herself. No sooner does she hear these cruelly prophetic words than the queen responds, 'It is because of this that my lord is the foremost of the Rāghava dynasty'.[58]

If censure of Rāma for his renunciation of his faithful wife is implicit in the *Rāmāyaṇa* in the hero's alacrity in stifling any protest on the part of his brother, it is unrestrained in the *Uttararāmacarita*. In the play the action, as suggested above, is regarded not so much as the stern duty of a self-sacrificing monarch as the rash gesture of a royal neophyte which is carried through only because of the unusual circumstance that those older and wiser than the young king are away from the capital. This point is re-emphasized in the *viṣkambhaka*, or prologue, to the second act of the play when the account of Sītā's banishment is recounted during the course of a conversation between the forest divinity Vāsantī and an ascetic woman, Ātreyī, who has lately come from the hermitage of

Vālmīki. Deeply shocked at the dreadful news, the goddess asks Ātreyī how such a thing could have been allowed to happen when the House of Raghu was under the guidance of Arundhatī and Vasiṣṭha and while the elder queens, Daśaratha's widows, were still alive. Ātreyī replies that all of the elders were away at the ashram of Ṛṣyaśṛṅga at the time.[59] Here, clearly, Sītā's banishment is regarded as a rash act and a wrong that would have been avoided with proper guidance. The sense of rashness is stressed again later in the play when the aged king Janaka, whose happiness has been shattered by the repudiation of his daughter, denounces Rāma's actions in menacing terms:

Oh! What cruelty on the part of the depraved populace! Oh! What rashness on the part of King Rāma! As I brood constantly on the dreadful impact of this thunderbolt of torment, it seems appropriate for me to let my rage blaze forth at once either with my bow or with a curse.[60]

However, in the main, this line of representation has the effect of exculpating Rāma or at least providing the mitigating circumstances of youth and inexperience.

That Rāma has acted improperly is also suggested by the play's claim that upon learning of his treatment of Sītā, Arundhatī, the wife of his guru, resolves not to return to an Ayodhyā that is devoid of Sītā, a resolution endorsed by Rāma's mothers and the sage Vasiṣṭha.[61] Then, too, the goddess Vāsantī, an old and dear friend of the royal couple, upon learning of the catastrophic turn of events begins her lamentation as follows:

Alas, my dear friend! Alas, noble lady! Such is the fate of your birth, Alas, dear Rāma! Or, rather, I have no further use for you![62]

But all of these comments pale before the repeated, bitter, and lacerating denunciations to which Bhavabhūti has Rāma subject himself throughout the course of this emotionally wrought drama. It is here, in the words placed in the hero's mouth by the poet that we see the first fully developed ethical critique of Rāma's abandonment of Sītā.

Rāma's self-reproaches form a kind of leitmotif of the play, and it would be cumbersome to adduce every such passage. One can, however, get a clear sense of the harshness of his self-judgement from the speech Rāma delivers immediately upon resolving to banish Sītā without a hearing and without even informing her of his intentions. As soon as the messenger who had brought him the evil tidings of the scandal among the populace has left the stage, Rāma begins his self-castigation, sparing himself no abuse:

Rāma:

> Oh! What a calamity! I have become so cruel, capable of such a despicable act.
>
> > Like a butcher slaughtering a pet bird, I am giving over to death through deception her, who, from childhood, has been indulged with every dainty thing and who, out of love, cannot bear to be apart from me. [45]
>
> Then why should an untouchable sinner like me defile the lady?

(So saying he gently lifts Sītā's head and withdraws his arm.)

> > Ah, innocent girl, renounce me, who have become a filthy outcaste through an act of unprecedented vileness; for, under the delusion that you had found a sandalwood tree, you have been clinging to a poisonous tree with noxious fruit. [46]

[He invokes a long list of his elders, friends, and allies.]

> You have all been robbed and cheated by wretched Rāma. But then, who am I now even to name them?
>
> > For I realize that these great people are as if tainted with sin even to be named by such a wicked ingrate as I. [48]
>
> I, who
>
> > have abandoned my beloved wife—the ornament of my house, who laid her head upon my breast and trustingly fell asleep, heavy though she was with the burden of the child moving uneasily in her womb—and cruelly tossed her away like a sacrificial offering to carnivorous beasts. [49]

(Placing Sītā's feet on his head)

> My lady! My lady! This is the last time Rāma's head will touch your lotus feet.[63]

The rhetorical effect here is a powerful one. Through Rāma's histrionic self-vilification, the playwright is able to fully articulate his (and his audience's?) disquiet at the ethical implications of the scene. Yet, by placing this censure in the mouth of Rāma himself, he avoids the difficulty of having another character indulge in such fierce denunciation. This, coupled with Sītā's prior sanction of her own abandonment, and the careful shifting of part of the blame for Rāma's lapse to his inexperience and the absence of his moral and ethical guides, enables the poet to voice a powerful critique of the hero's actions without making himself guilty of what, from another mouth, would amount not merely to *lèse majesté* but to actual blasphemy.[64]

The question of blasphemy calls attention to the further complexity introduced by theological concerns into the traditional culture's ethical evaluation of the problem of the renunciation of Sītā, an evaluation that, as we see, has itself undergone revision in response to shifting social values.

Although Pollock has persuasively argued that Rāma's identity as a principal *avatāra* of Viṣṇu is constitutive of the *Rāmāyaṇa* in its oldest surviving version, that of Vālmīki,[65] it has also been shown by the same author and others that the figure does not appear to have become established as a central receptacle of *bhakti* worship much before the twelfth century AD.[66] Whatever may be the reasons for the relatively slow development of a Rāma cult, it is clear that the devotional energy focused on the god and, perforce, his divine consort (Śrī-Sītā)—manifested in texts such as the *Rāmcaritmānas* of Tulsīdās and its Sanskrit exemplar the *Adhyātma Rāmāyaṇa*—is far more intense and immediate than that represented in Vālmīki.

This intensification and the concomitant increase of the eschatological significance of Sītā to the individual worshipper[67] infuses the suffering she endures during the course of the epic story with a set of problems not directly confronted by Vālmīki. Thus, in the *Adhyātma Rāmāyaṇa* and the *Mānas,* her defilement at the hands of the monster Rāvaṇa is avoided through the device of having her enter the fire for the duration of the period of her abduction while a phantom, *chāyāsītā,* is carried off by the *rākṣasa* in her stead. Her ill-treatment at the hands of Rāma, however, presents a theological problem of a far more serious order, a problem with which these texts choose not to deal. Thus the poem of Tulsīdās concludes with the coronation of Rāma and omits the problematic portions of Vālmīki's *Uttarakāṇḍa* which form the entire substance of Bhavabhūti's play.

The issue of how the repudiation of Sītā would be received by the immense contemporary audience of Tulsī's *Rāmāyaṇ* was played out recently in connection with the handling of this episode—sensitive both theologically and ethically—by the producer of the enormously popular television serial, *Rāmāyaṇ*. This, in turn, provided a further insight into how changing systems of values cope with potentially disquieting episodes in traditional texts that are regarded as possessed of great sanctity.

For all his elaborate claim to have synthesized the premier literary *Rāmāyaṇas* of India,[68] it is evident that the producer's, Ramanand Sagar's, principal inspiration was the *Rāmāyaṇ* of Tulsīdās. Accordingly, it was his intention to end the series with the restoration of Rāma to his ancestral throne. This decision, however, involving as it did the excision of the *Uttarakāṇḍa*, had unforeseen consequences. A sweeper caste of north India that calls itself the Vālmīki and identifies itself with the Ādikavi[69] took offence at the exclusion of the portion of the epic narrative in which their guru figures most prominently as a character. The Vālmīkis then agitated for the inclusion of the final section. This, in turn, was vigorously resisted by high-caste Hindu groups organized around the Vishwa Hindu Parishad (VHP) who objected to the depiction of Rāma's humiliating treatment of Sītā. After the matter reached the courts and was referred to a board of *Rāmāyaṇa* experts, it was decided that the disputed episodes would indeed be filmed and broadcast.[70]

What is of interest in the present connection is the way in which this popular contemporary version of the epic reconstructs the episode of Sītā's repudiation so as to avoid to the greatest extent possible causing offence to its intended audience by representing the ideal god–king Rāma as engaging in an action that is so markedly out of keeping with contemporary public standards of proper behaviour. For, however justified Sītā's banishment may appear by the canons of *realpolitik* current in the ancient *janapada* of Kosala and however powerfully the *Rāmāyaṇa's* ethos in which a man's honour as determined by the perceived sexual purity of his womenfolk is instantiated in the North Indian culture of *izzat*,[71] such brutal treatment of a wife whom one knows to be faithful can hardly be endorsed by the progressive elements in contemporary India's economic and political elites.

Sagar's revisions of the episode are illuminating in the ways they draw from suggestions in the traditional literature on the subject and yet play to a contemporary popular morality that favours a hero's spirited defence of his beloved. The filmmaker's solution to the problem is to place the initiative for Sītā's exile with the queen herself, thus freeing both Rāma and Lakṣmaṇa of any appearance of impropriety. This is how the episode is described by the British journalist Mark Tully who interviewed Sagar during the period when he was scripting and shooting the relevant scenes.

> When I was in Umargaon, Ramanand Sagar was wrestling with feminist problems in the last section of the *Ramayan.* He was thinking of placating the feminists by making Sita take the decision to go into exile herself.
>
> 'Sita notices that Ram can't take any decision, and so she asks him what is wrong. He replies that people are saying that she is not pure. She tells him that the people are fools, but Ram says, "They are our people." Then Sita says, "I will not stay and bring disrespect on this great family." Ram's brother Lakshman then tries to pacify her by saying, "We will cut their tongues out." So Sita then replies, "That's the men's way of doing things. We have ways of making people feel ashamed so that there is a permanent change in them." Eventually, after her exile, the people of Ayodhya go out and beg her to come back. She tells them, "You do not deserve me," and asks Mother Earth to swallow her up. So she is the winner, not the loser as the feminists seem to think.'[72]

Although Sagar is at pains to represent this strategy to a foreign journalist as a sop to feminists, it is difficult to see how a revision that makes no substantive change in the story other than to construct Sītā as the author of her own misery can be construed as rendering it much more palatable to those who speak for women's interests.[73] Far more likely, in the context of the very real social and political confrontation that developed over the question of whether or not to include the events of *Uttarakāṇḍa*, the group Sagar was really trying to appease was high-caste Hindu men whose feelings were articulated by the VHP. He does so by absolving Rāma of all initiative and virtually all blame for Sītā's mistreatment and by constructing 'the people' rather than Rāma as the ones who do not deserve Sītā.[74] This device also has the effect of relieving both Rāma and Lakṣmaṇa of the onus of deceiving Sītā into thinking she is being taken on a pleasurable excursion when she is, in fact, to be abandoned to her fate in the wilderness.

A historical survey of the theme indicates that the idea of placing the burden of these events on the woman is not wholly out of keeping with the spirit of the earlier versions of the tale. Vālmīki's Sītā, it should be recalled, never protests or criticizes Rāma's decision while, more to the point, Bhavabhūti's heroine actually commends Rāma prior to her repudiation for the attitude that will inevitably lead to it. Nonetheless, Sagar carries the motif to the limit in having Sītā come up with the plan and insist upon it over the objections of her husband and brother-in-law.[75]

Of course the ultimate exemplar for a shifting scale of ethical behaviour in response to changing situations is Kṛṣṇa, the Machiavellian counsellor of the Pāṇḍavas in the *Mahābhārata.* It is in his advice and the actions of his protégés, especially Arjuna, that we see that for the epic characters, as for most people, self-interest is more compelling than any rigid code of behaviour and that no matter how powerful an ethical norm may be, a countervailing argument of necessity can be successfully made. In this way, Arjuna is able to overcome the taboo on assaulting one's guru to shoot down the unresisting patriarch, Bhīṣma,[76] while even the earthly incarnation of Dharma himself, Yudhiṣṭhira, is persuaded to fudge the truth enough to encompass the destruction of the brahmin martial arts master, Droṇācārya.[77]

One well-known incident in particular, however, merits a brief examination in connection with the present discussion. This is the killing of Karṇa. Here the issue is stripped of all the complications of age, status, and deference that are brought to the fore by the cases of Bhīṣma and Droṇa. The final battle of Arjuna and Karṇa is chivalrous epic single combat in its purest form; for the principals are rivals and equals, both sworn to the strict code of the *kṣatriya.*

After their spectacular duel has proceeded for some time, one of Karṇa's chariot wheels is caught in the mud of the battlefield. He asks for a time out while he tries to free it, reminding Arjuna of his noble upbringing and of the rules of chivalrous conduct which prohibit striking at an opponent at such a moment. Karṇa argues that he should have nothing to fear from either Arjuna or Kṛṣṇa while engaged in freeing his wheel.[78]

Before Arjuna can respond, Kṛṣṇa replies for him; and his response is as typical as it is illuminating. Instead of addressing the specific rules of honourable engagement, he begins to taunt Karṇa, telling him that the lowly take refuge in *dharma* only when they are in trouble. He then ironically inquires of the immobilized warrior where his concern for righteousness was when he stood by and watched the humiliation of Draupadī and other atrocities committed by the followers of Duryodhana. The recital of all the Pāṇḍavas' sufferings serves its purpose by infuriating Arjuna, and the battle continues with no allowance made for Karṇa's disadvantage.[79]

The episode is noteworthy in providing yet another example of the shifting moral grounds that are exploited by the epic heroes. The argument here is that Karṇa, having himself condoned the violation of *dharma*, has no moral right to demand that his enemies treat him in accordance with it. In other words, the *kṣatriyadharma* that is so central an animating principle in the actions of the epic protagonists has no absolute value nor is it always binding on those who profess it. In rehearsing all the 'criminal' acts of Karṇa, especially his tacit participation in the violation of Draupadī, Kṛṣṇa makes a case not dissimilar to that made by Rāma in defending himself against the charges of impropriety levelled against him by Vālin: that his opponent is not, in fact, a worthy warrior rival, but a miscreant who can make no legitimate claim on the code of righteous combat.

One final example of the epic's handling of manifestly unethical behaviour on the part of its principal heroes is the one that marks the culmination of the epic conflict in the *Mahābhārata*, Bhīma's infamous 'low blow' that shatters Duryodhana's thighs.[80] Not only does Bhīma employ a blow forbidden by the rules of combat, he compounds his breach of propriety by placing his left foot on the head of his helpless, fallen enemy.[81]

Bhīma's departure from the *kṣatriya* code of conduct towards equals is shocking, so much so that Yudhiṣṭhira himself, Duryodhana's principal enemy, castigates his brother sharply, almost pleading his enemy's case:

> You must not press his head with your foot! Do not let righteousness escape you completely! It is a king and a kinsman that you have struck.

It is not right of you, blameless one. He is destroyed. His ministers, brothers, and subjects have been slain. He is our brother and even his funeral offerings have been cut off. It is not right what you have done. People used to say, 'Bhīmasena is a righteous man.' Why then, Bhīmasena, do you trample upon the king?[82]

Bhīma's actions are reviled still more forcefully by Kṛṣṇa's older brother, Balarāma, who is a particular authority on the niceties of combat with the mace. He flies into a rage at Bhīma's violation of the rules of combat, condemns him as a transgressor of the authoritative texts, and rushes upon him with his own weapon, the plough, raised, to kill him.[83]

Kṛṣṇa, however, intercepts his brother's charge and partially allays his anger by arguing: (1) that Bhīma was only fulfilling the vow he had earlier made,[84] (2) that Bhīma was fulfilling a curse on Duryodhana that he should have his thighs broken in this way,[85] (3) that Balarāma and Kṛṣṇa should, in any case, side with the Pāṇḍavas, and (4) that, at all events, the degenerate Kali Age is at hand.[86]

This casuistic line of argument is in direct contradiction to the reasoning employed by Kṛṣṇa in his efforts to dissuade Arjuna from killing Yudhiṣṭhira. Surely just vowing to do something wrong cannot make it right. Balarāma is not persuaded by his brother's arguments and, indeed, the epic poet Vyāsa himself describes it as specious righteousness *(dharmacchala)*.[87] Balarāma is, however, dissuaded from assaulting Bhīma, and instead he leaves in disgust saying that Bhīma's reputation as an honourable warrior is ruined while Duryodhana's will live forever.[88]

The fact of the matter is that even Kṛṣṇa is not persuaded by his own rhetoric, for only a few lines later he himself asks Yudhiṣṭhira how he can tolerate Bhīma's unrighteous actions in trampling the king's head.[89] Nonetheless, Kṛṣṇa's rhetoric has been successful in that it has accomplished his immediate pragmatic goal, the protection of Bhīma from the wrath of Balarāma. This pragmatism, it appears, is the principle that underlies most of Kṛṣṇa's ethical and moral argumentation in the poem. He adduces whatever argument will serve to accomplish his larger purpose, the victory of the Pāṇḍavas, the forces of a 'higher *dharma*'.

Yudhiṣṭhira's response to Kṛṣṇa's question is revealing in the present connection in that it shows that even the Dharmarāja can be moved by partiality and a desire for vengeance to overlook such wrongdoing:

I don't like it, Kṛṣṇa, that Wolf-belly, in his fury, has placed his foot on the king's head. Nor do I take any pleasure in the destruction of my family. But we were constantly being humiliated with every kind of abuse by the sons of Dhṛtarāṣṭra who, with many abusive words banished us to the wilderness. The pain of all this is still keenly felt in Bhīmasena's heart. Taking this into account, Vārṣṇeya, I have tolerated this. Therefore, having slain the foolish, greedy king, a slave to his desires, let the Pāṇḍava do as he likes whether it be righteous or unrighteous.'[90]

In a way the last and most poignant exchange on the issue of Kṛṣṇa and the Pāṇḍavas' unfailing recourse to unrighteous forms of combat is the final debate between the dying Duryodhana and Kṛṣṇa, the architect of his downfall. Ignoring the excruciating pain of his injuries, Duryodhana reviles Kṛṣṇa, taking him to task for the foul means he employed to destroy such honourable warriors as Bhīṣma, Droṇa, Bhūriśravas, Karṇa, and himself, concluding:[91]

If you had engaged me, Karṇa, Bhīṣma, and Droṇa fairly in battle, it is certain that you would not have won. It is only through the most devious and unworthy means that you have slain me and these other kings who would not deviate from the proper code of warrior conduct.[92]

Kṛṣṇa's response is interesting. He makes no effort to deny the unethical nature of his actions and those done by his protégés at his urging. He argues instead that it is Duryodhana himself, through his own many violations of *dharma*, who has forced Kṛṣṇa to adopt the dirty tricks needed to undo him. Thus, he claims, the onus for these acts lies with Duryodhana, and not with him.[93] But when bands of celestial beings appear in the sky to laud the fallen Duryodhana and shower him with flowers, the Pāṇḍavas are deeply ashamed at their having slain so many virtuous heroes through deceit and give way to grief and depression.[94]

In order to cheer them up Kṛṣṇa tells them frankly that Duryodhana was right; there was, in fact, no way that they could have defeated him and his allies in a fair fight. Therefore, he concludes, it was necessary to resort to trickery, a mode of behaviour sanctioned through its use by the gods in their battles with the *asuras*.[95] This brutally pragmatic exhortation, whose fundamental message is simply that the ends justify the means, allays, for the moment, the Pāṇḍavas' guilt and sorrow, and they retire for the evening rejoicing in their victory.[96]

In conclusion it is to be observed that although the great Sanskrit epic poems do largely live up to their reputations as Indian culture's great repositories of moral and ethical values, in short, of *dharma*, a close synchronic analysis of the various moral and ethical crises and dilemmas the works contain accompanied by the diachronic examination of the culture's shifting response to these dilemmas reveals a number of interesting transformations as the epic stories are subtly revalorized over time.[97] Thus, while the epics and the culture in which they are embedded make frequent rhetorical appeals to '*sanātana dharma*' a theoretically single, eternal, and immutable code of values, in reality the texts and their interpreters have always understood *dharma* to consist of a complex set of codes from which the justification for a wide variety of actions could be drawn. It is the genius of these poems that so marvellously supple and 'context sensitive' a mechanism devised in antiquity can still remain adaptive and useful in the modern age.

NOTES

Abbreviations

ABORI	*Annals of the Bhandarkar Oriental Research Institute*
JAOS	*Journal of the American Oriental Society*
JAS	*Journal of Asian Studies*
JIP	*Journal of Indian Philosophy*

1. Not all such religious or theological texts are necessarily written. Along with such written examples as the Sahajiyā Vaiṣṇava treatises on the life of Kṛṣṇa (see Edward C. Dimock, Jr., *The Place of the Hidden Moon: Erotic Mysticism in the Vaiṣṇava-sahajiyā Cult of Bengal,* Chicago and London: University of Chicago Press, 1966, Paperback Edn, 1989), one may include the innumerable sermons and lectures delivered orally during *satsaṅga* and the like by Hindu religious teachers in India and elsewhere, and renderings of essentially oral, but often published, forms such as the *śāṅkāvalīs*, or lists of 'doubts' or questions submitted to teachers by the

faithful in connection with the actions of characters in the *Rāmcaritmānas.* See Philip Lutgendorf, *The Life of a Text: Performing the Rāmacaritmānas of Tulsidas,* Berkeley and Los Angeles: University of California Press, 1991, pp. 210–11, 392–401, esp. 393–5.

2. For a recent example of scholarly discussion of such issues, see Bimal Krishna Matilal, ed., *Moral Dilemmas in the Mahābhārata,* Delhi: Motilal Banarsidass, 1989.
3. Note, for example, the *Adhyātma Rāmāyaṇa's* and *Rāmcaritmānas's* omission of the Sītatyāga and the public controversy over the omission of the episode from Sagar's Doordarshan serialization. See 'The Second Coming', *India Today,* 31 August 1988, p. 81.
4. I have been struck, in recent years, by the degree to which this assumption underlies the thinking of a large number of even highly educated members of India's urban middle class, especially those representing traditional social elites. Recently, during the course of a conference on *Rāmāyaṇa* Studies that I jointly organized at the University of Hyderabad, an event that took place in the midst of terrible communal violence in that city, I was surprised to hear many of the Indian participants express astonishment that any segment of contemporary Indian society would be distressed to hear politicians speak about the goal of bringing back the idealized society of the 'Rāmrājya'. To them the term conjured up a social vision so unambiguously virtuous that they were genuinely perplexed that there could be segments of the society who would view it with apprehension.
5. On this point, see T.S. Rukmani, 'Moral Dilemmas in the Mahābhārata', in Matilal, *Moral Dilemmas,* pp. 20–34.
6. See Matilal, *Moral Dilemmas,* pp. 7–19. Truthfulness is of particular importance, according to the normative texts of ancient India, in the case of kings. For in its breach the very foundations of the state are threatened. This follows from the well-established ideas that the king is both the moral exemplar for his people and the maintainer—through his moral perfection—of the social, political, and even natural orders. On this, see Sheldon I. Pollock, trans., *The Rāmāyaṇa of Vālmīki: An Epic of Ancient India, Volume 2: Ayodhyākāṇḍa,* ed. by Robert P. Goldman, Introduction and Annotation by Sheldon Pollock, Princeton: Princeton University Press, 1986, pp. 28–9; Pandurang V. Kane, *History of Dharma- śāstra,* 8 vols, Poona: Bhandarkar Oriental Research Institute (Government Oriental Series, Class B, no. 6), 1962–75, vol. 3, pp. 3–6, 28–9, 44–6, 52–5, 96–8; Kane 1962–75, vol. 3; and *Manusmṛti,* 10th Edn, Bombay: Nirṇayasāgar Press, with Commentary *Manvarthamuktāvalī* of Kullūka, ed. by N.R. Acharya, 1946, 9.243 (hereafter *Manu*).
7. These events form the substance of virtually the entire lengthy *Ayodhyākāṇḍa,* the second of the epic's seven books. For a discussion, see *The Rāmāyaṇa, vol. 2, Ayodhyākāṇḍa,* p. 10.
8. Matilal, *Moral Dilemmas,* pp. 7–8; *Mahābhārata: Critical Edition,* 1933–71, 19 vols, Poona: Bhandarkar Oriental Research Institute, with *Harivaṃśa,* critically edited by V.S. Sukthankar et al. (hereafter *MBh*), 8.48–9.

Following Kṛṣṇa's advice, Arjuna addresses his elder with the familiar pronoun *tvam* which, according to Kṛṣṇa, is tantamount to death for a man of honour such as Yudhiṣṭhira (*MBh* 8.49.67ff.). Although Matilal describes this kind of solution as 'face-saving' and even 'childish', he argues that the semantic issue it raises cannot be totally ignored, for it effects a compromise whereby both the intention of the speaker and his moral adherence to truth-keeping are respected (*Moral Dilemmas*, p. 16). The catastrophic consequences of blind adherence to one's word are, as is typical in the epic, averted through the intervention of Kṛṣṇa. It is Kṛṣṇa who characterizes Arjuna's youthful vow as childish, telling him that it would be stupid to adhere to it when the consequences would be so grave (*MBh* 8.49.23). He narrates to Arjuna the story of a sage, Kauśika, who in his obsession with always telling the truth divulges the whereabouts of some people to a band of dacoits who have been pursuing them. The innocents are slain by the robbers and the pious sage goes to hell for his sin (*MBh* 8.49.41–6). Here Kṛṣṇa argues for a moral relativism that eschews absolutes and quite sensibly weighs the consequences of even so virtuous an act as telling the truth. By the same token he argues that untruthfulness is necessary in the aid of the preservation of life and wealth and at the time of marriage, a sentiment cited by the *Rām* commentators in exculpation of Daśaratha's broken promise to Kaikeyī to make her son the heir to the throne. See below and footnotes 12 and 13.

9. *Vālmīki Rāmāyaṇa: Critical Edition*, 1960–75, 7 vols, Baroda: Oriental Research Institute, General Editors, G.H. Bhatt and U.P. Shah (hereafter *Rām*), 7.96.4–13.
10. *Rām* 2.99.3.
11. For a thorough and learned exposition of the problem of Daśaratha's various promises and its treatment by the *Rām* commentators, see Pollock, *The Rāmāyaṇa, vol. 2, Ayodhyākāṇḍa*, pp. 507–8.
12. For the remarks of the commentators and the specific citations from *śāstra*, see ibid., pp. 507–8. The conditions under which untruthfulness is permissible or even mandated are quite similar to those mentioned by Kṛṣṇa in his exhortation to Arjuna at *MBh* 8.49.28–9; 53. See footnote 8 above.
13. *Bṛhaddharmapurāṇa*, ed. by Haraprasād Śāstrī, Calcutta: Asiatic Society of Bengal (Bibliotheca India, no. 120), 1888–97, Reprint Varanasi: Chaukhamba Amarabharati Prakashan, 1974, 1.47.69. See, too, *Bhāgavatapurāṇam*, Bombay: Veṅkateśvara Steam Press, 1908, 8.19.43; and Pollock *The Rāmāyaṇa, vol. 2, Ayodhyākāṇḍa*, pp. 507–8, note to verses 3 and 5.
14. *Rāmāyaṇa of Vālmīki*, 7 vols, Bombay: Gujarat Printing Press, 1914–20, with three commentaries called *Tilaka*, *Shiromani* and *Bhooshana* (hereafter GPP), *Tilakaṭīkā* 2.107.4.
15. It is perhaps only in the somewhat hermetic world of Indology that one would have to make so obvious a statement.

16. An interesting discussion of the somewhat protean nature of the legal texts is to be found in Richard W. Lariviere, 'Justices and *Paṇḍitas*: Some Ironies in Contemporary Readings of the Hindu Legal Past', *JAOS* 48: 4, 1989, pp. 757–69. Also relevant here is the pronounced degree to which traditional Indian culture exhibits what Ramanujan has termed 'context sensitivity' as opposed to some other cultures which may demonstrate a predilection for universalizing. See A.K. Ramanujan, 'Is There an Indian Way of Thinking? An Informal Essay', *Contributions to Indian Sociology* 23: 1, 1989.
17. *Uttaramacaritam* of Bhavabhūti, ed. by S.K. Belvalkar (Harvard Oriental Series, no. 22), Cambridge: Harvard University Press, 1918 (hereafter *URC*), V.34.
18. See J.L. Masson, 'Fratricide among the Monkeys: Psychoanalytic Observations on an Episode in the *Vālmīkirāmāyaṇa*', *JAOS* 95, 1975; and *The Oceanic Feeling: The Origins of Religious Sentiment in Ancient India*, Dordrecht: D. Reidel (Studies of Classical India, vol. 3), 1980, pp. 75–95. In the latter (p. 95), Masson cites a verse found in the vulgate version of the *Mahābhārata* (equals critical edition 7.1375) which compares the infamy of the slaying of Droṇa by Arjuna to that of Vālin by Rāma: *MBh* 7.1375:

    From the killing of Droṇa will arise the same infamy that Rāma gained from the murder of Vālin: an infamy that will long endure in the triple world with its moving and unmoving contents.

    For a thorough discussion of the various moral issues raised by the episode and the various justifications offered by commentators and scholars for Rāma's actions, see Rosalind Lefeber (forthcoming), ch. 7, 'The Death of Vālin'.
19. *Rām* 4.17.12–45.
20. *Rām* 4.18.2–39.
21. *Rām* 4.18.39.
22. One such question involves the fairness of the judicial procedure through which Rāma finds Vālin guilty, condemns him, and puts him to death. For in the event, Rāma, purporting to act in the juridical role of the king, executes Vālin after hearing only the unsubstantiated charges of a clearly interested party. He makes no effort to determine the facts of the affair, and he permits Vālin no opportunity to present his side of the dispute. Sugrīva's story, it should be noted, has a number of inconsistencies while his own action in usurping Vālin's wife and kingdom without making any attempt to verify his suspicion of Vālin's death has drawn the notice of some scholars. See Masson, 'Fratricide among the Monkeys'.
23. *Rām* 4.18.40–4.
24. Some commentators provide additional rationales for Rāma's ambush of the monkey king over and above those placed in Rāma's mouth. Thus Govindarāja (Lefeber, 1994, p. 245, note on verse 4.18.39) articulates a theological rationale, arguing that if Rāma had permitted Vālin to see

him, the former would have immediately fallen at this feet as a supplicant and devotee. In this case, the commentator urges, Rāma could not have killed him. Moreover, Vālin's ally Rāvaṇa might also have come to Rāma as a supplicant thus forcing Rāma to spare him and so thwarting the purpose of the gods and the *raison d'etre* of the *Rāmāvatāra*. (Lefeber, 1994, p. 245). The commentator Tryambakarāya Makhī, somewhat in the spirit of the commentators' rationalization of Daśaratha's apparent breaking of his promise to Kaikeyī's father, justifies what he acknowledges as Rāma's improper conduct on the grounds that recourse to such conduct is, in fact, just when one is faced with adversity (Lefeber, 1994, p. 47). As Lefeber remarks, at least one scholar of the *Rāmāyaṇa*, Srinivas Shastri, has argued that Rāma may have had a very pragmatic reason for killing Vālin from ambush in that he might not have been powerful enough to have defeated him in open combat (Lefeber, pp. 45–50).

25. For an insightful discussion of Valmīki's characterization of Rāma as a unique prince who incorporates features of both *kṣatriya* and brahmin *dharmas*, of violence tempered by compassion and of sovereignty tempered by renunciation, see Pollock, *The Rāmāyaṇa, vol. 2, Ayodhyākāṇḍa*, pp. 64–73.

26. *Rām* 1.24.15; 1.25.10–12. The story of the slaying of Tāṭakā is told at 1.23.11–1.25.22. The vulgate version of the episode is much more elaborately related. See Robert P. Goldman, trans. and ed., annotation with Sally J. Sutherland, *The Rāmāyaṇa of Vālmīki: An Epic of Ancient India, Volume 1: Bālakāṇḍa*, Princeton: Princeton University Press, 1985, p. 338, note to verse 1.25.13, for a more detailed discussion of the passage.

27. *Rām* 1.23.28–30; 1.24.3–19.

28. *Rām* 1.24.13, compare *Rām* 1.25.5.

29. The moral prohibition on killing a female is clearly stated by many of the *dharmaśāstras* whose authors generally classify the act, *strīhatyā*, as an *upapātaka*, or lesser sin relative to the five *mahāpātakas* (*Manu* 11.59–66, esp. verse 66; Kane, *History of Dharmaśāstra*, vol. 4, pp. 33, 96; Robert P. Goldman, *Gods, Priests, and Warriors: The Bhārgavas of Mahābhārata*, New York: Columbia University Press, 1977; 'Matricide, Renunciation, and Compensation in the Legends of Two Warrior Heroes of the Sanskrit Epics', in *Proceedings of the Stockholm Conference Seminar in Indological Studies. Indologica Taurinensia* 10; and *The Rāmāyaṇa, vol. 1, Bālakāṇḍa*, p. 336, note to verse 15). On fighting with women as a breach of the ethical code of a *kṣatriya*, compare the well-known *MBh* episode dealing with the death of that great paragon of the code of chivalry, Bhīṣma, who gives up his life rather than fight the princess turned male warrior, Śikhaṇḍin (Robert P. Goldman, 'Transsexualism, Gender, and Anxiety in Traditional India', *JAOS* 113: 3, 1993; *MBh*, 6.103ff).

30. *Rām* 1.25.2–3; Robert P. Goldman, 'Fathers, Sons, and Gurus: Oedipal Conflicts in the Sanskrit Epics', *JIP* 6, 1978.

31. *Rām* 1.24.17–19.

32. It is worth noting that in the *Rām*, Rāma insists up until the very last

moment that he will not actually kill Tāṭakā, but merely disfigure her by cutting off her nose and ears (*Rām* 1.25.11–12). In other words, in keeping with his later treatment of Vālin, he is willing to apply the appropriate civil punishment to a woman who has transgressed the bounds of propriety, but he is resistant to the notion of engaging her in battle.

33. See Masson, *The Oceanic Feeling*, pp. 114–16; Robert P. Goldman and J.L. Masson, 'Who Knows Rāvaṇa: A Narrative Difficulty in the *Vālmīki Rāmāyaṇa*', *ABORI* 50, 1969; Kathleen Erndl, 'The Mutilation of Śūrpaṇakhā', in Paula Richman, ed., *Many Rāmāyaṇas: The Diversity of a Narrative Tradition in South Asia*, Berkeley and Los Angeles: University of California Press, 1991, pp. 66–8.
34. *Rām* 7.64–7 narrates the episode of Śambūka. See especially 7.65.8–23 for a description of who is eligible to perform *tapaḥ* in given *yugas*. The episode marks an unusually rigid elite response to the practice of *tapaḥ* and other attributes of the renunciant and spiritual life on the part of members of the lower social orders. Elsewhere both epics provide examples of *śūdra* or other low caste ascetics who appear to live the life of *vanavāsīs* without censure or penalty. Two such figures, killed by Daśaratha and Pāṇḍu respectively, as a result of 'hunting accidents' play significant roles in the epic narratives as it is the curses they place upon their killers that lead to the deaths of these kings and, respectively, to the exile of Rāma and the divine parentage of the Pāṇḍavas (*Rām* 2.57–8; *MBh* 1.109.5–30). Another example of an approved appropriation of an elite or brahmanical function (teaching of *dharma*) by a representative of the 'subaltern' groups can be seen in the story of the *dharmavyādha* (*MBh* 3.198–206). It should be noted, however, that the text explains the anomaly in this case as a result of the fact that the low-born hunter was in an earlier life a brahmin who, like Pāṇḍava and Daśaratha, accidentally killed a sage while hunting. In this case the curse involves a loss of ritual status, but not a loss of brahmanical wisdom (*MBh* 3.205.22–9; 3.206.1–5).
35. *Rām* 3.16–17. Śūrpaṇakhā, it appears, is punished for what she *is* as much as for what she *does* or attempts to do. The only hesitation the epic culture seems to have had in such a case related to the extent of the punishment and not its justification. Rāma's instinct, when confronted with Tāṭakā, it should be recalled, was to mutilate her in the same way he does Śūrpaṇakhā. It is the idea of killing a woman that arouses—at least temporarily—his resistance.
36. *MBh* 1.123.
37. *MBh* 1.123.6.
38. *MBh* 1.123.10.
39. *MBh* 1.123.26–8.
40. *MBh* 1.123.29–39.
41. The only moral lapse that the epic attributes to Arjuna, and the one that precludes his entry into heaven with his earthly body, is his own failure

to fulfil a vow, his boast that he would defeat the Kaurava armies in a single day (*MBh* 17.2.21–2).

42. *MBh* 1.135–7.
43. *MBh* 1.136.4.
44. *MBh* 1.136.5–9.
45. *MBh* 1.137.7–9.
46. Nīlakaṇṭha in his comment on *MBh* (vulgate) 1.148.4 (equals critical edition, 1.136.4) explains Yudhiṣṭhira's decision to lure the six to their death as follows, 'The sense of the verse that begins, "If we put six people in here . . . " is "Otherwise, suspecting that we have fled [our enemies] may resolve to search for us. That must not happen".' The text's only hint of an explanation of why the Niṣāda woman is selected for murder is that, in the common and all but empty epic cliché, she is 'doomed' (*kālacoditā*).
47. *MBh* 18.3.12–16. Note the curious eschatology expounded in the passage, in which a person whose deeds during life have been preponderantly virtuous must first spend some time in hell before proceeding to heaven, while a predominantly evil person goes first to heaven before reaching hell.
48. Kane, *History of the Dharmaśāstras*, vol. 3, pp. 391–406.
49. At *Rām* 7.44.6–8, he recounts Sītā's vindication by Agni and the other gods in the presence of the *ṛṣis* going on to state at *Rām* 7.44.9ab:

    Moreover my own heart knows that glorious Sītā is pure.

50. Addressing his brothers, he states at *Rām* 7.44.13:

    For fear of scandal I would renounce you and even my life itself. How much more readily would I give up Janaka's daughter.

51. *Rām* 7.44.18:

    And you must not talk back to me regarding Sītā under any circumstances. For, any attempt to dissuade me would incur my most severe displeasure.

52. Thus, for example, at 7.44.22 his eyes are filled with tears.
53. The citizens of Ayodhyā, it is clear, take Rāma's perceived lapse in taking Sītā back as a sign that they too, in emulation of the king, will have to tolerate promiscuity on the part of their own wives:
    Rām 7.42.19:

    asmākam api dāreṣu sahanīyaṃ bhaviṣyati /
    yathā hi kurute rājā prajā tam anuvartate //

    Now we shall have to tolerate [such behaviour] on the part of our own wives. For the people follow the lead of their king.

54. *Rām* 7.87.13–20; 7.88.1–4. This kind of situation is repeated elsewhere in the epics. Earlier in the *Rām*, Rāma subjects Sītā to highly abusive language, rejection, and the humiliation of being paraded before the

assembled monkeys and *rākṣasas* despite the fact that he is certain of her innocence. He accepts her only when she has publicly demonstrated her purity through an ordeal by fire and the fire god, an unimpeachable witness, has publicly testified to her innocence (*Rām* 6.102–6). Much the same scenario is played out in Duḥṣanta's brutal rejection of Śakuntalā and subsequent acceptance after the truth of her account of her marriage to the king is corroborated by a disembodied but heavenly and authoritative voice at *MBh* 1.68.71–80; 1.69.28–42.

55. *Rām* 7.88.3:

 Please forgive me, brahmin, for having renounced Sītā out of fear of the people even though I knew that she was innocent.

56. Several of the Śrīvaiṣṇava commentators on the *Rām* are aware of Rāma's painful ambivalence. In their comments to *Rām* 6.102.5 and 7 (vulgate 1.114.5 and 7) which describes the hero's tearful brooding as he awaits his first meeting with Sītā since her abduction, they observe that he knows that there will be censure of Sītā for having lived in Rāvaṇa's house but that it is also a grave wrong to abandon an innocent woman.
57. *URC* 1.11–12.
58. *URC* immediately following 1.12:
 Sītā:

 ado jevva rāhavakuladhuraṃdharo ajjautto /

 It is noteworthy that although Vālmīki's Sītā at no point protests her banishment or utters any criticism of Rāma for his treatment of her, it is not until Bhavabhūti that we see her giving her consent to such an action before the fact. This is a point to which I will return in connection with the development of this idea in the recent television rendition of the epic.
59. *URC* after 2.6.
60. *URC* 4.24.
61. *URC* after 2.6.
62. *URC* after 2.6:
 Vāsantī:

 hā priyasakhi hā mahābhāge īdṛśas te nirmāṇabhāgaḥ / hā rāmabhadra / athavālaṃ tvayā /

63. *URC* 1.45–6; 48–9:
64. It is to be kept in mind, however, that with reference to the traditional Hindu construction of royalty as in fact divinity, a construction that bears special force in the case of Rāma, there is little if any effective difference between the two. See Pollock, *The Rāmāyaṇa of Vālmīki: An Epic of Ancient India, Volume 3: Araṇyakāṇḍa*, ed. by Robert P. Goldman, Introduction and annotation by Sheldon Pollock, Princeton: Princeton University Press, 1991, pp. 15–54; 'Ramayana and Political Imagination in India', *JAS* 52: 2, 1993.

65. Pollock 1991: 15–54.
66. R.G. Bhandarkar, *Vaiṣṇavism, Śaivism and Minor Religious Systems*, 1913, Reprint, Varanasi Indological Bookhouse, 1965, p. 47; Pollock, 'Ramayana and Political Imagination'.
67. This phenomenon is by no means restricted to the North Indian devotionalism that centres on Tulsī's poem and the institution of the Rāmlīlā. It is even more pronounced, perhaps, in the case of the Śrīvaiṣṇavas of South India. For a discussion of the theological significance of Sītā among the Rāmanandī monks of Ayodhyā, see Peter van der Veer, *Gods on Earth: The Management of Religious Experience and Identity in a North Indian Pilgrimage Centre*, London: The Athlone Press, 1983; Philip Lutgendorf, 'The Secret Life of Rāmcandra of Ayodhya', in Richman, *Many Ramayanas*. For the Śrīvaiṣṇavas, see Patricia Y. Mumme, '*Rāmāyaṇa* Exegesis in Teṉkalai Śrīvaiṣṇavism' in Richman, *Many Ramayanas*.
68. The series' prologue depicts actors made up to represent Vālmīki, Tulsīdas, Kṛttibās, Kambaṉ, etc.
69. This sort of Sanskritization by association is doubtless derived from the widespread legend that the author of the *Rāmāyaṇa* was himself of low caste. For a discussion, see Goldman, 1975.
70. For a more detailed account of this fascinating set of transactions, see 'The Second Coming', *India Today*, 31 August 1988, p. 81; and Richman, *Many Ramayanas*, p. 3.
71. Explicit connection between the 'Sītātyāga' and the brutalization of women is not difficult to find. It has been observed that the refusal on the part of thousands of Hindu (largely Sindhi) men to accept wives and daughters who had been abducted during the violence accompanying the partition of the subcontinent, a refusal that forced many of these women into beggary and prostitution, was often justified through reference to the *Rāmāyaṇa*. In discussing his interaction with Hindu men who had let their wives, sisters, and daughters enter brothels rather than accept them after their abduction, Sadhale remarks, 'I had a dialogue with some of the menfolk concerned. They all said the same thing. "Rama had abandoned his wife. They were all following him" ' (Anand Sadhale 'Epics on Ramakatha in Marathi', Unpublished paper presented at the International Conference on Ramayana Text and Tradition, University of Hyderabad, 3–5 Jan. 1991). This connection has often been noted by Indian feminists. See the eloquent passage in Kamla Bhasin and Ritu Menon's article on Sagar's *Ramayan* published in *Seminar* February 1988 and quoted in Mark Tully, *No Full Stops in India*, London: Viking, 1991, p. 132.
72. Tully, *No Full Stops in India*, pp. 132–3.
73. Then too, serious feminists are so marginalized in India that it seems unlikely that Sagar would take them seriously as a body of critics of his enormously popular work. After all, the reputation of feminism in India is such that one of the country's most articulate and effective advocates

of women's rights has found it politic to disclaim the title of feminist in part 'because of its overclose association with the western woman's movement'. See Madhu Kishwar, 'Why I do not Call Myself a Feminist', *Manushi* 61, November–December, 1990, p. 3.

74. It should also be noted that the time slot during which the *Uttarakāṇḍa* episodes were broadcast was shifted from Doordarshan's 'prime time' for such national programmes on Sunday mornings to a time with a smaller anticipated audience. This was perhaps also an effort to limit the feared negative response.

75. It is not only with regard to this central incident that Sagar attempts to shift the responsibility for an ethically questionable action from prominent male characters to women. Another example is to be found in his treatment of the moral dilemma presented by Vibhīṣaṇa's betrayal of his older brother and king, Rāvaṇa, in violation of two of the most powerful norms of traditional Indian social life in assisting his enemy Rāma to destroy him. Despite the fact that Vibhīṣaṇa is foregrounded in the Rāma cult as one of the great exemplary *bhaktas* of the Lord, his treachery, even though done in the name of *dharma* and out of devotion to God, has been traditionally regarded with ambivalence by a society that places tremendous emphasis on deference and obedience to one's elders. Consider, for example, such proverbial jibes as the Bengali, *ghore śotru bibhīṣoṇ*, 'An enemy in one's own house, a Vibhīṣaṇa'. In the television serialization, Vibhīṣaṇa, although brutally treated by his brother, is irresolute and appears unable to muster the will to abandon him. He confides in his aged mother who instructs him to shift his allegiance to Rāma. It is interesting to contrast Vibhīṣaṇa's defection to the cause of *dharma* with the steadfast adherence of such *Mahābhārata* stalwarts as Bhīṣma, Droṇa, and Vidura to the unrighteous Duryodhana. On this point, see Rukmani, 'Moral Dilemmas in the *Mahābhārata*', p. 30.

76. *MBh* 6.114.

77. *MBh* 7.164.105ff.

78. *MBh* 8.66.60–5.

79. *MBh* 8.67.1ff.

80. *MBh* 9.57.40–5.

81. *MBh* 9.58.5.

82. *MBh* 9.58.15–17.

83. *MBh* 9.59.3–7.

84. *MBh* 9.59.14.

85. *MBh* 9.59.15.

86. *MBh* 9.59.21. And so, presumably, such breaches of *dharma* are only to be expected.

87. *MBh* 9.59.22.

88. *MBh* 9.59.23–6.

89. *MBh* 9.59.29–30.

90. *MBh* 9.59.31–4. The ultimate triumph of pragmatism over righteousness

is made very clear by Bhīma himself when he defends his actions before Gāndhārī at *MBh* 11.14.2–3.

91. *MBh* 9.60.26–38.
92. *MBh* 9.60.37–8.
93. *MBh* 9.60.39–46.
94. *MBh* 9.60.54–5.
95. *MBh* 9.60.57–62.
96. *MBh* 9.60.64–5. In this light Kṛṣṇa's (and Vyāsa's) often repeated cry, *yato dharmas tato jayaḥ,* 'Where there is righteousness, there is victory', might be inverted to mean, 'Might makes right'. For a further discussion of the conflicting values of Kṛṣṇaite devotionalism and the unmodified tradition of *kṣatriyadharma* as represented in the *MBh,* see David Gitomer, 'King Duryodhana: The *Mahābhārata* Discourse of Sinning and Virtue in Epic and Drama', *JAOS* 112: 2.
97. See Rukmani, 'Moral Dilemmas in the *Mahābhārata*', p. 25.

REFERENCES

Bhāgavatapurāṇam (Veṅkaṭeśvara Steam Press, Bombay, 1908).

BHANDARKAR, R.G., *Vaiṣṇavism, Śaivism, and Minor Religious Systems* (Indological Bookhouse, Varanasi, 1913; rpt, 1965).

Bṛhaddharmapurāṇa, edited by Haraprasād Śāstrī, *Bibliotheca Indica,* no. 120 (Asiatic Society of Bengal, Calcutta, 1888–97; rpt, 1974, Chaukhamba Amarabharati Prakashan, Varanasi).

DIMOCK, Edward C., Jr., *The Place of the Hidden Moon: Erotic Mysticism in the Vaiṣṇava-sahajiyā Cult of Bengal* (University of Chicago Press, Chicago and London, 1966), Paperback edition, 1989.

ERNDL, Kathleen, 'The Mutilation of Śūrpaṇakhā', pp. 67–88, in Richman, 1991.

GITOMER, David, 'King Duryodhana: The *Mahābhārata* Discourse of Sinning and Virtue in Epic and Drama', *JAOS* 112:2, 1992, pp. 222–32.

GOLDMAN, Robert P., 'Vālmīki and the Bhṛgu Connection', *JAOS* 96:1, 1976, pp. 97–101.

——, *Gods, Priests, and Warriors: The Bhārgavas of the Mahābhārata,* New York: Columbia University Press, 1977.

——, 'Fathers, Sons, and Gurus: Oedipal Conflicts in the Sanskrit Epics', *JIP* 6: 1978, pp. 325–92.

——, 'Rāmaḥ Sahalakṣmaṇaḥ: Psychological and Literary Aspects of the Composite Hero of Vālmīki's *Rāmāyaṇa*', *JIP* 8: 1980, pp. 11–51.

GOLDMAN, Robert P., 'Matricide, Renunciation, and Compensation in the Legends of Two Warrior Heroes of the Sanskrit Epics', in *Proceedings of the Stockholm Conference Seminar in Indological Studies, Indologica Taurinensia* 10, 1982, pp. 117–31.

——, *The Rāmāyaṇa of Vālmīki: An Epic of Ancient India, Vol. I: Bālakāṇḍa*, Translator and editor, annotation with Sally J. Sutherland (Princeton University Press, Princeton, 1985).

——, 'Transsexualism, Gender, and Anxiety in Traditional India', *JAOS* 113:3, 1993.

——, and J.L. Masson, 'Who Knows Rāvaṇa: A Narrative Difficulty in the *Vālmīki Rāmāyaṇa*', *ABORI* 50, 1969, pp. 95–100.

India Today, 'The Second Coming', 31 August 1988, p. 81.

KANE, Pandurang V., *History of Dharmaśāstra*, 8 vols (Bhandarkar Oriental Research Institute, Poona, 1962–75 [Government Oriental Series, class B, no. 6]).

KISHWAR, Madhu, 'Why I do not Call Myself a Feminist', *Manushi* 61, November–December, 1990, pp. 2–8.

LARIVIERE, Richard W., 'Justices and *Paṇḍitas*: Some Ironies in Contemporary Readings of the Hindu Legal Past', *JAOS* 48:4, 1989, pp. 757–69.

LEFEBER, Rosalind, edited by Robert P. Goldman, Introduction and annotation by Rosalind Lefeber, *The Rāmāyaṇa of Vālmīki: An Epic of Ancient India*, vol. 4, *Kiṣkindhākāṇḍa* (Princeton University Press, Princeton, 1994).

LUTGENDORF, Philip, *The Life of a Text: Performing the Rāmcaritmānas of Tulsidas* 1991a (University of California Press, Berkeley and Los Angeles).

——, 'The Secret Life of Rāmcandra of Ayodhya', 1991b, pp. 21–234, in Richman, 1991.

Mahābhārata, 6 vols, with the commentary of Nīlakaṇṭha (Chitrashala Press, Poona, 1929).

Mahābhārata: Critical Edition, 19 vols, with *Harivaṃśa*, critically edited by V.S. Sukthankar et al. [*MBh*] (Bhandarkar Oriental Research Institute, Poona, 1933–71).

Manusmṛti, edited by N.R. Acharya, with commentary *Manavarthamuktāvalī* of Kullūka (Nirṇayasāgar Press, Bombay, 1946), 10th edition.

MASSON, J.L. (J. Moussaieff Masson), 'Fratricide among the Monkeys: Psychoanalytic Observations on an Episode in the Vālmīkirāmāyaṇa', *JAOS* 95, 1975, pp. 672–78.

——, *The Oceanic Feeling: The Origins of Religious Sentiment in Ancient*

India (D. Reidel, Studies of Classical India, Dordrecht, 1980), vol. 3.

MATILAL, Bimal Krishna, edited by, *Moral Dilemmas in the Mahābhārata* (Motilal Banarsidass, Delhi, 1989).

MUMME, Patricia Y., '*Rāmāyaṇa* Exegesis in Te͟nkalai Śrīvaiṣṇavism', pp. 202–16, in Richman, 1991.

POLLOCK, Sheldon I., trans., edited by Robert P. Goldman, Introduction and annotation by Sheldon Pollock, *The Rāmāyaṇa of Vālmīki: An Epic of Ancient India*, vol. 2, *Ayodhyākāṇḍa* (Princeton University Press, Princeton, 1986).

——, edited by Robert P. Goldman, Introduction and annotation by Sheldon Pollock, *The Rāmāyaṇa of Vālmīki: An Epic of Ancient India*, vol. 3, *Araṇyakāṇḍa* (Princeton University Press, Princeton, 1991).

——, 'Rāmāyaṇa and Political Imagination in India', *JAS* 52:2, 1993, pp. 261–97.

RAMANUJAN, A.K., 'Is There an Indian Way of Thinking? An Informal Essay', *Contributions to Indian Sociology* 23:1, 1989, pp. 41–58.

RICHMAN, Paula, edited by, *Many Rāmāyaṇas: The Diversity of a Narrative Tradition in South Asia* (University of California Press, Berkeley and Los Angeles, 1991).

RUKMANI, T.S., 'Moral Dilemmas in the Mahābhārata', pp. 20–34, in Matilal, 1989.

Rāmāyaṇa of Vālmīki, 7 vols, with three commentaries called Tilaka, Shiromani, and Bhooshana [GPP] (Gujarati Printing Press, Bombay, 1914–20).

SADHALE, Anand, 'Epics on Ramakatha in Marathi', unpublished paper presented at the International Conference on Ramayana Text and Tradition, University of Hyderabad, January 3–5, 1991.

TULLY, Mark, *No Full Stops in India* (Viking, London, 1991).

Uttararāmacarita of Bhavabhūti, with the commentary of Ghanśyāma, Notes and Introduction by P.V. Kane and Translation by C.N. Joshi (Motilal Banarsidass, Delhi, 1971), 5th edition.

Uttararāmacarita of Bhavabhūti, edited by S.K. Belvalkar, Harvard Oriental Series, no. 22 [URC] (Harvard University Press, Cambridge, 1918).

Vālmīki Rāmāyaṇa: Critical Edition, 7 vols, General Editors G.H. Bhatt and U.P. Shah [*Rām*] (Oriental Research Institute, Baroda, 1960–75).

VAN DER VEER, Peter, *Gods on Earth: The Management of Religious Experience and Identity in a North Indian Pilgrimage Centre* (The Athlone Press, London, 1988).

Chapter 12

A Note on Identity and Mutual Absence in Navya-nyāya

Kamaleswar Bhattacharya

In his first major work (1968), Bimal K. Matilal gave, among others, a lucid exposition of the notion of identity in Navya-nyāya.

In India, as in the West,[1] identity as a relation posed a problem. So some schools of Indian thought conceived *tādātmya* not as 'identity' but as 'identity-cum-difference'.[2]

In Nyāya, *tādātmya* is conceived as pure 'identity'. But a problem arose—a problem of a different kind from that we are familiar with (namely a relation obtains between *two* things); and this problem does not seem to be well known.

Traditionally, in Navya-nyāya, 'mutual absence' (*anyonyābhāva*) or 'difference' (*bheda*)—one of the four types of 'absence' (*abhāva*) recognized in the system—is defined as an absence the counterpositive-ness to which is delimited by the relation of identity (*tādātmyasaṃbandhāvacchinnapratiyogitāko 'bhāvo nyonyābhāvaḥ*).[3]

But the word *tādātmya qua* 'identity', in this context, gives rise to an interesting problem. Since the 'identity' (*tādātmya*) of an 'object having contact' (*saṃyogin*) is nothing but 'contact' (*saṃyoga*), this definition of mutual absence covers also the 'constant absence' (*atyantābhāva*) of an 'object having contact', the counterpositive-ness to which is delimited by the relation of contact (*saṃyoga*). The definition thus suffers from the technical defect of 'overpervasion', (*ativyāpti*). So far as I am aware, this problem was raised for the first time by Raghunātha Śiromaṇi in his *Dīdhiti* on Udayana's *Ātmatattvaviveka*;[4] but its solution is to be found only in Mathurānātha Tarkavāgīśa's *Rahasya* (still unpublished) on Raghunātha's *Dīdhiti* on the *Siddhāntalakṣaṇaprakaraṇa* of Gaṅgeśa's *Tattvacintāmaṇi* (*Anumānakhaṇḍa*),[5] and in the *Dinakarī* commentary on the *Siddhāntamuktāvalī*, while full discussions are to be found in the *Ramarudrī* commentary on the *Dinakarī*, as well as in Śrīkaṇṭha Dīkṣita's commentary on Janakīnātha Bhaṭṭācāryacūḍāmaṇi's *Nyāyasiddhāntamañjarī*.

Here is what the *Dinakarī*[6] states: *tādātmyatvena tādātmyāvacchinnatvaṃ pratiyogitāyāṃ vivakṣitam. tena saṃyogāvacchinnapratiyogitāke saṃyogyatyantābhāve nātivyāptiḥ* 'What is meant is that the counterpositive-ness is delimited by identity as identity. Thus there is no overpervasion [of the definition of mutual absence] to include the constant absence of an object having contact, the counterpositive-ness to which absence is delimited by contact.' This terse expression of the problem and its solution is made fully explicit by the *Rāmarudrī*[7] as follows:

tādātmyam tadgato 'sādhāraṇo dharmaḥ. tādātmyatvena tādātmyāvacchinnatvavivakṣāyāḥ phalam āha: teneti, saṃyogyatyantābhāva iti. saṃyogitādātmyasya saṃyogarūpatayā saṃyogasaṃbandhāvacchinnapratiyogitākātyantābhāve yathāśrute 'tivyāptiḥ.[8] *idānīṃ tu tādātmyatvāvacchinnā tādātmyaniṣṭhā yā saṃsargavidhayāvacchedakatā tannirūpakapratiyogitākatvalābhena atyantābhāvīyasaṃyoganiṣṭhapratiyogitāvacchedakatāyāḥ saṃyogatvenaivāvacchinnatvāt nātivyāptir iti bhāvaḥ. yady api saṃyogitādātmyasya saṃyogarūpatvena tattādātmyatvam api saṃyogatvam evety ativyāptinirāso durghaṭa eva, tathāpi saṃsargatāvacchedakam tādātmyatvaṃ na saṃyogatvādirūpaṃ kiṃ tu saṃyogyādivṛttyasādhāraṇadharmatvam eva. ato 'tyantābhāvīyapratiyogitāyāṃ saṃyogasya saṃyogatvenaivāvacchedakatvāt asādhāraṇadharmatvenāvacchedakatvavirahān na tatrātivyāptiḥ. bhedapratiyogitāvacchedakasaṃsargasyaiva asādhāraṇadharmatvena bhānābhyupagamāt; anyathātyan tābhāvānyonyābhāvayor bhedāsaṃbhavād iti bhāvaḥ.*

'Identity' means the uncommon property residing in a certain object. [The author of the *Dinakarī*] explains with what end in view it is meant that [the counterpositive-ness] is delimited by identity as identity . . . Since the identity of an object having contact is identical with contact, there follows an overpervasion [of the definition of mutual absence] to include the constant absence the counterpositive-ness to which is delimited by the relation of contact—if we adopt the usual meaning of the word. But now is obtained the property of having a counterpositive-ness describing the property of being a delimitor as a relation, which resides in identity and is delimited by identity-ness. There is, therefore, no overpervasion, because the property of being the delimitor of the counterpositive-ness to the constant absence, which resides in contact, is delimited only by contact-ness. That is the idea. Although the overpervasion is hard to avoid because the identity of an object having contact being identical with contact, its identity-ness also is nothing but contact-ness, still the identity-ness which is the delimitor of the property of being a relation is not identical with contact-ness etc., but is only the property of being an uncommon property residing in an object having contact etc. Thus it is as contact that contact is the delimitor of the counterpositive-ness to the constant absence, it is not its delimitor as an uncommon property. There is therefore no overpervasion there. It is indeed only of the relation delimiting the counterpositive-ness to a difference that is admitted the appearance as an uncommon property; otherwise, there would be no difference between a constant absence and a mutual absence. That is the idea.

The explanation is clear and is in need of no comment. To avoid all ambiguity, the Navya-naiyāyikas, when they talk about mutual absence or difference, express themselves as follows: A mutual absence or difference is an absence the counterpositive-ness to which is delimited by identity *as a relation*.[9]

But is it necessary to assume a relation as delimitor of the counterpositive-ness to a mutual absence? Raghunātha, in his commentary on the *Ātmatattvaviveka*, says that there is no proof in favour of the assumption that the counterpositive-ness to a destruction (*dhvaṃsa*), *a prior* absence (*prāgabhāva*) and a mutual absence (*anyonyābhāva*) is delimited by a particular relation.[10] This view of Raghunātha is quoted by Jagadīśa Tarkālaṃkāra in his commentary on the *Siddhāntalakṣaṇadīdhiti*: . . . *sādhanavanniṣṭhānyonyābhāvapratiyogitāyāḥ sādhyatāvacchedakatādātmyasambandhāvacchinnatve mānābhāvasya svayam eva Bauddhādhikāraṭippaṇyāṃ uktatvāt*[11]—a passage that is diversely interpreted nowadays in India.[12] As Śrīkaṇṭha Dīkṣita explains, in the case of a constant absence (*atyantābhāva*), it is necessary to assume a relation delimiting the counterpositive-ness, as it can be delimited by different relations. Otherwise, it would be impossible to explain why, e.g. the cognition 'There is no pot on the ground by inherence' (*samavāyena bhūtale ghaṭo nāsti*) is compatible with the cognition 'There is a pot on the ground by contact' (*saṃyogena bhūtale ghaṭo 'sti*). In the case of destruction, prior absence and mutual absence, however, it is not necessary to assume such a relation, since the counterpositive-ness to these absences cannot be delimited by various relations.[13]

This seems to be the older view, as reflected in Nyāya literature from Udayana onwards.[14] As Śaśadhara clearly states, the counterpositive-ness to a constant absence (in the form 'It is not here' —*idam iha nāsti*) is delimited by a relation, while the counterpositive-ness to a mutual absence (in the form 'It is not this'—*idam idaṃ na bhavati*) is delimited by identity.[15] Evidently, 'identity' (*tādātmya*) is here to be understood, not as a relation (*saṃbandha, saṃsarga*), but as a property (*dharma*).[16]

NOTES

1. Cf. Willard Van Orman Quine, *Word and Object*, Cambridge, Mass.: MIT Press, pp. 114ff.
2. Cf. Nāgeśa (Nāgoji) bhaṭṭa, *Paramalaghumañjūṣā*, ed. by Kālikāprasād Shukla, Baroda: Baroda Sanskrit Mahavidyalaya, 1961, p. 42: *tādātmyaṃ ca tadbhinnatve sati tadabhedena pratīyamānatvam iti bhedābheda samaniyatam.*
3. 'Difference or otherness is exemplified in denials such as 'A pot is not a cloth.' Nyāya explains that in such expressions a cloth, called the counterpositive (*pratiyogin*), is declared not to be related to a pot by identity.' Bimal Krishna Matilal, *The Navya-Nyāya Doctrine of Negation: The Semantics and Ontology of Negative Statements in Navya-Nyāya Philosophy*, Cambridge, Mass.: Harvard University Press (Harvard Oriental Series, 46), 1968, p. 46.
4. Raghunātha Śiromaṇi, *Ātmatattvavivekadīdhiti* (*Bauddhādhikāravivṛti* or *-ṭippaṇī*), p. 580.
5. *Anumānadīdhitirahasya*, MS, in the collection of Dr Prabal Kumar Sen of Calcutta, fol. 112b.
6. *Siddhāntamuktāvalī*, pp. 71–2 (on *Siddhāntamuktāvalī* on *Kārikā* 12).
7. Ibid.—Similarly, Śrīkaṇṭha Dīkṣita on Janakīnātha Bhaṭṭācāryacūḍāmaṇi, *Nyāyasiddhāntamañjarī*, p. 54.
8. Cf. Raghunātha, loc. cit.: *tādātmyaṃ hi tadvṛttidharmaviśeṣaḥ. saṃyoginaś ca tādātmyaṃ saṃyoga eva, tathā ca saṃyogasaṃbandhena saṃyogino 'tyantābhāve tivyāptiḥ.*
9. Thus Jagadīśa Tarkālaṃkāra on *Siddhāntalakṣaṇadīdhiti*, p. 108: *bhedatvaṃ ca saṃsargavidhayā tādātmyāvacchinnapratiyogitākābhāvatvam.* Cf. Mathurānātha, loc. cit., line 2.
10. *dhvaṃsaprāgabhāvānyonyābhāvapratiyogitāyāṃ saṃbandhabhedānupraveśe mānābhāvāc ca. Ātmatattvavivekadīdhiti*, loc. cit.

11. *Siddhāntalakṣaṇa-Jāgadīśī*, pp. 156–7.
12. See *Indologica Taurinensia*, V, 1977, pp. 40–1 and n. 14.
13. See Śrīkaṇṭha Dīkṣita, loc. cit.
14. Thus Udayana, *Kiraṇāvalī*, p. 220: *saṃsṛjyamānapratiyoginirūpyo 'bhāvaḥ saṃsargābhāvaḥ. tadātmatvābhimānapratiyoginirūpyaś ca tādātmyābhāvaḥ*. See also Śaṅkaramiśra, *Upaskāra* on *Vaiśeṣikasūtra* IX. 1. 4 (pp. 376–7). Vardhamāna thus reproduces the opinion of his illustrious father, Gaṅgeśa: *pratiyoginam āropya yatra niṣedhadhīḥ sa saṃsargābhāvaḥ, pratiyogitāvacchedakam āropya yatra niṣedhajñānaṃ sa tādātmyābhāva ity asmatpitṛcaraṇāḥ. Kiraṇāvalīprakāśa*, p. 601. Or: *asmatpitṛcaraṇās tu pratiyogyadhikaraṇayoḥ saṃsargam āropya yo niṣedhaḥ sa saṃsargābhāvaḥ, ghaṭasaṃsargaḥ paṭo nety atra tu saṃsargo nāropyate kiṃ tu tādātmyam ity anayor bhedaḥ. Nyāyalīlāvatīprakāśa*, pp. 573–4. Cf. Gaṅgeśa, *Śaktivāda*: 2013 (Note that, according to the 'Ancients', like the counterpositive-ness to constant absence, that to destruction and prior absence,—the two other varieties of 'relational absence', *saṃsargābhāva*,—is also delimited by a relation; while, according to the 'Moderns', it is not. See *Journal Asiatique* CCLXXII, 1984, p. 75n. 62; 82.) From Udayana onwards, mutual absence is also defined as *tādātmyapratiyogiko 'bhāvaḥ. Lakṣaṇāvalī*, p. 283 (71); Śrīvallabha, *Nyāyalīlāvatī*, p. 576; Keśavamiśra, *Tarkabhāṣā*: 148. *Tādātmyapratiyogika* 'having identity as the counterpositive' is explained here by the commentators as *tādātmyāvacchinnapratiyogika* 'having a counterpositive determined by identity', in order to distinguish neatly a mutual absence from the constant absence of identity (*tādātmyātyantābhāva*). But there is no 'carelessness' on the part of the authors as has sometimes been assumed in our time: what is denied, in mutual absence, is the identity of one thing with another (whereas, in relational absence, it is the relation of one thing with another that is denied). 'Mutual absence' is therefore called *tādātmyābhāva*, 'absence of identity', and it is also widely held in Navya-nyāya that the constant absence of a property (*dharmātyantābhāva*) is identical with the mutual absence with the possessor of that property (*dharmyanyonyābhāva*). It is this view of mutual absence that Śrī Harṣa had before him—a fact that has not been always perceived. See *Khaṇḍanakhaṇḍakhādya*, p. 119.
15. *idam iha nāstīti saṃsargeṇa pratiyogitāvacchidyate, idam idaṃ na bhavati ity atra tādātmyena. Nyāyasiddhāntadīpa*, p. 123.—Cf. Śaṅkaramiśra, *Ātmatattvavivekakalpalatā*, p. 571.
16. Similarly, in late Advaita-Vedānta, *Gauḍabrahmānandī* (*Laghucandrikā*) on Madhusūdana Sarasvatī, *Advaitasiddhi*, p. 20.

REFERENCES

Gaṅgeśa, *Tattvacintāmaṇi, Anumānakhaṇḍa, Śaktivāda*: in *Gādādharī* (Chowkhambha, Varanasi, 1970), II, pp. 2001ff.

Jagadīśa Tarkālaṃkāra, *Ṭīkā* on Raghunātha Śiromaṇi's *Dīdhiti* on the *Siddhāntalakṣaṇaprakaraṇa* of Gaṅgeśa's *Tattvacintāmaṇi*, with the commentary of Vāmācaraṇa Bhaṭṭācārya, ed., Guru Prasad Shastri (Master Khelari Lal & Sons, Benares, 1933 [*Shri Rajasthan Sanskrit College Granthamala*], 1).

Jānakīnātha Bhaṭṭācāryacūḍāmaṇi, *Nyāyasiddhāntamañjarī*, with the commentary of Śrīkaṇṭha (not Nīlakaṇṭha!) Dīkṣita, ed., Gaurīnātha Śāstrī, Benares, 1884.

Keśavamiśra, *Tarkabhāṣā*, with the commentary of Viśvakarman, in *The Pandit*, New Series, vol. XXII, Benares, 1900.

Madhusūdana Sarasvatī, *Advaitasiddhi*, with the commentaries *Gauḍabrahmānandī* (*Laghucandrikā*), *Viṭṭhaleśopādhyāyī*, *Siddhivyākhyā* of Balabhadra . . . , ed., Mahāmahopādhyāya Anantakṛṣṇa Śāstrī (Nirṇayasāgar Press, Bombay, 1937), 2nd edition.

MATILAL, Bimal Krishna, *The Navya-Nyāya Doctrine of Negation: The Semantics and Ontology of Negative Statements in Navya-Nyāya Philosophy* (Harvard University Press, Cambridge, Mass., 1968) (Harvard Oriental Series, 46).

Nāgeśa (Nāgoji) bhaṭṭa, *Paramalaghumañjūṣā*, ed., Kālikāprasād Shukla (Baroda Sanskrit Mahavidyalaya, Baroda, 1961).

QUINE, Willard Van Orman, *Word and Object* (MIT Press, Cambridge, Mass., 1960).

Raghunātha Śiromaṇi, *Ātmatattvavivekadīdhiti* (*Bauddhādhikāravivṛti* or–*ṭippaṇī*) See Udayana, *Ātmatattvaviveka*.

Śaṅkaramiśra, *Upaskāra* on *Vaiśeṣikasūtra. The Vaiśeṣika Darśana with the Commentaries of Śaṅkara Miśra and Jayanārāyaṇa Tarkapañcānana*, ed., Paṇḍita Jayanārāyaṇa Tarkapañcānana, Calcutta, 1861 (*Bibliotheca Indica*).

——, *Ātmatattvavivekakalpalatā*. See Udayana, *Ātmatattvaviveka*.

Śaśādhara, *Nyāyasiddhāntadīpa*, ed., B.K. Matilal (LD Institute of Indology, Ahmedabad, 1976).

Siddhāntamuktāvalī: *Nyāyasiddhāntamuktāvalī of Viśvanātha Pañcānana Bhaṭṭācārya with Dinakarī* (*Prakāśa*) *Commentary by Mahādeva Bhaṭṭa & Dinakara Bhaṭṭa and Rāmarudrī* (*Taraṅgiṇī*) *Commentary by Rāmarudra Bhaṭṭācārya & Pt. Rājeśvara Śāstrī*, ed., Pt. Harirāma Śukla Śāstrī, Varanasi, 1972 (Kashi Sanskrit Series, 6; Nyāya Section, 1).

Śrī Harṣa, *Khaṇḍanakhaṇḍakhādya. Khaṇḍanakhaṇḍakhādyam . . . Tārkikaśiromaṇiśrīmacchaṅkaramiśrapraṇītavyākhyāsanātham . . . Kulayaśasviśāstriṇā saṃskṛtam,* Kāśī, Varanasi, 1988.

Śrīvallabha, *Nyāyalīlāvatī,* with the commentaries of Vardhamāna, Śaṅkaramiśra and Bhagīratha Ṭhakkura, eds, Pt. Harihara Śāstrī and Pt. Dhundhirāja Śāstrī (Chowkhambha, Benares, 1927–34).

Udayana, *Ātmatattvaviveka,* with the commentaries of Śaṇkaramiśra, Bhagīratha Ṭhakkura and Raghunātha Tārkikaśiromaṇi, eds, Mahamahopadhyaya Vindhyesvariprasada Dvivedin and Pt. Lakshmana Sastri Dravida (*Bibliotheca Indica*), Calcutta, 1907–39; rpt, 1986.

——, *Kiraṇāvalī,* ed., Jitendra S. Jetly (Gaekwad's Oriental Series, 154), Baroda, 1971, with the commentary of Vardhamāna, part II, ed., Narendra Chandra Vedantatirtha, Calcutta, 1956 (*Bibliotheca Indica*).

——, *Lakṣaṇāvalī:* Appendix to *Kiraṇāvalī,* ed., Jetly, pp. 276–83, with the commentary of Śeṣaśārṅgadhara, in *The Pandit,* New Series, vol. XXII, Benares, 1900.

Vardhamāna, *Kiraṇāvalīprakāśa, Nyāyalīlāvatīprakāśa,* see under Udayana, Śrīvallabha.

Chapter 13

Emotions as Judgements of Value and Importance*

Martha Nussbaum

Nun will die Sonn' so hell aufgeh'n
Als sei kein Unglück die Nacht gescheh'n.
Das Unglück geschah nur mir allein.
Die Sonne, sie scheinet allgemein.

Du musst nicht die Nacht in dir verschränken,
Musst sie ins ew'ge Licht versenken.
Ein Lämplein verlosch in meinem Zelt,
Heil sie dem Freudenlicht der Welt.

[Now the sun is going to rise, as bright
as if no misfortune had happened during the night.
The misfortune happened only to me.
The sun sends light out neutrally.

You must not fold the night into yourself.
You must drown in eternal light.
In my tent a small lamp went out.
Greetings to the joyful light of the world.]

Friedrich Rückert
(text of the first of Mahler's *Kindertotenlieder*)

It is almost impossible to understand the extent to which this disturbance agitated, and by that very fact had temporarily enriched, the mind of M. de Charlus. Love in this way produces real geological upheavals of thought. In the mind of M. de Charlus, which only several days before

> resembled a plane so flat that even from a good vantage point one could not have discerned an idea sticking up above the ground, a mountain range had abruptly thrust itself into view, hard as rock—but mountains scul[p]ted as if an artist, instead of taking the marble away, had worked it on the spot, and where there twisted about the another, in giant and swollen groupings, Rage, Jealousy, Curiosity, Envy, Hate, Suffering, Pride, Astonishment, and Love.
>
> Marcel Proust, *A la recherche du temps perdu*

The story of an emotion, I shall argue, is the story of judgements about important things, judgements in which we acknowledge our neediness and incompleteness before those elements that we do not fully control. I therefore begin with such a story, a story of fear, and hope, and grief, and anger, and love.

I

Last April I was lecturing at Trinity College, Dublin. As my mother was in hospital convalescing after a serious but routine operation, I phoned at regular intervals to get reports on her progress. One of these phone calls brought the news that she had had a serious complication during the night, a rupture of the surgical incision between her oesophagus and her stomach. She had developed a massive internal infection and fever, and, though she was receiving the best care in a fine hospital, her life was in jeopardy. This news felt like a nail suddenly driven into my stomach. With the help of my hosts I arranged to return on the next flight, which was not until the following day. That evening I delivered my scheduled lecture, on the subject of emotions. I was not then the same exuberant self-sufficient philosopher delivering a lecture, but rather, a person barely able to restrain tears. That night in my room in Trinity College, I had a dream in which my mother appeared emaciated and curled into a foetal position in her hospital bed. I looked at her with a surge of tremendous love and said, 'Beautiful Mommy'. Suddenly she stood up, looking as young and beautiful as in the photographs of the time when I was two or three years old. She smiled at me with her characteristic wit and said that others might call her wonderful, but she preferred to be called beautiful. I woke up and wept, knowing that things were not so.

During the transatlantic flight the next day, I saw, with hope, that image of health before me. But I also saw, and more frequently, the image of her death, and my body wanted to interpose itself before that image, to negate it. My blood wanted to move faster than the plane. With shaking hands I typed out paragraphs of a lecture on mercy, and the narrative understanding of criminal offenders. And I felt, all the while, a vague and powerful anger—at the doctors, for allowing this crisis to occur, at the flight attendants, for smiling as if everything were normal, and above all, at myself for not having been able to stop this event from happening, or for not having been there with her when it did. On arriving in Philadelphia I called the hospital's intensive care unit and was told by the nurse that my mother had died twenty minutes before. My sister, who lived there, had been with her and had told her that I was on my way. The nurse asked me to come and see her laid out. I ran through the littered downtown streets as if something could be done. At the end of a maze of corridors, beyond the cafeteria where hospital workers were laughing and talking, I found the surgical intensive care unit. There, behind a curtain, I saw my mother in bed, lying on her back, as I had so often seen her lying asleep at home. She was dressed in her best robe, the one with the lace collar. Her make-up was impeccable. (The nurses, who had been very fond of her, told me that they knew how important it had been to her to always have her lipstick on right.) A barely visible tube went into her nose, but it was no longer hooked up to anything. Her hands were yellow. She was looking intensely beautiful. My body felt as if pierced by so many slivers of glass, fragmented, as if it had exploded and scattered in pieces round the room. I wept uncontrollably. An hour later I was on my way to my hotel, carrying my mother's red overnight bag with her clothes and the books I had given her to read in the hospital—strange relics that seemed to me not to belong to this world any more, as if they should have vanished with her life.

II

This story embodies several features of the emotions which it is my endeavour to explain here: their urgency and heat; their tendency to take over the personality and move one to action with overwhelming force; their connection with important attachments, in terms of which one defines one's life; one's sense of passivity before them; their apparently adversarial relation to 'rationality' in terms of cool calculation or cost–benefit analysis, or their occasionally adversarial relation to reasoning of any sort; their close connections with one another, as hope alternates uneasily with fear, as a single event transforms hope into grief, as grief, looking for a cause, expresses itself as anger, as all of these can be the vehicles of an underlying love.

In the light of all these features, it might seem very strange to suggest that emotions are forms of judgement. And yet it is this thesis that I shall defend. I shall argue that all these features are not only not incompatible with, but are actually best explained by, a version of the ancient Greek Stoic view, according to which emotions are forms of evaluative judgement that ascribe great importance to things and persons outside one's control. Emotions are thus, in effect, acknowledgements of neediness and lack of self-sufficiency.[1] The aim is to examine this view and the arguments that support it, showing how the original Stoic picture needs to be modified in order to be philosophically adequate. In this way I hope to restore to the philosophical and political discussion of emotion a dimension that has too frequently been overlooked in debates about whether emotions are 'rational' or 'irrational'.[2]

My focus will be on developing an adequate philosophical account. But since any adequate account in this area must respond not only to the data of one's own experience and to stories of the experience of others, but also to the work done to systematize and account for emotional experience in the disciplines of psychology and anthropology, I draw on those disciplines as well. Neo-Stoic views have recently been gaining ascendancy in cognitive psychology, in work on helplessness and control,[3] and on emotion as 'appraisal' of that which pertains to a creature's 'thriving';[4] and in anthropology, in work on emotion as an evaluative 'social construction'.[5] Since the Stoic view needs to be connected to a plausible developmental account of the genesis of emotion in infancy, I also draw on pertinent material from the object-relations school of psychoanalysis,[6] which converges with the findings of cognitive psychology and enriches the account of the complexity of human history.[7]

Throughout, the explananda will be the genus of which grief, fear, love, joy, hope, anger, gratitude, hatred, envy, jealousy, pity, guilt, and other relatives are the species. The members of this family are distinct, both from bodily appetites such as hunger and thirst as well as from objectless moods such as irritation or endogenous depression. Though there are numerous internal distinctions among the members of the family, they have enough in common to be analysed together; and a long tradition in philosophy, beginning from Aristotle, has so grouped them.[8]

III

The Stoic view of emotion has an adversary: the view that emotions are 'non-reasoning movements', unthinking energies that simply push the person around, and do not relate to conscious perceptions. Like gusts of wind or the currents of the sea, they move, and move the person, but obtusely, without vision of an object or beliefs about it. In this sense they are 'pushes' rather than 'pulls'. This view is connected with the idea that emotions derive from the 'animal' part of our nature, rather than from a specifically human part—usually by thinkers who do not have a high regard for animal intelligence. Sometimes, too, the adversary's view is connected with the idea that emotions are 'bodily' rather than 'mental', as if this were sufficient to make them unintelligent rather than intelligent.[9]

The adversary's view is grossly inadequate and, in that sense, it might seem to be a waste of time to consider it. The fact, however, that it has until recently been very influential, both in empiricist-derived philosophy and in cognitive psychology,[10] and through both of these in fields such as law and public policy,[11] gives reason to reflect on it.[12] A stronger reason for reflecting upon this view lies in the fact that the view, though inadequate, does capture some important aspects of emotional experience, aspects that need to figure in any adequate account. If we first understand why this view has the power that it undeniably does, and then see why and how further reflection moves us away from it, it will lead to an understanding of what we must not ignore or efface in so moving away.

Turning back to my account of my mother's death, we now find that the 'unthinking movements' view does appear to capture at least some of what went on: my feeling of a terrible tumultuousness, of being at the mercy of currents that swept over me without my consent or complete understanding; the feeling of being buffeted between hope and fear, as if between two warring winds; the feeling that very powerful forces were pulling my self apart, or tearing it limb from limb; in short—the terrible power or urgency of the emotions, their problematic relationship with one's sense of self, the sense of one's passivity and powerlessness before them. It comes as no surprise that even philosophers who argue for a cognitive view of emotion should speak of them this way: Seneca, for example, is fond of comparing emotions to fire, to the currents of the sea, to fierce gales, to intruding forces that hurl the self about, cause it to explode, cut it up, tear it limb from limb.[13] It seems easy for the adversary's view to explain these phenomena: for if emotions are just unthinking forces that have no connection with our thoughts, evaluations, or plans, then they really are just like the invading currents of some ocean. And they really are, in a sense, non-self; and we really are passive before them. It seems easy, furthermore, for the adversary to explain their urgency: for once we imagine them as unthinking forces we can without difficulty imagine these forces as extremely strong.

By contrast, the neo-Stoic view appears to be in trouble in all these points. For if emotions are a kind of judgement or thought, it would be difficult to account for their urgency and heat; thoughts are usually imagined as detached and calm. Also it is difficult to find in them the passivity that we undoubtedly experience: for judgements are actively made, not just suffered. Their ability to dismember the self is also overlooked: for thoughts are paradigmatic, as it were, of what we control, and of the most securely managed parts of our identity. Let us now see what would cause us to move away from the adversary's view and how the neo-Stoic view responds to our worries.

What, then, makes the emotions in my example unlike the thoughtless natural energies I have described? First of all, they are *about* something; they have an object. My fear, my hope, my ultimate grief, all are about my mother and directed at her and her life. A wind may hit against something, a current may pound against something: but these are not *about* the things they strike in their way. My fear's very identity as fear depends on its having an object: take that away and it becomes a mere trembling or heart-leaping. In the same way, the identity of the wind as wind does not depend on the particular object against which it may pound.

Second, the object is an *intentional* object: that is, it figures in the emotion as it is seen or interpreted by the person whose emotion it is. Emotions are not *about* their objects merely in the sense of being pointed at them and let go, the way an arrow is let go against its target. Their aboutness is more internal and embodies a way of seeing. My fear perceived my mother both as tremendously important and as threatened; my grief saw her as valuable and as irrevocably cut off from me. (Both, we might add—beginning to approach the adversary's point about the self—contain a corresponding perception of myself and my life, as threatened in the one case, as bereft in the other.) This aboutness comes from my active way of seeing and interpreting: it is not like being given a snapshot of the object, but requires looking at it, so to speak, through one's own window. This perception might contain an accurate view of the object or it might not. (And, indeed, it might take as its target a real and present object, or be directed at an object that is no longer in existence, or that never existed at all. In this way too, intentionality is distinct from a more mechanical directedness.) It is to be stressed that this aboutness is part of the identity of the emotions. What distinguishes fear from hope, fear from grief, love from hate—is not so much the identity of the object, which might not change, but the way the object is perceived: in fear, as a threat, but with some chance for escape, in hope, as in some uncertainty, but with a chance for a good outcome,[14] in grief as lost, in love as invested with a special sort of radiance. Again, the adversary's view is unable to account for the ways in which we actually identify and individuate emotions, and for a prominent feature of our experience of them.

Third, these emotions embody not simply ways of seeing an object, but beliefs—often very complex—about the object.[15] It is not always easy, or even desirable, to distinguish between an instance of *seeing x as y*, such as I have described above, from the belief that x is y. In order to have fear—as Aristotle already saw it[16]—I must believe that bad events are impending; that they are not trivially, but seriously bad; that I am not in a position to ward them off; that, on the other hand, my doom is not sealed, but there is still some uncertainty about what may befall.[17] In order to have anger, I must have an even more complex set of beliefs: that there has been some damage to me or to something or someone close to me;[18] that the damage is not trivial but significant; that it was done by someone; that it was done willingly; that it would be right for the perpetrator of the damage to be punished.[19] It is plausible to assume that each element of this set of beliefs is necessary in order for anger to be present: if I should discover that not x but y had done the damage, or that it was not done willingly, or that it was not serious, we could expect my anger to modify itself accordingly, or recede.[20] My anger at the smiling flight attendants was quickly dissipated by the thought that they had done so without any thought of disturbing me or giving me offence.[21] Similarly, my fear would have turned to relief—as fear so often does—had the medical news changed, or proven to be mistaken. Again, these beliefs are essential to the identity of the emotion: the feeling of agitation by itself will not reveal to me whether what I am feeling is fear or grief or pity. Only an inspection of the thoughts will help discriminate. Here again, then, the adversary's view is too simplistic: severing emotion from belief, it severs emotion from what is not only a necessary condition of itself, but a part of its very identity.

Finally, there is something marked in the intentional perceptions and the beliefs characteristic of the emotions: they are all concerned with *value*, they see their object as invested with value. Suppose that I did not love my mother or consider her a person of great importance; suppose I considered her about as important as the branch on a tree near my house. Then (unless I had invested the branch itself with an unusual degree of value) I would not fear her death, or hope so passionately for her recovery. My experience records this in many ways—not least in my dream, in which I saw her as beautiful and wonderful and, seeing her that way, wished her restored to health and wit. And of course in the grief itself there was the same perception—of enormous significance, permanently lost. This indeed is why the sight of the dead body of someone one loves is so painful: because the same sight that is a reminder of value is also an evidence of irrevocable loss.

The value perceived in an object appears to be of a particular sort—although here I must be more tentative since I am approaching an issue that is my central preoccupation. The object of the emotion is seen as *important* for some role it plays in the person's own life. I do not fear just any and every catastrophe anywhere in the world, nor (so it seems) any and every catastrophe that I know to be bad in important ways. What inspires fear is the thought of the impending damage that threatens my cherished relationships and projects. What inspires grief is the death of a beloved, someone who has been an important part of one's life. This does not mean that the emotions view these objects simply as tools or instruments of the agent's own satisfactions: they may be invested with intrinsic worth or value, as indeed my mother had been. They may be loved for their own sake, and their good sought for its own sake. But what makes the emotion centre around her, from among all the many wonderful people and mothers in the world, is that she is *my* mother, a part of my life. The emotions are in this sense localized: as in the Rückert poem in the epigraph, they take up their stand 'in my tent', and focus on the 'small lamp' that goes out there, rather than on the general distribution of light and darkness in the universe as a whole.

Another way of putting this point is that the emotions appear to be eudaimonistic, that is, concerned with the agent's flourishing. And thinking about ancient Greek eudaimonistic moral theories will help us to start thinking about the geography of the emotional life. In a eudaimonistic ethical theory, the central question asked by a person is, 'How should I live?' The answer lies in the person's conception of eudaimonia, or human flourishing. The conception of eudaimonia includes all that to which the agent ascribes intrinsic value; for instance, if one can show that there is something missing without which one's life would not be complete, then that is sufficient argument for its inclusion.[22] The important point is this: in a eudaimonistic theory, the actions, relations, and persons that are included in the conception are not all valued simply on account of some instrumental relation they bear to the agent's satisfaction. This is a mistake commonly made about such theories under the influence of utilitarianism and the misleading use of 'happiness' as a translation for eudaimonia.[23] Not just actions but also mutual relations of civic or personal *philia*, in which the object is loved and benefited for his or her own sake, can qualify as constituent parts of eudaimonia.[24] On the other hand, they are valued as constituents of a life that is my life and not someone else's, as my actions, as people who have some relation with me.[25] This, it seems, is what emotions are like, and this is why, in negative cases, they are felt as tearing the self apart: because they have to do with[26] damage to me and to my own, to my plans and goals, to what is most urgent in my conception of what it is for me to live well.

We have now gone a long way towards answering the adversary, for it has been established that his view, while picking out certain features of emotional life that are real and important, has omitted others of equal and greater importance, central to the identity of an emotion and to discriminating between one emotion and another: their aboutness, their intentionality, their basis in beliefs, their connection with evaluation. All this makes them look very much like thoughts after all; and we have even begun to see how a cognitive view might itself explain some of the phenomena the adversary claimed on his side—the intimate relationship to selfhood, the urgency. But this is far removed from the neo-Stoic view, according to which emotions are just a certain type of evaluative judgement. For the considerations we have brought forward might be satisfied by a weaker or more hybrid view, according to which beliefs and perceptions play a large role in emotions, but are not identical with them.

We can imagine, in fact, three such weaker views, each with its historical antecedents:[27]

(1) The relevant beliefs and perceptions are *necessary conditions* for the emotion.
(2) They are *constituent parts of* the emotion (which has non-belief parts as well).
(3) They are *sufficient conditions* for the emotion, which are not identical with it.

The logical relations among these options are complex and need scrutiny. (1) does not imply, but is compatible with (3). (3) does not imply, but is compatible with (1). (1) is compatible with (2)—the beliefs may be necessary as constituent elements in the emotion; but we might also hold (1) in an external-cause form, in which the beliefs are necessary conditions for a very different sort of thing that is not itself a belief. The same can be said for (3): a sufficient cause may be external or internal. (2) is compatible with (3), since even if the belief is just a part of the emotion, and not the whole, it may be a part whose presence guarantees the presence of the other parts.

We have gone far enough, I think, to rule out the external-cause form of (1) and of (3), for we have argued that the cognitive elements are an essential part of the emotion's identity, and of what differentiates it from other emotions. So we are left, it appears, with (2)—whether in a form in which the belief part suffices for the presence of the other parts, or in a form in which it is merely necessary for their presence. What are those other parts? The adversary is ready with a fall-back answer: non-thinking movements of some sort, or perhaps (shifting over to the point of view of experience) objectless feelings of pain and/or pleasure. A number of questions immediately come to mind about these feelings: What are they like if they are not *about* anything? What is the pleasure *in*, or the pain *at*? How are they connected with the beliefs, if they do not themselves contain any thought or cognition?[28] These questions will shortly be reviewed.

IV

I must begin a fuller elaboration and defence of the neo-Stoic view by saying something about judgement. To understand the case for the view that emotions are judgements, one needs to understand exactly what a Stoic means when he or she says that; I think we will find the picture intuitively appealing, and a valuable basis (ultimately) for a critique of the familiar belief–desire framework for explaining action.[29] According to the Stoics, then, a judgement is an assent to an appearance.[30] In other words, it is a process that has two stages. First, it occurs to me or strikes me that such and such is the case. (Stoic appearances are usually propositional, although I shall later argue that this aspect of their view needs some modification.) It looks to me that way, I see things that way[31]—but so far I haven't really accepted it. Now there are three possibilities. I can accept or embrace the appearance, take it into me as the way things are: in this case it has become my judgement, and that act of acceptance is what judging is. I can repudiate it as not the way things are: in that case I am judging the contradictory. Or I can let it be there without committing myself to it one way or another. In that case I have no belief or judgement about the matter one way or the other.[32] Consider a simple perceptual case introduced by Aristotle.[33] The sun strikes me as being about a foot wide. (That's the way it looks to me, that is what I see it *as*.) Now I might embrace this appearance and talk and act accordingly; most children do so. If I am confused about astronomy, I may refuse to make any cognitive commitment on the matter. But if I hold a confident belief that the sun is in fact tremendously large, and that its appearance is deceptive, I will repudiate the appearance and embrace a contradictory appearance. There seems nothing odd here about saying both that the way of seeing the world is the work of my cognitive faculties and that its acceptance or rejection is the activity of those faculties. Assenting to or embracing a way of seeing the world, acknowledging it as true *requires* the discriminating power of cognition. Cognition need not be imagined as inert. In this case, it is reason itself that reaches out and accepts that appearance, saying, so to speak, 'Yes, that's the one I'll have. That's the way things are'. We might even say that this is a good way of thinking about what reason *is:* an ability by virtue of which we commit ourselves to viewing things the way they are.

Let us now return to my central example. My mother has died. It strikes me, it appears to me, that a person of enormous value, who was central to my life, is no longer there. It feels as if a nail has entered my insides; as if life has suddenly a large rip or tear in it, a gaping hole. I see, as well, her wonderful face—both as tremendously loved and as forever lost to me. The appearance, in however many ways we picture it, is propositional: it combines the thought of importance with the thought of loss, its content is that this importance is lost. And, as I have said, it is evaluative: it does not just assert, 'Betty Craven is dead'. Central to the propositional content is my mother's enormous importance, both to herself as well as to me as an element in my life.

So far we are still at the stage of appearing—and notice that I was in this stage throughout the night before her death, throughout the long transatlantic plane ride, haunted by that value-laden picture, but powerless to accept or reject it; for it was sitting in the hands of the world. I might have had reason to reject it if, for example, I had awakened and found that the whole experience of getting the bad news and planning my return trip home had been just a nightmare. Or, I might have rejected it if the outcome had been good and she was no longer threatened. I did accept that she was endangered—so I did have fear. But whether or not she was or would be *lost*, I could not say. But now I am in the hospital room with her body before me. I embrace the appearance as the way things are. Can I assent to the idea that someone tremendously beloved is forever lost to me, and yet preserve emotional equanimity? The neo-Stoic claims that I cannot. Not if what I am recognizing is that very set of propositions, with all their evaluative elements. Suppose I had said to the nurses, 'Yes, I see that a person I love deeply is dead and that I'll never see her again. But I am fine: I am not disturbed at all'. If we put aside considerations about reticence before strangers and take the utterance to be non-deceptive, we will have to say, I think, that this person is in a state of denial. She is not really assenting to *that* proposition. She may be saying those words, but there is something that she is withholding. Or, if she is assenting, it is not to that same proposition but perhaps to the proposition 'Betty Craven is dead'. Or even (if we suppose that 'my mother' could possibly lack eudaimonistic evaluative content) to the proposition 'My mother is dead'. What I could not be fully acknowledging or realizing is the thought: 'A person whom I deeply love, who is central to my life, had died', for to recognize this is to be deeply disturbed.

It is of crucial importance to be clear about what proposition or propositions we have in mind. For, if we were to make the salient proposition one with no evaluative content, say, 'Betty Craven is dead',[34] we would be right in thinking that the acceptance of that proposition could be at most a cause of grief, not identical with grief itself. The neo-Stoic claims that grief is identical with the acceptance of a proposition that is both evaluative and eudaimonistic, that is, concerned with one or more of the person's most important goals and ends. The case for equating this (or these) proposition(s) with emotion has not yet been fully made, but so far it appears far more plausible that such a judgement could in itself be an upheaval. Another element must now be added. The judgements that the neo-Stoic identifies with emotions all have a common subject matter: all are concerned with vulnerable externalities: those that can be affected by events beyond one's control, those that are unexpected, those that can be destroyed or removed even when one does not wish it. This implies that the acceptance of such propositions reveals something about the person: that she allows herself and her good to depend upon things beyond her control, that she acknowledges a certain passivity before the world. This emerges in the complex combination of circumstantial and evaluative considerations that must be present in the relevant propositions.

At this point, it can be concluded not only that the judgements described are necessary constituent elements of the emotion, but that they are sufficient as well. It has been argued that if there is no upheaval the emotion itself is not fully or really present. The previous arguments suggest that this sufficiency should be viewed internally: as that of a constituent part which itself causes whatever other parts there may be. I have spoken of the way in which the relevant judgements are a part of the identity conditions of the emotion; however, there is need for further analysis since it may still appear counter-intuitive to make the emotion itself a function of reason, rather than a non-rational, cognitive movement.

Well, what element in me *is* it that experiences the terrible shock of grief? I think of my mother; I embrace in my mind the fact that she will never be with me again—and I am shaken. But how and where? Does one imagine the thought as causing a trembling in my hands, or a fluttering in my stomach? And if so, does one really want to say that this fluttering or trembling *is* my grief about my mother's death? The movement seems to lack the aboutness and the capacity for recognition that must be part of an emotion. Internal to the grief must be the perception of the beloved object and of her importance; the grief itself must quantify the richness of the love between us, its centrality to my life. It must contain the thought of her irrevocable deadness. Of course, one could now say that there is a separate emotional part of the soul that has all these abilities. But, having seemingly lost one's grip on the reason for housing grief in a separate non-cognitive part, reason looks like just the place to house it.

The adversary might now object that this is not yet clear. Even if one concedes that the seat of emotion of must be capable of many cognitive operations, there also seems to be a kinetic and affective aspect to emotion that does not look like a judgement or any part of it. There are rapid movements, feelings of pain and tumult: are we really to equate these with some part of judging that such and such is the case? Why should we not make the judgement a cause of emotion, but identify emotion itself with these movements? Or, we might even grant that judgement is a constituent element in the emotion, and, as a constituent element, a sufficient cause of the other elements as well, and yet insist that there are other elements, feelings and movements, that are not parts of the judgement. I have begun to respond to this point by stressing the fact that we are conceiving of judging as dynamic, not static. Reason here moves, embraces, refuses; it moves rapidly or slowly, surely or hesitantly. I have imagined it entertaining the appearance of my mother's death and then, so to speak, rushing towards it, opening itself to absorb it. So why would such a dynamic faculty be unable to house, as well, the disorderly motions of grief? And this is not just an illusion: I am not infusing into thought kinetic properties that properly belong to the arms and legs, or imagining reason as accidentally coloured by kinetic properties of the bloodstream. The movement towards my mother was a movement of my thought about what is most important in the world; that is all that needs to be said about it. If anything, the movement of my arms and legs, as I ran to University Hospital, was a vain mimesis of the movement of my thought towards her. It was my thought that was receiving, and being shaken by, the knowledge of her death. I think that if anything else is said it will sever the close connection between the recognition and the being-shaken of that experience. The recognizing and the upheaval belong to one and the same part of me, the part with which I make sense of the world.

Moreover, it appears that the adversary is wrong in thinking of the judgement as an event that temporally precedes the grieving—as some of the causal language suggests. When I grieve, I do not first of all coolly embrace the proposition, 'My wonderful mother is dead', and then set about grieving. No, the real, complete, recognition of that terrible event (as many times as I recognize it) *is* the upheaval. It is as I described it: like driving a nail into the stomach. The thought that she is dead sits there (as it sat before me during my plane ride) asking me what I am going to do about it. Perhaps, if I am still uncertain, the image of her restored to health sits there too. If I embrace the death image, if I take it into myself as the way things are, it is at that very moment, in that cognitive act itself, that I am putting the world's nail into my own insides. That is not preparation for upheaval, that is upheaval itself. That very act of assent is itself a tearing of my self-sufficient condition. Knowing can be violent, given the truths that are there to be known.

Are there other constituent parts to the grief that are not themselves parts of the judgement? In any particular instance of grieving there is so much going on that it is very difficult to answer this question if one remains at the level of token identities between instances of grieving and instances of judging. We have a more powerful argument—and also a deeper understanding of the phenomena—if we inquire instead about the general identity conditions for grief, and whether there are elements necessary for grief in general that are not elements of judgement. In other words, would be withdraw our ascription of grief if these elements were missing? I believe that the answer is that there are no such elements. There usually will be bodily sensations and changes involved in grieving; but if we discovered that my blood pressure was quite low during this whole episode, or that my pulse rate never went above 60 there would not, I think, be the slightest reason to conclude that I was not grieving. If my hands and feet were cold or warm, sweaty or dry, again this would be of no criterial value. Although psychologists have developed sophisticated measures based on brain activity, it is perhaps intuitively wrong to use these as definitive indicators of emotional states. We do not withdraw emotion-ascriptions otherwise grounded if we discover that the subject is not in a certain brain-state. (Indeed, the only way the brain-state assumed apparent importance was through a putative correlation with instances of emotion identified on other grounds.)

More plausible, perhaps, would be certain feelings characteristically associated with emotion. But here we should distinguish 'feelings' of two sorts. On the one hand, there are feelings with a rich intentional content—feelings of the emptiness of one's life without a certain person, feelings of unrequited love for that person, and so on. Such feelings may enter the identity conditions for some emotion; but the word 'feeling' now does not contrast with the cognitive words 'perception' and 'judgement', it is merely a terminological variant for them. As already mentioned, the judgement itself possesses many of the kinetic properties that the 'feeling' is presumably intended to explain. On the other hand, there are feelings without rich intentionality or cognitive content, for instance, feelings of fatigue, of extra energy. As with bodily states, they may accompany emotion or they may not—but they are not necessary for it. (In my own case, feelings of crushing fatigue alternated in a bewildering way with periods when I felt preternaturally wide awake and active; but it seemed wrong to say that either of these was a necessary condition of my grief.) So there appear to be type-identities between emotions and judgements; emotions can be defined in terms of judgement alone

NOTES

* This article is based on the first of my Gifford Lectures delivered at the University of Edinburgh, spring 1993, and to be published by Cambridge University Press; the provisional title is 'Need and Recognition: A Theory of the Emotions'. The subsequent lectures not only offer further arguments for the theory and extend it to the analysis of other emotions, but also argue that the theory as stated here needs to be modified in certain ways in order to yield an adequate account of the development of emotion and of the emotions of non-human animals. I address various normative questions about the place of emotions, so defined, in an account of public and private rationality. I cannot hope here to provide more than a sketch of those further developments, and hope that the reader will understand that some questions that may arise about this theory are questions that are addressed later. Despite these drawbacks, I did want to put forward this particular essay as my attempt to honour the memory of Bimal

Matilal, not only for its subject matter, but because it is at the core of my work, rather than a peripheral addendum. Matilal was a scholar of profound insight and intellectual courage, whose contribution to philosophy is *sui generis*, a paradigm of cross-cultural historical and philosophical inquiry. I also knew him as a person possessing great warmth, grace, and wit, whose particularity these abstract terms do not go very far towards conveying.

1. I discuss the Stoic view historically in my *The Therapy of Desire: Theory and Practice in Hellenistic Ethics*, Princeton: Princeton University Press, 1994, ch. 10. Some parts of the argument of this lecture, especially in Section IV, are closely related to that argument; but I have added new distinctions and refinements at every point in the argument, and, in Sections V and VII, have substantially modified my position. Further modifications occur subsequent to the material of this article.
2. Some elements of a related philosophical position are in William Lyons, *Emotion*, Cambridge: Cambridge University Press, 1980; Robert Solomon, *The Passions*, New York: Doubleday, 1976; Robert Gordon, *The Structure of Emotion*, Cambridge: Cambridge University Press, 1987; and Ronald de Sousa, *The Rationality of Emotion*, Cambridge, MA: MIT Press, 1987. None of these, however, stresses the evaluative nature of the emotions' cognitive content.
3. See especially, Martin Seligman, *Helplessness: On Development, Depression, and Death*, New York: W.H. Freeman, 1975.
4. See especially, Richard Lazarus, *Emotion and Adaptation*, New York: Oxford University Press, 1991; A Ortony, G. Clore, and G. Collins, *The Cognitive Structure of Emotions*, New York: Cambridge University Press, 1988; and Keith Oatley, *Best Laid Schemes: The Psychology of Emotions*, Cambridge: Cambridge University Press, 1992. For a related view, with greater emphasis on the social aspects of emotion, see James Averill, *Anger and Aggression: An Essay on Emotion*, New York: Springer, 1982.
5. See, for example, Catherine Lutz, *Unnatural Emotions: Everyday Sentiments on a Micronesian Atoll and their Challenge to Western Theory*, Chicago: University of Chicago Press, 1988. See also Jean Briggs, *Never in Anger*, Cambridge, MA: Harvard University Press, 1970.
6. Above all, the work of W.R.D. Fairbairn, *Psychoanalytic Studies of the Personality*, London: Tavistock, 1952; Christopher Bollas, *The Shadow of the Object: Psychoanalysis of the Unthought Known*, London: Tavistock, 1987; and Nancy Chodorow, *The Reproduction of Mothering*, Berkeley: University of California Press, 1980; with much reservation and criticism, the work of Melanie Klein, *Love, Guilt, and Reparation and Other Works, 1921–45*, and *Envy and Gratitude and Other Works, 1946–63*, London: Hogarth, 1984, 1985. Experimental psychology, anthropology, and psychoanalysis are brought together in an illuminating way in John Bowlby, *Attachment, Separation: Anxiety and Anger*, and *Los: Sadness and Depression*, New York: Basic Books, 1982, 1973, 1980.
7. Most of the detailed discussion of all this material is in parts of the project

subsequent to this paper; I include the references to convey an idea of my larger design.

8. The word I shall use for the explananda is 'emotions'. The Stoic view used the term *pathe*—previously a general word for 'affect'—in order to demarcate this class and to isolate it from the class of bodily appetites. For this reason, the philosophical tradition influenced by Stoicism has tended to use the word 'passions' and its Latin and French cognates. To contemporary ears, this word denotes a particular intensity, especially erotic intensity, as the more inclusive Greek term *pathe* did not. I therefore use 'emotion' as the best translation and the best generic term—although I shall comment both on the kinetic element that led to the original introduction of that word and also on the element of passivity that is stressed in the Greek term.
9. I believe, and argue subsequently, that emotions, like other mental processes, are bodily, but that this does not give us reason to reduce their intentional/cognitive components to non-intentional bodily movements. For my general position on mind/body reduction, see Martha Nussbaum and H. Putnam, 'Changing Aristotle's Mind', in *Essays on Aristotle's De Anima.* eds, Nussbaum and A. Rorty, Oxford: Clarendon Press, 1992.
10. See the illuminating criticisms of both in Anthony Kenny, *Action, Emotion, and Will*, London: Routledge, 1963, who shows that there is a close kinship between Humean philosophy and behaviourist psychology.
11. We see such views, for example, in the behaviourist psychology of Richard Posner, *The Problems of Jurisprudence*, Cambridge, MA: Harvard University Press, 1990, and to some extent in Richard Posner, *Sex and Reason*, Cambridge, MA: Harvard University Press, 1992. Even many defences of emotion in the law begin by conceding some such view of them—for documentation of this point, see my 'Equity and Mercy', *Philosophy and Public Affairs* 22, 1993.
12. The Stoics had similar reasons: the adversary's view was represented, for them, by some parts of Plato, or at least some ancient interpretations of Plato.
13. See my *Therapy*, ch. 12.
14. This difference of probabilities is not the whole story about the difference between fear and hope. In my case, where there was both a serious danger and a robust chance of escape, both were possible, and the shift from the one to the other depended on whether one focused on the possible good outcome or on the impending danger.
15. Subsequently, I argue that in the case of animal emotions, and in the case of some human emotions as well, the presence of a certain kind of *seeing as*, which will always involve some sort of a combination or predication, is sufficient for emotion.
16. *Rhetoric* II.5.
17. One might argue with this one, thinking of the way in which one fears death even when one knows not only that it will occur but when it will occur. There is much to be said here: does even the man on death row

ever know for sure that he will not get a reprieve? Does anyone ever know for sure what death consists in?

18. Aristotle insists that the damage must take the form of a 'slight', suggesting that what is wrong with wrongdoing is that it shows a lack of respect. This is a valuable and, I think, ultimately very plausible position, but I am not going to defend it here.
19. See *Rhetoric* II.2–3.
20. In my case, however, one can see that the very magnitude of accidental grief sometimes prompts a search for someone to blame, even in the absence of any compelling evidence that there is an agent involved. One reason for our society's focus on anger associated with medical malpractice may be that there is no way of proving that medical malpractice did not occur—so it becomes a useful target for those unwilling to blame hostile deities, or the cosmos.
21. Anger at self is a more intractable phenomenon, since it is rarely only about the events at hand; I discuss this elsewhere in my project.
22. On this, see Aristotle, *Nicomachean Ethics*, I; and for a particular case, IX.9, on the value of *philia*.
23. For the misreading, and a brilliant correction, see H.A. Prichard, 'The Meaning of *agathon* in the Ethics of Aristotle', *Philosophy* 10, 1935, pp. 27–39, and J.L. Austin, '*Agathon* and *eudaimonia* in the Ethics of Aristotle' in Austin, *Philosophical Papers*, Oxford: Clarendon Press, 1961, pp. 1–31.
24. For a good account of this, where *philia* is concerned, see John Cooper, 'Aristotle on Friendship' in *Essays on Aristotle's Ethics*, ed., A. Rorty, Berkeley: University of California Press, 1980, pp. 301–40.
25. The contrast between such eudaimonistic and more impartialist views is brought out and distinguished from the contrast between egoism and altruism, Bernard Williams, 'Egoism and Altruism', in *Problems of the Self*, Cambridge: Cambridge University Press, 1973.
26. As we shall see, 'have to do with' should not be construed as implying that the emotions take the conception of *eudaimonia* as their *object*. If that were so, they would be in error only if they were wrong about what conception of value I actually hold. On the neo-Stoic view they are about the *world*, in both its evaluative and its circumstantial aspect. If I grieve because I falsely ascribe to a thing or person outside myself a value he or she does not really possess (Stoics think of all grief as such), I am still really grieving, and it is true to say of me that I am grieving, but the grief is false in the sense that it involves the acceptance of propositions that are false.
27. See my *Therapy*, ch. 10.
28. By 'cognitive' processes I mean processes that deliver information (whether reliable or not) about the world; thus, I include not only thinking, but also perception and certain sorts of imagination.
29. I discuss this issue in a subsequent chapter of my project.
30. See my *Therapy*, ch. 10, with references to texts and literature.

31. It should be stressed that despite the usage of the terms 'taking in' and 'acknowledging', this notion of appearing is not committed to internal representations, and it is fully compatible with a philosophy of mind that eschews appeal to internal representations. It seems that neither Aristotle nor the Stoics had an internal/representationalist picture of the mind; nor do I. What is at issue is *seeing x as y:* the world strikes the animal a certain way, it sees it *as* such-and-such. Thus the object of the creature's activity is the world, not something in its head (or heart). In this essay I proceed as if all these ways of seeing can be formulated in linguistically expressible propositions. Subsequently I argue that this is too narrow a view to accommodate the emotional life of children and other animals, as well as many of the emotions of human adults. And it neglects the fact that other forms of symbolism—music, for example—are not simply reducible to language, but have expressive power in their own right.
32. Aristotle points out that such an unaccepted 'appearance' may still have some motivating power, but only in a limited way: as when a sudden sight causes one to be startled (but not yet really afraid), see *De Anima* III.9, *De Motu Animalium* II. Seneca makes a similar point concerning the so-called 'pre-emotions' or *propatheiai:* see *De Ira* II.3; it is remarkable that Richard Lazarus reinvents, apparently independently, the very same term, 'pre-emotions', to describe the same phenomenon in the animals he observes, see Lazarus, *Emotion and Adaptation.* The Greek sceptics suggest that one might live one's entire life motivated by appearances alone, without any beliefs—pointing to the alleged fact that animals are so moved. But their case is dubious, since, for one thing, it seems to misdescribe the cognitive equipment of animals.
33. *De Anima* III.3.
34. Of course the moment we insert the name of a human being, there is some evaluative content; and some moral theories would urge that this is all the value there should properly be, in any response to any death.

Chapter 14

On Śaṅkara's Attempted Reconciliation of 'You' and 'I': *Yuṣmadasmatsamanvaya*

Purushottama Bilimoria

True dialectic is not a monologue of the solitary thinker with himself, it is a dialogue between *I* and *Thou*

Feuerbach, 1843

PREAMBLE

The concept of *ātman*, as the pervasive, infinite principle of existence, without admitting to the substantive categories implicated in, say, the concept of matter on the one hand and the idea of the 'soul' or spirit self (*jīva, dehātman, ksetrajña, bhūtaprāṇa*) on the other hand, was perhaps the brahmanical answer to the critique of the very notion of the substantial self, which had been around in India (with the *srāmaṇas*, Jains, Ājīvākās, Cārvākas and Lokāyatas or materialists) long before the Buddha and his followers came upon the scene. The Buddhist doctrine of *anātman* (Pali *anattā*), when interpreted as a doctrine of no-self, may well be an attack on the notion of a substantial self even in the weaker sense, viz. as the *jīva*, etc. that does, according to the Hindu view, survive the death of the body; the *jīva* is not, however, the ultimate or supreme principle of reality in any sense of the term. The Buddhist rejection of the brahmanical conception of the self, although ostensibly directed against the *ātman*, could also be viewed as a more decisive rejection of the concept of the self as *jīva*. The argument is that because the brahmanical conception of *ātman* had a built-in rejection of substantiality (in the *vyāvahāric* or temporal modality), the Buddhist attack could well apply to one of its own alternative formulations, such as the 'mind-only' theory of the self developed by the Vijñānavāda or Yogācāra schools or, the *skandha-samūha* theory, or even the earlier *puggala* conception. However, insofar as the *ātman* remains a designation for the concept of the immutable, undifferentiated, unconditioned and autonomous principle of existence (and self-existent or *svabhāva* at that), regardless of whether it is substantial or not, the Buddhists were justified in calling into question the particular conception in their assertion of the *anātman* or no-self doctrine.

The way that the monistic tradition of brahminism resolved the problematic of *being* and the sense that people have of an individual and substantial self was to say that *ātman* was ultimately real and the phenomenal world (*jagat*) and the individual were illusory (*mithyā*; *māyā*) in nature. This is precisely the line of thinking that the eighth-century philosopher Śaṅkara championed, according to which also there can be no self, or no entity, that transmigrates. Śaṅkara expresses doubts that the nature of the self either as *jīva*, or as a stream of impermanent consciousness-moments, was a tenable view; he was all for giving up the idea of the discrete *jīva* altogether (*BS* II.2.10–20, 30–40, III.2.) But in so doing he at once relegates to the state of oblivion an entire tradition of richly developed conception of the self in its otherness. And why does he do that? My hypothesis is that Śaṅkara was fearful of the looming legacy of *asat* (non-existent or non-being) and its constant wager on *sat* (existent or being) in traditional philosophical development (notably in Buddhist dialectic), and if he could annihilate one side of the equation or identification (viz. of the origins with *asat* or non-being) which the *Ṛgveda* (10.72, 3; cf. *Nāsadīya-sūkta* 10.129) had in its moments of cynical wonder conceived of, *sat* would reign supreme. But in order to achieve this feat, Śaṅkara would almost have to empty *sat* of all its contents, other than its necessity (*astīti śāśvataḥ*), such that there could be no descriptions (i.e. naming) possible of it; its identification and re-identification must be entirely analytic, and so it remains emptied of all other modal possibilities (*parabhāva*). Moreover, if *sat* or being can at some deep level be said to be undifferentiable from *asat* (non-being) itself, i.e. difference as *différance* is itself swallowed up as it were—but a matter that must to the very end remain concealed in the theory—then the reality of *ātman* could be no more in doubt than the truth of the proposition A = A: A is necessarily A and cannot ever be non-A; that is its *svabhāva.*[1] In other words, Śaṅkara had discovered both (or perhaps voices its first acknowledgement within the Vedānta tradition) analyticity, and the alarming power and persuasiveness of the concept of nothingness of the self in the articulation of *asat.* In this essay, I shall endeavour to demonstrate this particular thesis by examining the various subtle moves and linguistic tropes Śaṅkara engineers in his commentarial work on the *Brahmasūtra*, the supposed aphoristic gem of the wisdom teachings of the vast and unwieldy Upaniṣads.

ŚANKARA'S CARTESIAN TURN

The Cartesian turn in India occurs with Śaṅkara in his celebrated *Adhyāsa* Meditation, or the introduction to the *Brahmasūtra*, where he is concerned with reconciling differing views through argument. It worries the youthful *sannyāsin* that there is a spate of views on the nature of the self, and he considers some candidates: the body alone with the attribute of intelligence; the inner organ; a momentary idea; nothingness; the transmigrating entity as agent and enjoyer; and the Lord as the inward enjoyer. This is about it. In an important way these reflect, in part, prevailing cultural conceptions, and they do not exhaust the philosophical range, actual and imagined. For instance, the Jains' conception is not canvassed; the *Sāṃkhya* conception comes in for attack much later; the *Nyāya* and *Mīmāṃsā* conceptions are bypassed, the conception of self as the other, conceived and constituted in an inseparable relation through the image of the other, or alterity, is glossed over. Indeed, the non-self, the 'other' in its utter *otherness, totiliter alter*, and the possibility of its non-being or inauthenticity echoing in one's self-image, is what appears to have troubled Śaṅkara most, and hence motivated him to retreat to the security of the self of some pristine, *sui generis* conception. Charging that these other views are based on fallacious arguments, he proceeds to offer what he considers the truly unassailable. So what is his argument? Śaṅkara develops this argument in his Introduction before he proceeds to comment directly on the texts of the *Brahmasūtra.* And so in this regard the argument stands apart from anything he might say later in his commentary. Indeed, that is what is unique about this piece of philosophizing, for this is one of the few places that Śaṅkara offers an argument he has thought through himself without substantive reference to the scriptural tracts of the tradition or to some other authority. He is obviously at pains to address a contention which the seers and sages perhaps missed or which might have arisen closer to his own era. One does not have to stretch one's imagination to guess who were his adversaries, who he is moved to tackle in these passages. Indeed, he gives us a clue or two as he goes on to consider possible (or perhaps actual) objections from the *Śūnyavādin* camp. And here he skilfully sets out an agenda for a whole gamut of speculation and theorizing that has gone on since in the Indian philosophical scene. Śaṅkara neither flinches nor minces any words; rather, he lays out his case in terse but lucid terms, in a style that is more elegant (in Sanskrit, that is) than Kant could boast about in his Critiques (in German). Let us then begin at the beginning, in Śaṅkara's own words. This is how Śaṅkara begins:

> The contents circumscribed by the dual concepts (*pratyayagocarayoḥ*) of 'you' (*yuṣmat*) and 'I' (*asmat*), namely, the object (*viṣaya*) and subject (*viṣayin*) respectively, being by their nature as contrary as light is to darkness, cannot reasonably have any identity (*itaretarabhāva* [or, *tādātmya*]: this being established, it is even less reasonable [*sutarā-mitretarabhāvānupapattiḥ*] that there be identity between their attributes [*dharma*].[2]

Already here two species of doubt are being entertained and responded to. The first has to do with the use of the personal pronouns in the first and second persons respectively, viz. *asmat* and *yuṣmat,* which we could render as 'I' (or occasionally, the royal 'we') and 'thou'. In literary trope, such as in Kālidāsa's *Śakuntalā,* such usage is acceptable, but what is its significance in the present context? Śaṅkara's implied response is that he has employed the two pronouns to mark absolute contrast and difference between two features of experience, which admittedly otherwise would be better served by the pronouns *aham* and *idam*, i.e. 'I' and 'that', which by definition exemplify two extremities. So why not use the latter juxtaposition? Or use '*bhavān*', the supplementary second person pronoun which, though it has the form of a third person pronoun (followed by a verb of the third person) is more honourable, and literally renders as 'his beingness'? The response is gleaned from the second line to the effect that a linguistic licence has been assumed for substitution of a third person pronoun with the second person, even though in practice we almost never conjoin first and second person pronouns; that is, we don't reply to the question: 'Who is there? Who are you?' by saying '*You* is I' or '*You* is me, Rāmu'; rather, we say, 'It is I', 'It is me, Rāmu'. For Śaṅkara immediately fills out the two pronouns taken as concepts with analogous contents, viz. subject and object. But notice the ingenious alignments: *asmatpratyaya* or *aham*, the first person designation is the subject, and there can be no dispute about this, for 'I' (or 'we') is self-referentially taken to be the subject; *yuṣmat*, on the other hand, is intended to stand for the object (*viṣaya*). Suppose we were analysing a simple sentence such as 'I hit you', clearly 'you' in the predicate position is the object as 'you' is the recipient of the action of my striking something. But Śaṅkara is intent on universalizing the *yuṣmatpratyaya* (you-concept) as objects-in-general, i.e. anything and everything that is other than the exclusive 'I'. Why else would 'you' be given the position of the object-concept in a subject–predicate relation? Indeed, one suspects that the model employed is not really a linguistic one at all, but a deeply ontological one, as when there is an attempt to forge or perhaps simply articulate a relationship between the 'I' of oneself and the 'thou' of the other (as, for example, in one's petition to a god, 'I beseech thee to come to me', or in an amorous relationship, 'I love you'). It would be a betrayal of one's emotional sentiment if the '*yuṣmat*' in these evocations could be substituted by *viṣaya* or 'object' in an inert sense, and one would be charged with mixing up levels of being, if not also making a category mistake. Nevertheless, I could be saying to my pet dog: 'I love you'. However, even here one takes it that the love for the sentient creature is not of the same order as one's appreciation of, say, an object of art, whatever its artistic merits. My dog achieves the status of '*bhavāt tatra eva*', or an honorary you, but I would more likely utter: '*bhavān mūrkhaḥ eva*' ('Ah, fool you are!').

True, even in such relationships there is no suggestion that an identity has been achieved, although people sometimes speak as if they have so achieved it in an exalted state of, say, an amorous, or sexual union. Again, one suspects that Śaṅkara's scepticism of the possibility of any kind of identity in worldly experience is rooted in the first instance in the absence of such an identity or non-differential identification in the very encounter of 'I' and 'you', which is a more serious encounter and often more shockingly uneasy than my running into a chair or a tree outside. So one might well be justified in pronouncing: Indeed, you and I are as far apart as light is from darkness. Is this image of the deep chasm between light and darkness a cutting metaphor or some sort of an analogy to dismantle the relation between 'I' and 'you', i.e. self and its other?

The imagined objection is that even if it is granted that there can be no identity between the *substance* (*dharmin*, or *viṣaya-bhūta*) in respect of object and subject, it could be that the identity is between their respective properties (*dharma*). Sometimes, when certain objects are placed near other objects, by dint of reflection or a process of emulation object *a* appears to acquire the property of object *b;* for instance, when an opaque crystal is placed over a strikingly yellow flower, one apprehends the crystal as having a diffused yellow texture, or the flower as somewhat opaque in its yellowness. Śaṅkara dismisses this alternative explanation for he avers that if the identity at the level of the substrates is not intelligible, then it is even less so at the level of the properties, because properties are no more capable of binding disparate entities together, than are individual fantasies capable of affording two lovers the same fulfilment as their physical togetherness can (unless, of course, they are avowed Freudians). And so the summation:

> Accordingly, the transference[3] (*adhyāsa*) of the object, represented by the concept 'you'—and of its properties on the subject, which is of the nature of consciousness (*cidātmaka*), represented by the concept 'I'—as well as the converse, the subject and its properties on the object can only be said to be an illusion (*mithyā iti bhavituṃ yuktam*).[4]

Notice here that the discussion has shifted from the issue of identification to that of transference, and the adversary (or the Fool) appears to be persuading Śaṅkara to forgo this possibility as well, for if identity is not possible then any other process which might lead to mutual identification should be rejected as well. Śaṅkara is cunning, for he admits this question in a rhetorical frame, only to press home the issue that the stated 'illusion' is what we might be looking for in explaining human apprehension of the world. It is, he says, a given in humankind's conventional praxis (*naisargika lokavyavahāra*) that there is false understanding owing to the inability to discriminate adequately between two entities (namely subject and object) and the properties in their very difference, which accounts for the ubiquitous tendency to mutually identify in the cause of transference of one set upon the other and to thereby forge a union (*mithunīkaraṇa*) of what is real (*satya*) with what is non-real (*anṛtá*), illustrated in speech-acts as 'I am this', and (more trivially) 'This is mine' or 'This is my Chicago hat'.[5] Now we see that the third person term is substitutable for the second person term, which conflation wipes out any distinctiveness that might be sedimented linguistically or, more significantly, ontologically between 'you' and 'this'. The 'I' in a given experience can have no real identifying reference to 'you' or to 'this' for all such references are to the non-self, while 'I' alone achieves reference to the self, and there can be no significant correlation between the two opposite referents, just as darkness is torn apart from light.

Śaṅkara therefore opines that the identification of self with non-self is an erroneous disposition as it is caused by ignorance (*avidyā*), the *tertium quid*, which conceals the self, the undisclosed witness as the transcendental self beneath the veil of the 'I' *qua* the embodied ego. (I prefer 'embodied ego' to 'empirical ego'.) The concealment prevents the apprehension of difference. *Différance* though it is, it is a significantly important concept the more modest among the Upaniṣads had set out to underscore in their own ways.

A comment or two on Śaṅkara's recourse to the all-important device of *adhyāsa* is apposite here too. I have said earlier that I understand the process being described as involving some kind of transference, rather than the simple-minded characterization of 'superimposition' (transimposition, involving layering of multiple images in three-dimensional lasergraphy to project a virtual reality into empty space, would do even better). There is a lengthy discussion in Śaṅkara on what is involved in *adhyāsa*, such as whether it is the substantive substrate, the (Lockean) primary properties, or the secondary qualities that are involved in the transference.

Two analogies are drawn from people's experience: viz. one moon appearing as two, and the regular illusion of a piece of nacre appearing as the absent silver. The several versions of the supposed prevailing theory that it is the properties of one object which are transferred upon the other (in either direction) are qualified and set aside because they are all prolix and make too many untenable assumptions. It looks more to be the case that the substrate of one, even though it is not given in the immediate environment of perception, is in its entirety projected and transferred upon the substrate of the object given but not properly cognized in perception. Thus 'silver' is, as it were, lifted and transferred upon 'nacre'.[6]

But where does 'silver' come from if it is not there outside? It comes from memory (*smṛti*). That is all very well as long as there is an objective substrate upon which the transference is possible (it is *pratyakṣavastu*, or *de res* object of perception); Śaṅkara is asked: but what if the substrate is a non-object and is outside the concept of 'you' as one's self says it is? In other words, how can the supposed 'inner' (*pratyak-*) self which is non-objective be the substrate for transference of the objective (*yuṣmatpratyayāpetasya ca pratyagātmano 'viṣayatvaṃ bravīṣi*)? To this Śaṅkara has two responses: First, it is not non-objective in the absolute sense, because it is the object of the concept 'I' and the inner-self is given in immediacy of the subject self-representation; additionally, there is no rule that says that two objects must be alike for there to be mutual transference (or any kind of identification). Even in respect of the object experienced as being outside (which in this case would be nacre), we can never be sure that the 'inner' (psychical) content has a corresponding 'outside' content. (This is Śaṅkara's general theory of perception in which he questions whether we only sometimes have false perceptions or whether all our perceptions are always already false, which is a bit like asking in the Rylean manner if Freedonia survives on counterfeit currency whether only just a few of the coins on its street are counterfeit?)[7] The second response is that there can be no *adhyāsa* without at least one substrate being involved; in other words, we have to assume that there is nacre in reality for the physically absent silver to be transferred. This response seems to run somewhat counter in spirit to the first response, in that Śaṅkara who couldn't a moment ago allow us to posit any substrate outside of the inner content of our regular perceptions, now wants a substrate for our illusory perceptions. Why is it so difficult to assume, the *Śūnya-vāda* adversary asks, that delusions can be without substrates, such as a *keṣoṇṅka* (a bright spot inside finger-stimulated eyelids, often mistaken for a psychedelic gyrating *śakti*-ball by quixotic devotees of gurumired Cats-killing *bhoginīs*)? Why not just say this: *śūnyasya śuktyātmanā, vivartamānasya rajatarūpeṇa vivartatā*:[8] nothingness first appears as nacre; then nacre is perceived as silver; the silver is the result of a mistaken transference (of the properties of silverness, splendour, etc.), which is *asat* (in both theories), onto nacre, which is *śūnya* or empty, that is the substrate, if one has to have one, and so is nothingness (emptiness) itself. Tempted to agree with this reply, for it is consistent with Śaṅkara's relativity theory of *māyā*, Śaṅkara is afraid that this might entail sacrificing the substrate which he ultimately needs for his absolute principle on the bottomless altar (or *altarity*) of nothingness. He could not countenance that defeat in argument, and so suggests that it is less prolix to assume that the process is inexplicable or indeterminate (*anirvacanīyata*).[9] If Śaṅkara holds this view then it is a case of disanalogy and he should not have invoked it in the first place, no more than the design argument in Western theodicy should have used the clock-maker's analogy, which Hume later skilfully ravaged.

Just on this thesis of *adhyāsa*, there is another conceptual fault which is smoothly glossed over by commentators on Śaṅkara alike. It is this that all along Śaṅkara has spoken of the transference of *asmat* on *yuṣmat* (that is the 'I' is falsely projected upon and identified with the body, mind, intellect, spouse's love, etc.) Here the principle of *sat*, presented in 'I' falls on the bewitching *asat*; the direction is clearly from *sat* to *asat*. However, in the analogy he uses of delusory perception, the projective identification is reversed, i.e. he draws on a counterprojective identification process, namely the transference of the silver (property, or substance), which is *asat* in the theory, onto an object, the substrate of nacre, which is *sat* (or at least admitted to be so under the principle of *anirvacīnya*, or indeterminacy); the direction is clearly from *asat* to *sat*. But non-being cannot be an effect of being! This is why the Buddhists (who score a point in their critique here) urge that if you allow *asat* the power to arise and prey upon something other than itself, then why stop at supposing that *asat* (non-being) does not have the power of settling upon another product of *asat* from a prior moment, and which too originated from another *asat* event-moment, and that too is dependent non-linearly on another, and so on *ad infinitum*? Dependent co-origination (*pratītya-samutpāda*) would be sufficient to explain the process, and we would not need to make spurious metaphysical assumptions like *svābhāvika* (i.e. things having inherent existence, *svabhāva*) or even the softer *anirvacīnyatā*. Alternatively, when the analogy is transposed to the actual issue at stake, and keeping to the direction of the analysis, and if it is the case, as Śaṅkara insists, that in *adhyāsa* there is transference of the 'I' on the non-self, would it not be consistent to regard the 'I' (*ātman*), the self, as the *asat* notion (the non-being) which is transferred upon the non-self, which is *sat* (or being)? If this reversal is permitted, then it would do irreparable damage to Śaṅkara's argument for which he seems to rely so heavily on analogies from everyday experience. He would still be left with some room to manoeuvre if he admitted the 'self's' nature to be *anirvacīnya* or something akin to the *anirvacīnyatā* principle he falls back on. But such a notion would be fraught with ambiguity, uncertainty, even insentience and devoid of the powerful presence (the *logos*) in which he wants to ground the self.

Another criticism, most forcefully articulated by the seventh-century Mādhyamika philosopher, Candrakīrti, but which fails to receive a decisive response in Śaṅkara, is that the transcendental (T-) self which the Upaniṣads and Hindu philosophers speak of is in the end an intellectually conceived notion of the self. Either this T-self is the same as or it is different from the psycho-physical complex that knows it. If it is the same, then it is not transcendental; if it is different, then its knowledge is not possible. Both possibilities lead to a logical *cul-de-sac.*[10] If, again, the self is said to be merely a constellation of parts of the psycho-physical complex, then, as King Milinda was reminded, it would be as odd as looking for the referent to the term 'cart' in the parts that make up the vehicle, individually or wholly. One is tempted to say that it is both same and different, but this would not satisfy Śaṅkara's stipulation of the purity of the self without being mixed up with the non-self of the psycho-physical complex. If we said that the T-self is neither the same as nor different from the pyscho-physical complex, then also Śaṅkara would have difficulties, not because this proposition might be seen to be violating the law of non-contradiction, which it does not, or because it rejects the excluded middle, for Śaṅkara not being a Naiyāyika would care less, but because this dialectic has the subtle force of persuading one not to take any decisive position or stance on the essential, non-dissimulable, nature of the self without difference or deferrals (*différance*) of any kind. It will be too *tear*-ful (as Mark C. Taylor might put it), and there would be no escape from the ubiquitous spiral of *duḥkha, saṃsāra,* re-death, for did not the Upanisadic seers whisper that the transcendental is blessedly blissful, truthfully real, and basking in its own consciousness?

Let us at this point recapitulate some problems we have raised with certain signalling terms in Śaṅkara's preamble. We began with the question: Why does Śaṅkara, in the first instance, create a pairing between *asmat* and *yuṣmat* as the correlatives of subject and object respectively? '*Asmat*' is strictly speaking, 'I', or the royal *we,* which encompasses in its broad sweep everyone in general and no one in particular, as when we say 'We will all die one day', 'As a citizen of Freedonia I believe in democracy', and is not unlike the general 'one' used in place of 'we' or the oblique 'I', as in 'One has to exercise caution', 'One cannot tell what the future holds for India'. As Heidegger pointed out, the oblique 'I' is as good, therefore, as not being a subject of any specificity, or as signifying a *being* of any ontological significance. That is one oddity.

Second, what makes Śaṅkara think that the logical opposite, the counter-correlate of 'I'/'we' is 'you', which is less oblique? Why is the self taken to be signified by 'I' and the non-self by 'you'? Suppose we lived in a community (of green Martians, if need be) where the (linguistic) convention reflected a commitment to a particular ontology wherein is acknowledged, honoured and perhaps even eulogized, the self in the other each time one's intentionality is turned to the other, while at the same time effacing one's self-presence, believing that one's own self is *in abscondi, is effaced*, or altogether non-existent. One therefore is never found to be saying, 'I will lead you', but instead, 'You will go ahead of me' (even if one intends to lead the way, as my humble Japanese guide put it, 'I will follow you' as he led the tour, always a pace behind his guests). In other words, the presupposition underlying all personal designations is that the first person term is never directly used, and if it is then one assumes that it is devoid of any substantial referential content unlike its second person, and perhaps even third person correlate, precisely because the purpose (*telos*) in each moment of relating with the other is to bask in the presencing of the self which is only ever fully solicited and awake in the face of other *qua* the Other as Levinas would put it. This is not also unlike Martin Buber's doctrine of the exaltation of the 'thou' over and above the deceited 'I'. Such a convention and the background world-view this imputes would forbid Śaṅkara from making the alignments he makes, i.e. of the self with *asmat* and the non-self with *yuṣmat*, even in the crudest metaphorical sense. The soteriology accordingly might point to the virtue of questioning the worth and realness of one self *vis-à-vis* the other, as the Protestant theologian Dietrich Bonhoeffer expressed his inner struggle under Nazi encampment: 'Who am I? This or the Other?'[11]

Third, what gives Śaṅkara such a self-assured confidence that there is no scope for any relationship or connection in reality (i.e. in the metaphysical sense) outside of the separation which the atomistic structure of language imposes upon us? Are *you* really what my self is not in every respect? (In the commentary, *Bhāmatī*, we are simply told that this is the counter-correlate, but so is the predicate term.)[12] Why does he set up this enormous chasm, a difference without remainder, between the concept-terms descriptive of first and second person referents, only to betray this sense in his example where he suddenly substitutes 'you' for 'that' (*idam*)? Should it not be that if 'I' does identify with what is not- 'I', its diametrically other, then in terms of *yuṣmat*, one should be saying '*I Am You*', not 'I am that'? What has he got against 'you'? Romantic usages aside, what is so utterly incommensurable and logically impossible about this phrasing? 'I am this' is acceptable, and so is 'you are that' in another context (as in the Upaniṣadic *tat-tvam-asi*, although the 'that' here assumes a very different identity, as does the 'you'), but it appears not to be acceptable to say 'I am you' or 'you are me'. This is the banishment of the 'you', which I want to suggest is indicative of Śaṅkara's endeavour to marginalize the world-as-being, or being-in-the world; indeed to deny *being* to the world, the entire cosmos, and to anything that is not capped by 'I' which alone privileges *sat* and therefore all else is *anṛtá* or *asat*. So the oppositional correlates he is setting up are essentially those of *sat* and *asat*, recalling the Ṛgvedic doubts about the existence of either, whose poor cousins Śaṅkara finds in the terms 'I' and 'you' respectively, with only one correlate capable of surviving the dissimulation wrought by an increasingly cynical philosophical development post-Nāgārjuna. I am puzzled that this disparity and uneasy conceptual trope has not worried commentators and even critics of Śaṅkara.

The fourth question is: Is Śaṅkara right in running together the 'object' (*viṣaya*) and the concept 'you' as its signifier, for this could imply that each time I refer to you, I take you to be an object (*viṣaya*), not simply as the other, that too of course, but as a *jada* ('inert' thing); or that 'you' have the same status as the *sarīra, manas, buddhi* (body, mind, intellect) tagged on to the 'I' in precisely those kinds of identifications as when I say, 'This is my body', or 'So my mind thinks'. For all these entities are ultimately dispensable and not necessary for my self-existence as *sat*, because the reality to which the 'I' really answers is not contained in any other entities or properties with which there is an apparent identification, albeit in speech-acts.

Further, if I say 'This is my body', 'I am Rāmu', 'I am mindful', 'I feel pain', where does *yuṣmat*, 'you', enter into the picture? Am I, even if falsely, thinking of the bodily subjectivity in some sense as being 'you'? If the body is not mine, is it yours then? Is that what this usage intends? Have I even tried to identify myself with some elusive 'you'? I suspect that either Śaṅkara is setting up a red-herring, or he has created a smoke-screen of a loose subject–object discourse, behind which lurks the 'I–You' dialectics even as he is intent upon annihilating the subject-bearer of the concept 'you', as the *other*, which I, on this side, in even the most casual encounter will perforce entertain or acknowledge. (Each species recognizes its own member as the 'other', 'you', even if they bump into each other only during the apocalypse, less lonely for that.) Is this some kind of a narcissistic denial of the *other* for being too close to one's own nature? Though sometimes we do playfully/ privately refer to our own introspected mood, behaviour, emotion, etc. as though their bearer was an *alter ego*, a second subjective badfellow, as it were; so we find ourselves admonishing the *alter* subject: 'Yaar, you're mad', or the voice of conscience bleats out with ellipses: '[You] . . . shouldn't have done that, mate.' But this is of no consequence, for this other self which one beholds, even if momentarily in reproach, is no better than the 'other' one needs to get away from. I am aware that a common linguistic conventions sanction the interchangeability of *idam* and *yuṣmat*, but its execution by Śaṅkara betrays a *'pas de deux'* (in the Derridean sense of the dance of duplicity and mutual erasure).

Even so, let us grant Śaṅkara his claim to the alleged category mistake in pairing the concept of 'I' with that of 'you'; but now consider the following paralogism: Since all arrangement of the world outside oneself proceeds through concepts, i.e. a set of linguistic signs doubling up on each other in an ordered sequence, whose loci is the mind or consciousness, there being no access to the world but through this intensional route, one then finds oneself in a closed world whose limits are the limits of the available store of concepts. One infers by certain observed behaviours that others might have a similar intensional structure, but there can be no real assurance, for this 'I' and *that* 'I' (if there is one) cannot possibly be identical; it is safer to relegate that other as some estranged 'you' and wish it luck. This is not to deny that overall there is a loose significative adjustment which allows regulative interactions between this 'you' and that 'you'; but beyond that, entry into each other's experiences, or into my 'I', is prohibited: you cannot look into my 'I'. So ultimately there is only this one 'I', the true subject; and therefore one can only speak of what is 'one's own', and 'whereof one cannot speak one maintains silence'.

This indeed smacks of extreme solipsism, a position which in our times only Wittgenstein has dared entertain (though not utter loudly), as J.N. Findlay points out, viz. a world in which 'all data and objects and interlocutors are necessarily one's own familiars'; or, alternatively, 'the self becomes an extensionless point, and all that remains over is the world of objects'.[13] The self's privileging, in this conception, can only be of it-self, and the seamless, colourless realm of the objective world simply breaks down into equally colourless concepts which are ideographs that neither feature meaningfully in our experience nor achieve complete reference to anything other than their own appearance held together in a web of symbols. We can construct meaning and interpretations out of them—as language tries to—but at a deeper level it has a hollow ring of emptiness, or outgrown tautology. Consider Wittgenstein: 'that *A* believes that *p, A* thinks *p, A* says *p*, are of the form '*p* says *p*';[14] confer Śaṅkara: 'I believe that *B* [= *Brahman*]', 'I pray *B*', 'I am *B*', is really about 'I is *I*'.

Put succinctly, the summation of all the questions we have asked so far is: Why does Śaṅkara believe that there is something logically quixotic about the subject–object relation which language makes transparent to our understanding? I have long pondered about this. There is probably a kernel of truth in Ganeśwar Miśra's sharp observation, which he attributes to Vidyāraṇya (a thirteenth-century prominent Vedānta writer), while invoking Strawson on the fundamental distinction between the logical subject and the logical predicate.[15] Let us consider this comment.

Miśra suggests that Śaṅkara is simply telling us something about the form of a proposition or judgement; he is not interested in making any factual point or claim at all, rather, his concern is purely logical and formal. Now in a proposition a predicate of a universal, which is an idea, is being related to a subject, which is a particular, such as the 'I'. But whenever a predicate is attributed to a subject of any kind there is a logical error: because subject and object are of different kinds; thus the ascription of a universal, an 'unsaturated' idea, say, brahmin, to the particular, 'saturated', designated by 'I' (in, e.g. 'I am a shoemaker') can only but be a logical error. According to Miśra, Śaṅkara is not making any factual claims about the existence, the nature, self-nature, or the extension of self; nor is he saying that the world is an illusion, or phenomenal (*pace* Radhakrishnan) or whatever. Rather, like Wittgenstein, whose philosophical cousin he appears to be in the

Indian tradition, the architect of Kāladī leaves the world exactly as it is (as he found it). At no point does Śaṅkara depart from revisionary metaphysics (i.e. the logical refinement of concepts) to fall back on the authority of 'revelation' (*śruti*): his task, as that of any philosopher, is purely elucidatory, and his *vairāgya* implies utter disinterest in the content of judgement.

This is a very interesting comment, although I think Miśra has overstressed the logical metaphor and the Wittgensteinian anti-metaphysical rhetoric. Nonetheless, his basic point stands, for there is as much problem for the *contents qua* signifieds of the predicate and subject to be so related as there is for their pre-reflective conceptual counterparts, for all such relations are contingent, temporal, and far from ideal. Miśra's critique points to this, particularly if the subject term is considered especially privileged, as 'I' is supposed to be. Be that is at may, what is the privileged content of 'I'? How is this known? We now shift to the epistemological and ontological problems.

To the question how do we know the self if our only mode of self-awareness is through the falsely identified non-self (this body, this name, as being mine), Śaṅkara replies: 'it is apprehended as the true content shimmering beneath the "I", and because the self, opposed to the non-self, is well-known in the world as the immediately perceived reality.' He gives the classical argument, with regard to the indubitability, self-certainty, of the reality of 'I': 'everyone is conscious of the existence of (his) self, and never thinks, "I am not". If the existence of the self were not known, everyone would think "I am not" (which is self-contradictory). This, then, is the substrate of "I-ness", the ground of being' (*pratyagātma, ātmanāmahaṃkāraspadam.*[16] Couple with this his earlier observation that the supposition that if the referent were the substrate of the body, there would be no continuity between the 'I' in the young days, the 'I' in current days, and the 'I' in old age.[17] The corollary is that, given the disparate cognitions from one moment to the next, since the substrate would undergo constant change, there can be no sense of a unified personal identity. So even a substrate is denied. These are serious claims about the life of self which need close scrutiny.

Śaṅkara is right in pointing out that personal identity cannot be understood in the same way as the identity of a thing, i.e. of material substance, and that the attempt to deny one's self leads to absurd consequences. As we know, self-denial led Descartes to conclude that the 'mind', *res cogitans*, a mental being, in its extendedness, exists, no more and no less; nothing, however, about the self as Śaṅkara is wont to establish follows from the Cartesian insight. There are serious logical flaws in Descartes' argument, as Chisholm and others have rightly pointed out,[18] but Śaṅkara is nowhere near providing the missing premises and analysis that would clinch the argument in his favour. An essential self-consciousness need not be the sufficient 'pre-reflective' remainder of the *cogito* (or even of the Husserlian *cogitationes* as Sartre argued): it could as well be a walking, loving, willing, singing, suffering self as the subjective pole of a self-constituting consciousness but which is only ever so in constant encounter with the world, the objective. There is no third, transcendent self-consciousness between the knower–known dualism. Indeed, Hume famously (and countless Buddhists infamously before him) failed to discover anything like an abiding self short of a bundle of perceptions and sensory impressions: 'I never catch *myself* without a perception.'[19] He was not too bothered by the questions of continuity and unity or self-reflexivity, although this lacunae made Hume's critique incomplete, which philosophers have since addressed with great force, notably Charles Taylor and Derek Parfit in their different ways, the former preferring a more hermeneutical–naturalistic route, the latter the analytical route.[20] The author of *Bhāmatī*, for reasons of his own, does not believe either that the self is ever disclosed in perception or in introspection, for the notion of 'I' (*ahaṃ-pratyaya*) disclosed herein is that of agency or enjoyer, etc.[21] This gives us about us much clue about the true/real self as the trickster psychoanalyst of our day can disclose the hidden subjectivity of his or her hopelessly ambiguated client, who will soon learn to deploy a device similar to Śaṅkara's imagery of *adhyāsa*. One may also wonder with Chuang-Tzu whether Chuang Chou dreaming that he was a butterfly was not the butterfly dreaming he was Chuang Chou, which in this context would lead one to ask whether the 'I' who thought she/he knew her/his deep and real self, was not the 'I' of the *alter ego* imagining that there is an 'I' to be known. The Hindu-hermetic circle is almost as vicious as it looks.

But on the certitude with which one knows oneself, for again

Śaṅkara is adamant that the sense of my individuated existence in the utterance 'I am' is a methodological criterion for indubitable knowledge and the possibility of transcendental experience, at a level which goes beyond the mere thought, or *cogito*, which is its starting point (verily the object of the concept of 'I', *asmat-pratyayaviṣayatā*). The denial of 'I exist' is said to be self-contradictory, in which case it has more than an existential status; the locus of a 'thought-form' (which Kant trades in for the empirical ego) it is there as a category without which there can be no conception of a thinker, a doubter, in short, of a being endowed with consciousness. Thus the 'I' is there of necessity. It is not just the *condition* of knowledge (as it is for Kant), it is knowable as pure undifferentiated consciousness. Its *knownness* is of a kind (*paravidyā*) not shareable with *knowing* in the phenomenal or empirical sense.

Second, the problem of continuity as Śaṅkara raises it, without any mention of the dimension of inner time-consciousness that I take him to be denying, is utterly misconceived, for there is nothing logically odd in supposing that consciousness possessed of memory, intentionality and imagination can retain a trace or semblance of a more or less unbroken personal life; the identity of a person throughout time is explicable in terms of the *beliefs* and *self-narratives, habits, style of existence*, coupled with hopes, expectations, appropriations and projections a being has about the states at various temporal stretches, and the thread that holds these together need be no less contingent, accidental and polymorphologic than, say, Gandhi's attempted autobiography reveals. Besides, as Strawson has shown, the notion of the individual embeds much more than just the body, or the mind, for personal subjectivity is conceivably much wider in scope, with interactive aspects, past histories and future trajectories, than traditional thinking, mind–body dualism in particular, has supposed it to be.[22] Also, agency involving rational choice, as Parfit has shown, requires no conception of a continuing re-identifiable substrate, much less a substance of any sortal kind.[23]

Third, adverting to the question of unity, Kant presented us with a brilliant and unsurpassable account of the unity of apperception for which an empirical concept answers to the *a priori* reflection on what is the subject of thinking. He does not deny an introspected self, but this substrate as a unifying factor is *presupposed* in perception rather than its accounting for synthetic unity of apperception, and there is, further, no direct experience of oneself as substance. Although reflecting on the nature of mental activity gives us the *idea* of a determinate substance, beyond the concept, however, it remains unknown, and Kant would not allow us to speculate as to its real nature. Kant's arguments are succinctly summed up for us by Sellars thus:

(i) the I is a being of unknown species which thinks;

(ii) the I doesn't simply 'have thoughts': *it thinks*—but in knowing *that* it thinks, and *what* thinks, we are not knowing what sort of being it is.

(iii) the I must have a nature—what it is we cannot know, though we *can* know that it is not material substance.[24]

Sellars comments: '[N]evertheless, although the I as an object of experience is not material substance, Kant insists . . . that *as noumenon* the I may be the same being as that which appears to us as our bodies.'[25] Kant goes a step further and suggests that the ultimate logical subject is the person, not unlike the Strawsonian person, and that the person might even be a composite being, a totality—a position, as I should urge again, resonates better the Upaniṣadic conception of the self as a dimensionally multivalent being whatever else self might be (which is exactly what I take Śaṅkara to be rejecting).[26]

But what about the suggestion that the secret lies transcendentally in the realm of the concealed consciousness as the abiding witness and the giver of the 'I-cognition' (*ahaṃ pratyaya*)? This position also has problems, not least for the reasons Rāmānuja in his rebuttal has drawn out, which is instructive to consider here.

Rāmānuja makes the now well-accepted phenomenological observation that consciousness is inherently intentional in its internal structure, that is to say, consciousness is always consciousness ething, that there is no such thing or being as a pure undifferentiated consciousness. Indeed, *différance* is the essential trait of consciousness; and I imagine that this also entails the essential temporality—or being-in-time—of consciousness. Rāmānuja's conention is that consciousness is a presence which is a distinguishing attribute of the I; indeed, *différance* runs right through and through all awareness.[27] All consciousness, Rāmānuja argues, implies difference—I take it in the sense of differing and deferring—not least the cognition 'I see this'. His argument turns on the intentionality of the eidetic structure, in which all presences are given as noetic correlates via noema or internal meaning states.[28] It follows that consciousness of the self does not stand unaffected by the internal difference; in other words, there is no escaping the fact that consciousness of the self, however refined and reflexively turned back upon itself from the stream of awareness states, is just another 'consciousness of . . . ' presencing; the knower here as anywhere else is the *ahaṃkāra*, the agent or subjective pole in all knowing, which in its effort to distance itself from the plethora of 'I-am-aware' cognitions sediments into *ahaṃ-pratyaya* (literally, 'I-concept'), rather close to the idea of 'transcendental ego' in Kant, or even in Husserl. But this is precisely the sense of 'I' which Śaṅkara wants to deny either as the true knower of the self (for this latter 'I' lodged in *ahaṃkāra*, i.e. inner common sense, is contingent, changeable, and ultimately destroyed by senility or death), or because the transcendental ego is not transcendent enough! But to Rāmānuja's thinking there cannot be any other subject which escapes this difference, and it need not be expected to. Rāmānuja has no difficulty, then, in accepting the logical identification of subject and object in the horizon of a consciousness marked by difference. However, he has a slight problem giving precise sense to the idea of the 'I-concept' that comes about as a result of the bracketed reflections of consciousness on the *ahaṃkāra*; he asks a pertinent question here: does consciousness become a reflection of the *ahaṃkāra*, or does the *ahaṃkāra* become a reflection of consciousness?[29] This is like asking: does the mirror reflect your image or do you project your image onto the mirror? What is the logical relation between the 'I' and consciousness and should this not be considered as involving the same opposition or contradictoriness (i.e. of transference)? I am not so sure that Rāmānuja's solution of identity-in-difference gets us out of the viciousness that he rightly charges Śaṅkara with (for identity-in-difference is as incoherent an idea as is total self-identity), but he has at least made us aware that non-difference in matters of consciousness (for example, in 'I am consciousness') would amount to as crude a form of tautology as it would be to say that 'Rāmu holds a stick' is about the stick only, or about Ramu only.[30]

CONCLUSION

The self that begins very early in the tradition as a nebulous mantric effect, that is, as a remnant of the primeval sacrifice of the *puruṣa* by the gods, against a background in which existence itself, human existence included, appears to be problematic in that there is a non-intelligible residue at the root of our experience which no inquiry may be able to answer, and where the idea of non-existent (*asat*) is entertained as the possible ground of all being, of *sat* (existent) as much as of the *Word* which came to be referred to as Brahman—this self gradually emerges into a powerful symbol for the unity of all being, consciousness and the ultimate. But in the Upaniṣads this conception of self is deepened in other ways: embodied existence and a notion of personhood is never discarded for the utterly abstract. *Ātman* as a unifying principle of personal life, encompassing traits of intelligence, affection, agency, self-awareness, for instance, in the *Mahānārāyaṇa Upaniṣad*, is located in a space within the heart, which is described as being thumb-sized, shaped like an inverted lotus and suspended somewhere between the Adam's apple and the navel. In this space is a subtle ball of fire which consumes food and warms the body, and which in the *Śvetāśvatara* measures a hundredth part of a hair's split-end, forming the basis also of the contingent consciousness of everyday life: Contingency is the key descriptive term here: that is the reality and impulse the Upaniṣads never run away from. It is this that gives one the identity of the *person* (*anguṣṭha-mātraḥ puruṣaḥ*).[31] This indwelling principle is also identified through other personal names, such as *paramātman*, *Śakti* and, indeed, 'I' and 'you'. This then is at the root of the transcendental ego and the embodied ego (*ahaṃkāra*). In a very real sense this self as represented in its various modalities is a constituted or constructed self. But this constituted and constituting person–nature has an enduring capacity for life which gives the bearer the sense of continuing identity and reflective self-consciousness. That makes for a sense of personal self, and it meets the essential conditions of what it is to be a person, namely intelligence, affectivity, agency, rationality, self-understanding, self-esteem, mutual recognizance[32] and, some would want to add to these, the capacity to *suffer*. Śaṅkara's denuded self robs the person of all the characteristics without which neither sound ethics nor an ontologically balanced and realistic world of community solidarity can be effected; realism is subverted: there are in his world, no individuals, and therefore no rights, interests, emotions, or authentic desires; there is no individuation, so there is no possibility of relations, of appreciation of beauty, strength, health, harmony, feelings, sexuality, etc., all of which were at the heart of the quest for the *prāṇic* elixir from the *Ṛgveda* to the Upaniṣads, the epics and Purāṇas (the *ānanda* of the Upaniṣads notwithstanding, for one wonders whether *mokṣa* is pleasurable). This blindfolded pursuit of the *logos* or logocentrism, of course, begins to be dismantled with Śrī Harṣa and other late Vedānta writers, only to be suspended by the silence following the disappearance of Buddhism from India, further arrested by the arrival of European-Orientalist romanticizers and, finally, nailed to the tomb with the gallant zealousness of twentieth-century harbingers of neo-Vedānta, notably Radhakrishnan and his theosophic pretenders or modern-day 'Advaitins'.

But the beginnings of a theory of *person* as descriptive of the self is extinguished by Śaṅkara and others, which is a great pity, for in the temptation to derive in pure abstraction an image of a self that would be absolute, Śaṅkara lost sight of the contingency of personal life and whatever else he thought ultimately mattered. Hegel was not far of the mark when he complained that Brahman, to which *ātman* is analytically equivalent (from Śaṅkara onwards), is a purely abstract, impersonal principle without self-consciousness (because it has no Other), and whose being is potential, not actual; and that because of this the world of particulars can have no part in it, rather they are entirely outside it, alien and independent of it, never in any true sense being created or sustained by Brahman. But both Hegel and Śaṅkara discard the painful Other by dissolving difference through their logic or principle of absolute identity, and making self-consciousness 'only [the] motionless tautology [of], Ego is Ego, I am I'.[33]

Finally, rather than declaring the death of the self of Vedānta, let me advert to one more possible motivation in Śaṅkara's project of ridding the experience of the ego of all the expressive categories of language and psychological atomism. By a series of reflective procedures of negative thinking (*neti neti*), as Zilberman had aptly pointed out, Śaṅkara wants to convince his audience that the paradoxes of description pertain wholly to the nature of linguistic structures themselves; the freedom that he wants is of another kind.[34] But he utilizes this philosophical foundationalism to ground a particular semantic of Indian culture, which by then was gaining hold, namely of the *sannyāsin* tradition, the ideal of the perfect renunciate life, which he hoped would become the absolute denominator of the Hindu civilization, as it was of the *śramaṇic* tradition, which brahmanism had by then successfully appropriated. This motivation perforce has to strike at the linguistic process and bracket out, in particular, the significative, denotative, symbolic and intentional functions of language, on which thinking relies so completely to make corresponding bridges with the objective world, as Professor Bimal Matilal has well alerted us to.[35] Deconstruct language and show, as later Wittgenstein did, that all discourse of the world proceeds through a collective conspiracy of 'language-games'; this will be sufficient to open up the possibility of a direct experience (*aparokṣānubhūti*) of the perfect ideal of *sannyāsa* (the experience is not of any *vastu* or objective state of affairs but of a supposedly self-transcendent subjective state of euphorically inward withdrawal). *Ātman* becomes but its meta-descriptive symbol, and a semiotic pointer to Brahman, the absolute meta-linguistic counterpart, indeed the abstracted Absolute. This linguistic design succeeds, as it was intended to, in exerting an enormous influence on the culture and organization of Indian society.[36]

NOTES

Abbreviations

Tait	*Taittirīya*
Chand.	*Chāndogya*
Brh.	*Bṛhadāraṇyaka*
Up.	*Upaniṣads.* Texts for the Upaniṣads are from *The Principal Upaniṣads,* ed. and trans. by S. Radhakrishnan, London: Allen & Unwin, 1975.

BS *Brahmasūtra*, and the reference is specifically to Śaṇkara's commentary or *Bhāṣya* (*BSB*) on the text with his own Introduction, variously called, *Samanvaya* or 'Reconciliation through Proper Interpretation', and the *Adhyāsa* Meditation, which I prefer. Texts for the *Bhāṣya* are from Śrī Sankaragranthavāliḥ *Complete Works of Śrī Śaṅkaracharya in the Original Sanskrit*, vol. VII, Vani Vilas ed., Samata Books, Madras, 1983. With interpolations of Vācaspati Miśra in his *Bhāmatī* commentary on the *BSB* (C. Kunhan Raja and S.S. Suryanarayana Sastri, editors/ translators, Adyar: Theosophical Publishers, 1933). Additional reference to, *Brahmasūtra-Catuḥsūtrī, The First Four Aphorisms of Brahmasūtra along with Śaṅkarācārya's Commentary*, by Pandit Hari Datta Sharma, Poona: Oriental Book Agency, 1967.

1. See Bimal K. Matilal's discussion on 'Necessity and Indian Logic' in his *Logical and Ethical Issues of Religious Belief* (Stephanos Nirmalendu Ghosh Lectures on Comparative Religion), Calcutta: University of Calcutta, 1982, pp. 144–51. Since the necessity of even the non-existent in Buddhist conception was conceded to (i.e. whatever is the *paramārtha* in the sense of the ultimate, whether as *sat* or indeed as *asat*, must of necessity be), Śaṅkara would not be too concerned about *necessity* but would rather want to register something about the nature of the ultimate; what better or greater achievement if both horns of the dilemma could by some transcendental deduction be brought into harmony or be reconciled under a single category. Kant would doubtless be in sympathy with this *samanvaya* move, though he would not accept its solution in Śaṅkara (or in Hegel for that matter), as I show in the text.
2. *yuṣmadasmatpratyayagocarayorviṣayaviṣayinostamaḥ prakāśadviruddhasvabhāvayoritaretarabhāvānupapattau siddhāyām; taddharmāṇāmapi sutarāmitaretarabhāvānupapattiḥ*. Ibid. p. 1.
3. *ityataḥ asmatpratyayagocare viṣayini cidātmake yuṣmatpratyayagocarasya viṣayasya taddharmāṇāṃ cādhyāsaḥ tadviparyayeṇa viṣayiṇastadharmāṇām ca viṣaye'dhyāso mithyeti bhavituṃ yuktam* (*BSB* p. 2; and p. 3 in Hari Sharma edition).
4. Contrary to established practice I find it less persuasive to translate *adhyāsa* as 'superimposition', I prefer 'transference' as this term used widely in common Western parlance, minus the psychoanalytical overtones, conveys better the sense of projecting and de-placing a psychical element from point A to point B.
5. *tathāpyanyo'yasminnanyo'nyātmakatāmanyo 'nyadharmanśca adhyāsya itaretarāvivekena atyantaviviktayordharmaṇoḥ, mithyājñānānimittaḥ, satyānṛte mithunīkṛtya, ahamidam mamedam iti naisargiko 'yaṃ lokavyavahāraḥ*. (loc. cit.)

Note that Śaṅkara uses the term '*mithunīkṛtya*', which has carnal overtones, especially in its common form *mithuna*, a famed path to enlightenment in Left-Hand Tantra.

6. *ucyate smṛtirūpaḥ pratra pūrvadṛṣṭavabhāsaḥ; tam kechid anyatra anyadharmādhyāsaḥ iti vadanti; kechittu yatra yaddhyāsaḥ tadvivekāgrahanibandhano bhramaḥ iti; anyate tu yatra yaddhyāsaḥ tasyaiva viparītadharmatvakalpanāma acakṣita iti; sarvathāpi tvanyasyānyadharmāvabhāśatām na vyabhicarati; tathā ca loka 'nubhavaḥ śuktikā hi rajatavadavabhāsate ekaścandraḥ sa dvitīyavad iti.*
7. I owe this observation to Arindam Chakrabarti, 'On the Purported Inseparability of Blue and the Awareness of Blue: An Examination of Sahopalambhaniyama', *Dialogue Series: I Mind-Only School of Buddhist Logic*, ed., Doboom Tulku, Delhi: Tibet House, 1990, p. 27.
8. Cited in Hari Datta Sharma, p. 14.
9. The term *anirvacanīya* is not Śaṅkara's, it is culled out of the texts on the definitional debate over *adhyāsa* in the *Bhāmatī* commentary on *adhyāsa*, under the *Sarvathāpi* text (see note 6), in Parimala Sanskrit Series No. 1. *The Brahmasūtra Śaṅkara Bhāsya with commentaries Bhāmatī, Kalpataru and Parimala*, Introduction by Ester Solomon, Ahmedabad: Parimal Publication, 1981, p. 32–4.
10. See the fine study on Candrakīrti by Peter Fenner, *The Ontology of the Middle Way*, Dordrecht: Kluwer Academic Publishers, 1990 (Studies in Classical India Series No. 10).
11. *The Cost of Discipleship* (poem trans. by J.B. Leishman), New York: Macmillan, 1959, p. 16.
12. Op. cit., pp. 6–7.
13. Wittgenstein, *Logico Tractatus*, 5.64; cf. J.N. Findlay, *Ascent to the Absolute*, London: Allen & Unwin, 1970, p. 156; 'My Encounters with Wittgenstein', in *Studies in the Philosophy of J.N. Findlay*, eds, Robert S. Cohen et al., Albany: SUNY Press, 1985, pp. 62ff.
14. *Tractatus*, 542; cf. discussion in J.N. Findlay, 'My Encounters with Wittgenstein', p. 64.
15. Ganeśwar Miśra, *Language, Reality and Analysis*, edited with Introduction by J.N. Mohanty, Leiden: E.J. Brill (Indian Thought No. 1), 1990, pp. 15–18, 61–7, and ch. 4.
16. *Bhāmatī*, pp. 35–6, Solomon/Parimal Edn.
17. But consider the different ways in which contemporary philosophers respond to this issue, from Swinburne to Parfit, to Putnam with his brain-in-vat and twin-earth thought experiments. (See next three notes.)
18. See Amelié Rorty, ed., *The Identity of Person*, Berkeley: University of California Press, 1976; Sydney Shoemaker and Richard Swinburne, *Personal Identity*, Blackwell, 1984; Bernard Williams, *Problems of the Self*, Cambridge: University Press, 1973, pp. 1–18; 64–81.
19. Hume, *Treatise of Human Nature*, Oxford University Press, 1978.
20. Derek Parfit, *Reasons and Persons*, Oxford: Clarendon Press, 1984; Charles Taylor, *Sources of the Self the Making of the Modern Identity*, 1989.

21. Op. cit.
22. Peter Strawson, *Individuals: An Essay in Descriptive Metaphysics*, London: Methuen, 1959.
23. Parfit (*Reasons and Persons*) develops the thesis that one can act morally or rationally but impersonally, i.e. without regard for others, because it is difficult, in real and imagined cases, to establish and know just who or what a person is (oneself included), how to distinguish between different persons, different lives, and so on.
24. Wilfrid Sellars, 'This I or He or It (the Thing) which Thinks', in his *Essays in Philosophy and History* (Philosophical Studies Series in Philosophy 2), Dordrecht: D. Reidel Publishing Company, 1974, p. 96.
25. Ibid.
26. See *Bhāmatī*, p. 5, in Kunhan Raja et al. Edn.
27. *Śrībhāṣya Rāmānujā's Commentary on Vedānta Sūtra*, Sacred Books of the East Series (Gen. ed. Max. Müller) trans. Thibaut translation, Delhi: Motilal Banarsidass, 1967, p. 61.
28. Ibid., p. 41.
29. Ibid., p. 63.
30. Ibid., p. 61; cf. Wittgenstein on the solipsistic charge that it is all about 'p is p', note 7 above.
31. See P. Bilimoria, *The Self and Its Destiny in Hinduism*, Deakin University, 1992, p. 17.
32. Nicholas Rescher, 'What is a Person?' *Human Interests*, Stanford University, 1990, pp. 6–7.
33. G.W.F. Hegel, *The Phenomenology of Mind*, trans. by J.B. Baillie, NY/London: Harper Torhcbooks, 1967, p. 219; see again note 1 above. Radhakrishnan attempting to redress the situation, or Śaṅkara's excesses, errs in the opposite direction, in that he makes Śaṅkara and the Upaniṣads conform uncompromisingly to Hegel's strictures for a muted transcendentalist position that has a historicized trajectory and temporal spiritual (self-) realization as the utopian goal. Radhakrishnan saw himself in part, as the philosopher-king, and the post-Gandhian nation-state as the other element pointing towards the *bhum*-ification of *Hiranyagarbha*, the manifest Absolute Spirit. See my 'Savings the Appearances in Plato's Academy', in S.S. Rama Rao Pappu, ed., *New Essays on Radhakrishnan*, Delhi: Indian Books Centre, 1995.
34. See excellent analysis in David B. Zilberman, *The Birth of Meaning in Hindu Thought*, ed. by Robert S. Cohen, Dordrecht: D. Reidel Publishing Company, 1988, pp. 222ff.
35. See Bimal Matilal's *The Word and the World*, Delhi: Oxford University Press, 1991.
36. See Max Weber and others, discussion in Zilberman.

Chapter 15

Two Truths, or One?

Radhika Herzberger and Hans G. Herzberger

Thomas Mann begins his essay on Schopenhauer by telling us that the pleasures of metaphysics are mainly aesthetic. Without sharing that high degree of philosophical detachment, we acknowledge that the present essay was motivated by aesthetic as well as historical concerns. Because it is part of an effort to understand Indian philosophers as particular individuals with distinctive problem situations and doctrines it is properly classified as historical. Because it aims to locate particular doctrines within larger philosophical visions, it might also be classified as aesthetic. Our essay develops a long perspective going back to the early origins of *pramāṇa* theory. Drawing the reader back in time puts us in a better position to trace historical sources for certain important ideas of Dharmakīrti and Diṅnāga, and to contrast the treatment of those ideas in their respective philosophical systems.

Traditional Indian thought identified four broad goals for human life, to be pursued through four stages in a canonical progression. The goals were *dharma, artha, kāma* and *mokṣa* (duty, wealth, pleasure and liberation); and the stages of life were *brahmācārya, gṛhastha, vānaprastha* and *sannyāsa* (student, householder, forest dweller and ascetic).

Earlier stages of life were primarily reserved for the pursuit of pleasure, wealth, and the moral duties of caste, while later stages were meant to prepare the ground for *mokṣa* or liberation from *saṁsāra*, the endless round of birth and rebirth in which all living things are caught. It was an outlook and a set of values which tended to stabilize the existing social order.

This general design for life admitted variations, as in the case of Śaṅkarācārya, who died at the age of thirty-two. Bypassing the stage of a householder, Śaṅkarācārya renounced this world and achieved *mokṣa* as a youth. Later King Janaka achieved liberation without having first renounced the worldly life. These are two prominent exceptions to the canonical four-stage progression.

Our essay explores relations between this framework of goals and the epistemology of *pramāṇa* theorists. At a different level, it also examines tensions between worldly pursuits and *mokṣa*, as they emerge from our texts.

It was Akṣapāda Gautama who first joined a system of values to *pramāṇa* theory. The system of values in question was older, part of the inheritance to which the Upaniṣads, Buddhists and Jains had contributed. Gautama considered a transcendent state of liberation to be the highest goal of human endeavour—as contrasted with worldly pursuits, which cease when they are fully understood.

Received doctrine presupposed two ontological domains, one of which was caught in the cycle of *saṃsāra.* For the rational framework of Nyāya philosophy, Gautama modified this ontology in a subtle way. Early Naiyāyikas held the goal of enquiry to be *apavarga* (final release), which was not directly the object of any *pramāṇa*, and was not strictly speaking an object at all. As a first step towards restructuring the relation between knowledge and the canonical goals, we suggest that Gautama held the proper object of philosophical enquiry to be the cycle of *saṃsāra*, and *apavarga* to be a state which comes into being when a certain kind of knowledge has been obtained, namely an adequate understanding of its own cyclical impediments.

The cycle posited by Akṣapāda Gautama, and enumerated in the second aphorism of the *Nyāyasūtra (NS)*, consists of sorrow, birth, activity, character faults, and false belief.[1] After enumerating these elements, the aphorism states that cessation of any element in this cyclically ordered set leads to the cessation of its predecessor. By iteration then each one in turn will cease.

What causes one of these elements to cease? Two incompatible answers to this question can be found in the *NS*: that *mokṣa* can be achieved cognitively—by knowledge gained through *pramāṇas* —and that it requires the practise of yoga. According to the first proposition, a kind of knowledge suffices for attaining the highest good (*niḥśreyas*). According to the second proposition, the cycle of rebirth cannot cease without the adoption of yogic practices.

To resolve this *prima facie* contradiction, we conjecture that the first alternative was the position originally advocated by Akṣapāda Gautama, and the second alternative was Vātsyāyana's later interpolation. We suspect that Vātsyāyana may have retreated to this position in response to Nāgārjuna's *Vigrahavyāvartanī (VV)* which had charged Gautama's *pramāṇa* doctrine with circularity.

Against this background, we raise the question whether Diṅnāga's doctrine of *pramāṇas* in the *PS* can be seen as a genuine attempt to solve the problem raised by Nāgārjuna. Do the definitions of both *pratyakṣa* and *anumāna* in the *PS* in fact meet the challenge posed by Nāgārjuna? By providing a formally adequate *pramāṇa* doctrine did Diṅnāga restore Gautama's link between *pramāṇas* and the highest good that Nāgārjuna had sought to break?

In the invocatory stanza of his *PS*, Diṅnāga praised the Buddha as *pramāṇabhūta* (one who is incarnated as the means of knowledge), and in his *Prajñāpāramitāpiṇḍārtha*—a Mahāyāna classic —Diṅnāga held the highest good to be non-dual gnosis (*prajñāpāramitā jñānam*).[2] Such passages are consistent with our view that he was restoring Gautama's link between knowledge and values, and they offer us a way of establishing a sharper contrast between Diṅnāga and Dharmakīrti.

Carrying Nāgārjuna's problem into the domain of Dharmakīrti's thought, additional questions arise. Why did Dharmakīrti discard the philosophical advances promised by Diṅnāga's response to Nāgārjuna? Why did he reject the intermediate universe Diṅnāga posited between mundane and transcendental reality? Why did he dissociate inference from perception; and, having done so, how could he retain the epithet *anumānabhūta* for the Buddha? In the framework of *NB* and *PV*, that epithet cannot easily carry the high praise it had earlier expressed, because in that framework it would seem to imply that the Buddha incarnate has *mithyājñāna* (false belief).[3]

Behind these questions lies our belief that Nāgārjuna's critique of earlier *pramāṇa* doctrine as circular had a crucial impact on the metaphysical visions implicit in *PS*, *NB* and *PV*. We believe that Nāgārjuna's criticism encouraged development of the two-truth doctrine in the *Nyāyabhāsya (NBh)* and in Dharmakīrti's system; indeed Nāgārjuna himself had laid the groundwork for a distinction between two kinds of truth.

Our argument has five parts, which unfold along chronological lines:

Section I attributes to Akṣapāda Gautama the view that the four items he lists in *NS*.I.1.1–2 are not strictly speaking objects or *prameyas*, in spite of the fact that *apavarga* appears on the list. The items in question are *tattvajñāna, mithyājñāna, niḥśreyas* and *apavarga* (essential knowledge, false belief, highest good and final release). In Gautama's epistemology, only objects (*prameyas*) can be known. We attribute to Gautama the position that *apavarga* and the highest good are by-products of the essential knowledge of *prameyas* in various categories. We also describe an indirect relation binding the *pramāṇas* to *apavarga, tattvajñāna* and *niḥśreyas*.

Section II traces changes in Akṣapāda Gautama's philosophical vision according to Chapter IV of *NS*. We draw attention to his revised doctrine of the cycle and the way in which Nyāya was brought under the tutelage of yoga.

Section III offers a diagnostic exposition of more general changes in the Nyāya perspective. We concentrate here on Nāgārjuna's critique of *pramāṇa* doctrines and Vātsyāyana's response to the charge that those doctrines involve circularity or infinite regress.

Section IV outlines some wider consequences of Diṅnāga's *pramāṇa* doctrine.

Section V considers the formal adequacy of Dharmakīrti's *pramāṇa* doctrine in *PV* and *NB*, describes the nature of *samvṛtti* in *PV*, and draws out connections with Nāgārjuna's two-truth doctrine.

I

Gautama's first *sūtra* enumerates sixteen categories intended to cover that essential knowledge (*tattvajñāna*) which leads to the highest good (*niḥśreyas*). These sixteen categories are: means of knowledge (*pramāṇa*), object of knowledge (*prameya*), doubt (*saṃśaya*), intention (*prayojana*), example (*dṛṣṭānta*), premise (*siddhānta*), argument (*avayava*), hypothetical reasoning (*tarka*), conclusion (*nirṇaya*), debate (*vāda*), disputation (*jalpa*), sophistry (*vitaṇḍa*), fallacious reason (*hetvābhāsa*), trickery (*chala*), casuistry (*jāti*), and the clincher (*nigrahasthāna*).[4]

This capacious list embeds various tools of the debater's trade within an epistemological framework. The most significant categories for our present purpose are the means of knowledge (*pramāṇa*) and the objects that are epistemologically accessible to those means (*prameya*). We will presently show that the list of the means and objects of knowledge is fairly exhaustive of the objects in Akṣapāda Gautama's universe of discourse. Things not listed therein are not objects within the cyclical universe, nor within any transcendental universe, nor within any universe occupying an indeterminate position between these two domains.

Four means of knowledge are described in *NS*.I.1.4: perception, inference, analogy and testimony.[5] The objects of knowledge (*prameya*) are enumerated in *NS*.I.1.9) as: the *ātman*, the body, the objects of the senses, intellect, mind, activity, character faults, reincarnation, fruits of action, sorrow and final release (*apavarga*).[6] Each of these listed *prameyas* is defined in separate sūtras.

With the exception of false belief (*mithyājñāna*), the elements of the cycle mentioned in *NS*.I.1.2 are included in the *prameyas* listed in *NS*.I.1.9 above. This should make every element in the cycle, excepting false belief, also an object of knowledge. Presumably false knowledge is not anything in itself; rather it is false knowledge regarding the objects listed in I.1.9. Thus it would seem to be on a par with *tattvajñāna* (mentioned in *NS*.I.1.1) which is not listed as an object at all.

Do the objects listed in *NS*.I.1.9 exhaust Akṣapāda Gautama's universe of discourse? Are there any objects which lie beyond the cognitive faculties and are in that sense transcendental? Obviously, the *ātman*, which is the first to appear in the list of *prameyas*, is not a transcendental object for Akṣapāda Gautama, and its definition suggests that it is known inferentially through a sign (*linga*). It would then appear to be the same sort of object as mind.[7] But what about *niḥśreyas* (the highest good) and *apavarga* (release)? Are they like the *ātman*, accessible to the ordinary means of cognition? The definition of *apavarga* (final release) sets it in a different category from the other *prameyas.* According to that definition, 'The complete cessation of that (*tad*) is final release'.[8] The demonstrative *tad* refers to the elements of the cycle enumerated in *NS.*I.1.2. *Apavarga* is release from sorrow, birth, activity, false belief. And that is how it is explained by Vātsyāyana. *Apavarga* is not thus a direct object of the *pramāṇas,* nor is it a transcendental object. Indeed though listed as one, it is not an object in the strict sense at all. It is more like a residual state that comes into being upon cessation of the cycle.

Given the above definition of *apavarga,* what is the relation between *niḥśreyas* (the highest good), mentioned in *NS.*I.1.1, and *apavarga* (final release), mentioned in *NS.*I.1.2? The highest good is a product of our complete and essential knowledge of the sixteen categories, of which the elements listed in *NS.*I.1.2 are a part. Essential knowledge of the elements in the cycle, based on a proper understanding of the framework of knowledge provided by the Nyāya system, brings about cessation of the cycle and eventually leads to the highest good.

The objects enumerated in *NS.*I.1.2 are thus not morally neutral. Being tied up with the cycle of rebirth, they are impediments to liberation, which is the highest good. Proper knowledge of these impediments, however, leads to their cessation. When they cease, the highest good is realized. This is the framework in which we connect *NS.*I.1.1 with *NS.*I.1.2.

Because the highest good is not an object of knowledge, and because the objects of the cycle, those objects which are not morally neutral, cease to be impediments upon being known, cognition based on *pramāṇas* is liberating. Neither *apavarga* nor *niḥśreyas* are objects in the sense that the *ātman* is; nor are they transcendental objects. *Apavarga* is an indirect product of the knowledge of the sixteen categories. The unarticulated premise here is that knowledge is liberating: to know the impediments is to be free of them. That freedom which is brought about by a liberating knowledge is the highest good. In this way *pramāṇas* are linked to freedom, and epistemology is integrated with a system of values.

We suggest that within a liberation–rebirth metaphysics, Akṣapāda Gautama's doctrine is cognitivist in the sense that according to it, cognitive means suffice for bringing about the highest ends. It teaches that knowledge, even without the benefit of yogic practice, can be liberating. For example, to know a given character fault, and to know what kinds of action it incites, is to be free from that fault and its consequences in action. At the end of life, those who are free from character faults and the actions they incite, are freed from rebirth and suffer no more pain; for them, the cycle has completely ceased.

Socrates had taught that virtue is knowledge—to know the good is to be good. Akṣapāda Gautama taught that freedom is knowledge—to know the impediments to the good is to be freed from those impediments, from rebirth, from pain, and from the cycle of life. On this doctrine, because the *pramāṇas* are necessary and sufficient means for knowing impediments to the good, they are also necessary and sufficient means to the highest good.

II

Chapter Four of the *Nyāyabhāṣya* redefines the *prameyas* including the objects of the cycle. In his introductory comments Vātsyāyana somewhat defensively states that his purpose is not to change the original doctrines of the founder.[9] However, he himself had set the stage for doing exactly that with a dilemma he had posed earlier:

> Sirs, does essential knowledge arise with reference to each of the several objects or does it arise in connection with only some? What then is the difference here [between occasional arising and arising in reference to each object]? It cannot arise simultaneously with reference to each object [to be known], because that which is to be known is infinite. Nor can it arise with reference to only some for in those cases where it does not arise delusion is not turned away; therefore the unfortunate possibility of a [permanent] remainder of delusion is faced. Indeed essential knowledge with reference to one object cannot negate delusion with reference to another.[10]

This dilemma challenges Akṣapāda Gautama's project of confining the impediments to liberation to those objects mentioned in the cycle: sorrow, birth, activity, character faults and so on (as stated in *NS*.I.1.2). Essential knowledge of each of these objects, based on formulations provided later in the first chapter, is to bring about liberation.

One approach to resolving the dilemma would be to distinguish between objects of knowledge and categories of those objects. That which is to be known may be infinite at the level of objects and finite at the level of categories. If so, one could claim that essential knowledge can be attained for some objects within each category of the Nyāya system, even though essential knowledge of all objects within each category could never be complete. On this basis the Nyāya doctrine might be defended as permitting, without guaranteeing, escape from residual delusion. At the level of objects, anyone who tries to follow an infinite path towards the ultimate end would find final release forever out of reach; but finite paths could also exist towards the same ultimate end. Akṣapāda Gautama's categories are finite in number, so liberation could be attainable in principle. Even if we were to question the definition of essential knowledge (*tattvajñāna*) as provided in the *sūtras*, the number of independent delusions from which freedom through knowledge is sought in a lifetime would in any case be limited by our own mortality, and could never be infinite. By suggesting that Gautama's chain must be infinite, Vātsyāyana cut at the heart of this early integration of epistemology with a system of values.

Having thus put aside the earlier vision, Vātsyāyana poses the question: 'How does essential knowledge (*tattvajñāna*) arise?', as a prelude to a *sūtra* that introduces a new element. That *sūtra*, followed by Vātsyāyana's elucidation, reads:

> [Essential knowledge proceeds] from the practice of a particular form of meditation (*samādhi*).
>
> When the mind having been abstracted (withdrawn) from the Sense-organs, is kept steady by an effort tending to concentration—the contact that takes place between this Mind and the Soul, and which is accompanied by a conscious eagerness to get at the truth, is what is called 'Meditation'. During this meditation no cognitions appear in regard to the object of the senses. From the practice of the said Meditation proceeds True Knowledge.[11]

Thus Nyāya is brought under the tutelage of Yoga:[12] There is the advice that Yoga should be practiced in forests, caves and on river banks.

On this doctrine, essential knowledge (*tattvajñāna*) of the sixteen categories would not be a sufficient condition for reaching the 'highest good' (*niḥśreyas*), but would require also the practise and retirement to forest and cave. Here Vātsyāyana reads a set of presuppositions into *NS*.I.1.2.[13]

This is not the place to pursue several interesting questions which inevitably arise, including the question whether Vātsyāyana held knowledge to be even a necessary condition for *apavarga*. If it were neither sufficient nor necessary, it would be entirely independent, without definite relevance to the transcendental good.[14] Whatever position one takes on that more difficult question, however, we have a diagnostic question: Why was Nyāya brought under the tutelage of Yoga? An answer to that question would illuminate the causal nexus between *pramāṇas*, the highest good, and worldly affairs (*vyavahāra*). The affairs of the world and their very direct links with the *pramāṇas* take on prominence in this context, and help to unravel the historical situation.

III

Chapter II of the *NB* relates an exchange between a Nāgārjunian figure and his Naiyāyika opponent on questions relating to justification of the *pramāṇas*. Are the means of cognition like a lamp which illumines both itself and the object? Or are they like a standard of weight, itself in need of calibration? The first analogy, which likens the *pramāṇa* to the light of a lamp, cannot account for perceptual errors. The second analogy, based on the assumption that a means of cognition need not always be accurate, lays the burden for correcting errors on yet another means of cognition. Vātsyāyana preferred the latter analogy and the epistemology it exemplified. The *NB*, however, records a dissenting point of view within the Nyāya framework. The spokesman for this view delivers an aphorism (*NS*.I.1.19) upholding the analogy between the *pramāṇas* and lamplight, and later, raises the problem of an infinite regress unless the *pramāṇas* are self-justifying.[15]

Vātsyāyana would have to concede, however tacitly, that in the event there may be no final justification for a *pramāṇa's* claim to truth, but he did not explicitly address this question nor discuss in detail the possibility of an infinite series of justifications. In fact he entirely bypassed the question of justification for the claims to truth of *pramāṇas* in favour of a different question: How can that which is a means (*sādhana*) or the cause (*hetu/nimitta*) of cognition become also an object (*viṣaya*) of cognition? In other words: How can a *pramāṇa* also be a *prameya?* Vātsyāyana's answer consists in describing cases taken from everyday usage to show that this does indeed happen and ordinary speech records such a process. Within the everyday context of action, pragmatic considerations can be brought to bear. Here an aim of knowledge is to help in pursuit of the four goals of life: happiness, duty, wealth and release; and *pramāṇas*, by giving right knowledge of an object, help us to gain it as a means to the basic aims of life. There can be no infinite regress in everyday affairs, for the objects of *pramāṇas* in practice can be reached and one or other of the four goals of life can be achieved. This focus on practice offers a new approach to the vexed problem of justifying the *pramāṇas*.

'If perception were to be apprehended on the basis of [yet another] perception, there would be an infinite regress'. If this objection is raised, we reply: This is not so, because worldly affairs (*vyavahāra*) are accomplished by apprehending [the distinction in each case between the *pramāṇa* as] the instrument and [in the other as] the object of cognition: 'I apprehend (or gain) the objects through perception; I apprehend (or gain) the object through inference; I apprehend (or gain) the object through analogy; I apprehend (or gain) the object through scripture; thus my knowledge is perceptual; my knowledge is inferential; my knowledge is analogical; my knowledge is scriptural.' For the person who apprehends (*upalabhamānasya*) as the object and instrument of cognition, worldly business—whose aim (*prayojana*) is duty, wealth, happiness, final release and the abandonment of everything contrary [to these four aims]—is achieved. The affairs of the world are accomplished only then. And in the affairs of the world there is nothing to be accomplished by an infinite regress, nothing which would require the postulation of an infinite regress.[16]

Efficacious action is possible for those who test one cognition against others, and base their actions—which are always aimed at wealth, happiness, duty or final release—on *pramāṇas*. This is so because success in action involves either getting hold of real objects or moving towards one of the four basic human goals. Neither objects nor goals are forever out of reach, which would be the case if infinite regress were inescapable. This pragmatic justification is given prominence in the opening lines of the *NBh.*

Because action is successful when the object [towards which it is directed] is cognized on the basis of a *pramāṇa*, the *pramāṇa* is endowed with a [real] object [or with a goal]. There is no cognition of an object apart from a *pramāṇa*. There is no successful activity apart from cognition of the object.[17]

Drawing the concept of *vyavahāra* into a discussion of *pramāṇas* offers Vātsyāyana a means of doing two things: to deny infinite regress in worldly affairs; and—perhaps more importantly—to use pragmatic grounds in justifying the *pramāṇas* as necessary for *vyavahāra.* Vātsyāyana represents a point of view quite distinct from that of the Naiyāyika who likens the *pramāṇas* to the light of a lamp that illumines both itself and its object.[18] This doctrine, despite its difficulties in accounting for error, had one advantage: it bypassed the threat of infinite regress in justifying *pramāṇas.* Another advantage was that it could stand without recourse to any novel set of premises recognizing as an ideal, success in the basic aims of human action.

Vātsyāyana's interpretation of the *NS*, which Uddyotakara followed, put aside Akṣapāda Gautama's original doctrine that understanding the cycle, through *pramāṇas*, releases one from it. Instead Vātsyāyana offered a new double ideal for Nyāya: to provide the basis for our understanding of the world, and the means for achieving goals. He replaced the ideal of knowledge as freedom, with the ideal of knowledge as power.

IV

To praise the Buddha as '*pramāṇa*-incarnate', is to elevate *pramāṇas* to a transcendental level. For a *pramāṇa* theorist to do this would be either turning a blind eye to history or marking a particular innovation—especially in view of Nāgārjuna's earlier critique of the *pramāṇa* doctrine and the modifications it brought forth in Nyāya philosophy. Diṅnāga's use of the particular epithet *pramāṇabhūta* suggests that he considered Nāgārjuna's problem of defining the *pramāṇas* in a formally adequate way, to have been solved in his system. And it seems to us a defensible proposition, that Diṅnāga's definitions of perception and inference in the *PS*, were formally adequate. By this we mean that those definitions are grounded within the indicated system: they are neither circular nor do they support any vicious regress.[19]

In response to Nāgārjuna's criticisms, Vātsyāyana drew the concept of worldly affairs (*vyavahāra*), with its fourfold goals of duty, wealth, happiness and release, into the explanatory framework of Nyāya. He offered a pragmatic justification for *pramāṇas*, making use of the role of belief in action. Because actions are based on beliefs acquired through *pramāṇas*, he reasoned that success in action demonstrates the truth of the underlying beliefs and the reliability of *pramāṇas* as sources of knowledge. He concluded his justification with the factual observation that successful action actually occurs without infinite regress.[20]

A challenge now hung in the air: a philosophical niche was open for *pramāṇa* theorists to fill, if only they could devise non-circular definitions for the instruments of knowledge. Diṅnāga's *PS* occupied this niche by defining *pratyakṣa* (perception) without recourse to *avyabhicārin* (non-deviating), *abhrāntam* (non-deluded) and *avisaṁvādakam* (non-contradicted), modifying qualifiers of perception used by Akṣapāda Gautama and Dharmakīrti in their systems. He also provided criteria for judging products of *anumāna*, the second faculty of knowledge.

Diṅnāga defined perception as *kalpanā poḍham* (that [faculty] which is without imagination). To be dissociated from the imagination is to be dissociated from the process of giving names. The imaginative faculty is shot through with language:

> In the case of proper names: an object qualified by a name is spoken of as 'this is Ḍittha'; in the case of genus names by a genus as 'this is a cow'; in the case of action names by an action as 'this is a cook'; in the case of substance names by a substance as 'this is a cane bearer', 'this is a horn bearer' and so on.[21]

According to Diṅnāga, a person using the faculty of imagination knows an object as blue, while the same person when using the faculty of perception knows blueness. Imagination involves propositional structure, division of experience into properties and their bearers. Any act of naming, or more properly of describing, associates an object with a class, defined by some predicate. To say that an object is blue is to ascribe it to the class of blue things. And that ascription is subject to error. The ascription is correct if the appearance (*ābhāsa*) of the object in question matches the property expressed by the predicate and exemplified by the class to which the object is assigned.

According to Diṅnāga, truth and falsity are properties of sentential acts of judgement; they are not properties of appearances, which belong in the field of perception. Appearances are neither true nor false, and the qualifying adjectives *abhrāntam* and so on are not required.

Diṅnāga explicitly stated that perceptual errors involve the mind or the constructive faculty.[22] Even though truth and falsity cannot properly belong to the appearances rendered by perception, they are *pramāṇa* because they help to validate perceptually based sentential judgements.[23]

When, on the other hand, an external object alone is what is cognized (*prameya*) then:

the shape of the object is itself the means of knowing it (pramāṇa)

for there, even though the cognition is self cognizable, the own form [of cognition] having been disregarded, the state of the appearance of the object is the means of knowing it. Because the object

is measured thereby.

Whatever the shape in which the object appears to cognition, as white or anything else, that is the form in which the object is cognized.

The object is measured by its appearance (*ābhāsa*), which is perceptual.

In Diṅnāga's philosophy, the two faculties of perception and imagination (*kalpanā*) encompass all of experience, worldly as well as transcendental. For the experience of the yogi is to be brought within the compass of perception if it is non-verbal and therefore direct, without any overlay of scriptural speculation.

Perception is not the only human faculty whose claims to truth are sustained by Diṅnāga, who enumerates inference (*anumāna*) as another *pramāṇa*. He uses the *Trairūpya* to divide the region of experience over which imagination holds sway into two parts. The *Trairūpya* selects out from all constructs of imagination those which fall within the field of inference. The definitions of both *pramāṇas,* perception and inference, are thus formulated by Diṅnāga in a formally adequate way, rendering them safe against Nāgārjuna's criticisms of circularity.

Before *anumāna* as a *pramāṇa* can be elevated to a transcendent status, and the epithet *pramāṇabhūta* justified, at least one more requirement has to be fulfilled, with respect to Diṅnāga's Buddhist commitments. Briefly stated, the gnosis which is the *prajñāpāramitā* experience,. has to be disconnected from language and not divided into subject and object. Universals, objects and relations, which enter into any inference, have to be reconciled with the pervasive Buddhist doctrine that ultimate reality is beyond the bounds of language. And yet these categories underlie even the simplest sentence recording sensory experience. Diṅnāga analyses 'This is blue' in terms of 'This belongs to the class of blue objects'. It becomes necessary therefore to discover in Diṅnāga's system a demonstration that universals, objects and relations can be reduced to perceptual units which alone are real; that they can be constructed on the basis of mere appearances beyond the subject–object distinction and prior to language. Diṅnāga himself explains this process of constructing complex wholes on the basis of smaller units with the help of an analogy:

> Therefore, the existence of the body as depending on the hand etc. could be one of the instances of '[the existence of something being] assumed in dependence on realities.' Thus if there were no such element of the universe as elements of the visible, [audible] etc., there would be no wholes constructed in them.[24]

Elsewhere one of us [RH] has argued at length that the Trairūpya has a reductionist programme built into it—that universals, objects and relations in the domain of *anumāna* are reducible to elements from the domain of perception.[25] That argument was supported by textual evidence from Diṅnāga providing details of the constructive process. Our belief that the entities involved in *anumāna* are reducible to *pratyakṣa* rests on the arguments of that work which are briefly outlined above. We go on to sketch some metaphysical implications that flow directly from this way of construing the *anumāna* doctrine in Diṅnāga's philosophy.

A religious composition attributed to Diṅnāga mentions a

threefold division of experience into *parinispanna*, *paratantra* and *parikalpita*.[26] *Parinispanna* is the level of the enlightened and *parikalpita* is the level of the unenlightened. *Paratantra* is a middle level which is indeterminate—it can be drawn into either one of the other two levels. We are inclined to assign *anumāna* to the *paratantra* level, as an instrument of our judgement about the external world accessible to those who are enlightened and also to those who are not. When used by the unenlightened, its strictly cognitive character is accompanied by pleasure and pain resulting from desire. The same instrument is available to those who are enlightened and know the external world without any accompanying desire, attraction or revulsion.

A passage from one of Diṅnāga's lost works indicates that the post-Nāgārjuna problem of how the Buddha could speak without commitment to fictions, was indeed a real concern of the time and not a modern, alien consideration:

> If you do not admit of the existence of the body, [the following will be concluded: Buddha,] the holder of the right view, should have preached [for example, the four methods of contemplation] in vain: moreover, there should be no person accused of holding nihilistic views of the universe and there should be no distinction in our deeds [with regard to their religious merits].[27]

V

The historical dialogue we have tried to reconstruct in this essay suggests a certain tension between two realms—the mundane world and that reality which transcends it. It also suggests a conceptual connection between that tension and the explanatory ideal of a *pramāṇa* theory: whether the theory was supposed to provide an account of transcendental experience, as Akṣapāda Gautama held, or of the affairs of the world, as Vātsyāyana believed. The tension arose, we suggest, as a result of Nāgārjuna's critique of the definition of *pramāṇas*. Having demolished various concepts underpinning existing explanations of the world, Nāgārjuna paid homage to 'dependent origination (*pratītyasamutpāda*) as pacified, silent and empty. However, the Mahāyāna ideal did not allow him to retreat into a noble silence. That ideal required peace, silence and emptiness to be brought to the attention of the world. Retreat into silence, however noble, had to make room for words. Thus Nāgārjuna declared:

dve satye samupāśritya buddhānāṁ dharmadeśanā
lokasamvṛttisatyaṁ ca satyaṁ ca paramārthataḥ.[28]

which could be translated this way:

The Buddhas' teaching of the doctrine takes simultaneous refuge in two truths: the veiling truth of the world and a transcendental truth.

So translated the stanza focuses on two levels of the doctrine—the noble silence and words about that silence. The words, which belong to a distinctive kind of meta-discourse, would include sermons of the Buddha and perhaps also Nāgārjuna's demonstration of the emptiness of all epistemic constructs. When Buddhas speak, their words cover the noble silence with a translucent veil.

Alternatively, Nāgārjuna's stanza could be translated into English without the word 'simultaneous':

The Buddhas' teaching of the doctrine takes refuge in two truths: the veiling truth of the world and the transcendental truth.

One word can make a world of difference; on the second translation, speech has for its subject neither silence nor the path to silence, but the world governed by causal laws and filled with mirages. Here the veil of words covers silence and negates it.

Candrakīrti's gloss on Nāgārjuna's stanza identifies three distinctive features of *samvṛtti*: it is ignorance (*ajñānam*); it is origination through mutual support; it is worldly usage based on language with its divisions of the real into knowledge and the known. His gloss leaves open the question of how ignorance (*ajñāna*) can be a guide to truth (*satya*).[29] This opens one more philosophical niche, into which Dharmakīrti's *pramāṇa* doctrine may be seen to fit.

The *Nyāyabindu (NB)* defines perception to include the modifying qualifier *abhrāntam* (non-deluded). This revisionist definition puts aside Diṅnāga's solution to the problem of Nāgārjuna, without offering a clearly viable alternative. Once again, Nāgārjuna's question looms: Is truth a property of *pramāṇas* or of the *prameya*? More picturesquely: Who is the father and who is the son? If the father begets the son who in turn begets the father they are caught in a regress not unlike Candrakīrti's account of *samvṛtti* as *paras-parasambhavam anyonyasamāśrayeṇa.*[30] It had been Candrakīrti's view that *anyonyāśraya* might indeed be a feature of a cyclical world covered with ignorance.

Vātsyāyana, faced with a similar philosophical problem, had made a similar response: while implicitly conceding the futility of a philosophical justification, he sought a pragmatic justification, declaring that actions based on right cognition met with success and not with infinite regress (*anavasthā*). Indeed *pramāṇas* were necessary conditions for achieving at least three of the four goals of life: duty, wealth and happiness.

The *Nyāyabindu* declares at the very beginning: *samyagjñāna-pūrvikā sarvapuruṣārthasiddhiḥ* (the realization of all human purpose is preceded by right knowledge).[31] The *PV* also attempts to connect *pramāṇas* and the actions they generate with the *artha*, or objects of those actions. Perception connects directly with objects, identifying appearances (*ābhāsa*) and their objects (*artha*) in a phenomenalist manner. Inferences are several steps removed from objects, and yet the information they impart can be a means to acquisition of the object.[32]

Faced with the problem of perceptual mistakes, Dharmakīrti declares that there is indeed no *artha* or real object in the case of a dark hair-like clump which clouds one's eyes. The appearance in this case is merely an appearance and not perception, for only perceptual knowledge of a real object results in perception.[33] Thus the *NB* definition of perception as non-deluded (*abhrāntam*) lurks within the folds of Dharmakīrti's pragmatic justification for *pramāṇas* in the *PV*.

The *PS* contains neither pragmatic considerations nor qualifying adjectives of perception such as *abhrāntam*, so it would seem that Dharmakīrti did not follow Diṅnāga's lead in this matter. He drew inspiration from another source—Vātsyāyana—and sought to concede the charge of philosophical circularity while offering a pragmatic justification for *pramāṇas*. In Section III we described Vātsyāyana's adoption of this position in full view of Nāgārjuna's criticism. In Section IV we sketched an interpretation of the *PS* as offering a formally adequate account of perceptual knowledge as well as epistemological criteria in the realm of the imagination. We have defended our view that both *PV* and *NB* fall short of Diṅnāga's treatment of these topics in *pramāṇa* theory.[34]

According to our reading, Dharmakīrti's justification of *pramāṇas* is circular: he holds an object to be real if it is cognized through a proper means of knowledge, and a means of knowledge to be proper if the objects it cognizes are real. The mutual support between *pramāṇa* and *prameyas* fits the mark of *samvṛtti* given by Candrakīrti as *anyonyāśraya.* And Dharmakīrti's *anumāna* shares two additional features that Candrakīrti ascribes to *samvṛtti:* from the ultimate standpoint it is false (*mithyā*), and it is connected with linguistic universals (*sāmānya*) that lie within the web of language. Language and the universals central to it, spring from a beginningless habit energy (*anādivāsanā*); but universals mask the nature of reality which consists of nothing but unique individuals.[35]

Dharmakīrti's account of *pramāṇas* addresses a question left open by Candrakīrti and his followers: how ignorance (*ajñānam*) can be characterized as *satya.* On this account, inference, and the universals which support it, mask the nature of reality. Therefore they spring from ignorance, even though they offer practical knowledge of objects which can be efficacious in action.

In the later *pramāṇa* doctrines, we sense a residual circularity which is not easy to accommodate, even on some lower level concerned with practice and mundane needs. Two-truth doctrines require truth on each level, and they require a philosophy meeting conditions of formal and material adequacy on each of those levels. One level may be thought of as lower than the other, but the same philosophical standards should apply uniformly to both. For this reason we have some doubt whether Nāgārjuna's problem of circularity can be put to rest by portioning discourse into a lower and a higher level of truth.

We have studied some ways, in which the particularities of several distinctive philosophers were accommodated into the generic doctrines of their schools. We have noted how some ancient commentators glossed over differences, blunting the thrust of counter-arguments. The consequences have been both historical and aesthetic—historical because this process may obscure the distinctive philosophical contributions of various members of the school; and aesthetic because it may distract us from the enterprise of seeing how details of each position fit together into a vision of the whole.

NOTES

1. *duḥkhajanmapravṛttidoṣamithyājñānānāmuttarottarapāye tadanantarāpāyād apavargaḥ. Nyāya-sūtra* of Gautama with the *Nyāyabhasya* of *Vātsyāyana*, ed. by Ganganath Jha, Poona (Poona Oriental Series), 1939 (hereafter *NBh* and *NS*), *NS*.I.i.2, p. 7.
2. *prajñāpāramitā jñānam adhayayaṃ sā tathāgataḥ sādhyātādarthyayogena tācchabdyaṁ dharmamārgayoḥ.* Frauwallner, 'Dignaga, sein werk und seine Entwicklung', *Weiner Zeitschrift für die Kunde Südasiens*, Band III, 1959, pp. 83–164.
3. See *Pramāṇavārttikam of Dharmakīrti*, ed. by Dwarikadas Shastri, Varanasi: Bauddha Bharati Series-3, 1968 (hereafter *PV*) II.72: *tasmān mithyāvikalpo'yam artheṣv ekātmatāgrahaḥ* ('Therefore, it follows that to grasp a unitariness among [diverse] objects is a false [act of the] imagining').
4. *pramāṇaprameyasaṁśayaprayojanadṛṣṭāntasiddhāntāvayavatarkanirṇayavādajalpavitaṇḍahetvābhāsachalajātinigrahasthānānām tattvajñānān niḥśreyasādhigamaḥ.* (*NS*.I.i.1, p. 2).
5. *pratyakṣānumānopamānaśabdāh pramāṇāni.* (*NS*.I.i.3, p. 10).
6. *ātmaśarīrendriyārthabuddhimanaḥpravṛttidoṣapretyabhāvaphaladuḥkhāpavargās tu prameyam.* (*NS*.I.i.9, p. 22).
7. *icchādveṣaprayatnasukhaduḥkhajñānāny ātmano lingam iti.* (*NS*.I.i.10, p. 24).
8. *tadatyantavimokṣo'pavargaḥ.* (*NS*.I.i.22, p. 32).
9. To identify the successive authors of the various sūtras that make up the *NS* is a complex task which has vexed scholars since S.C. Vidyabhusan, *A History of Indian Logic*, Calcutta, 1921.
10. *kim na khalu bhoḥ yāvanto viṣayās tāvatsu pratyekaṁ tattvajñānam utpadyate atha kvacid utpadyata iti. kaś cātra viśeṣah? na tāvad ekaikatra yāvad viṣayam utpadyate jñeyānām ānantyāt. nāpi kvacid utpadyate—yatra notpadyate tatrānivṛtto moha iti mohaśeṣaprasangah. na cānyaviṣayeṇa tattvajñānenānyaviṣayo mohaḥ śakyaḥ pratiṣeddhum iti.* (*NBh*.IV.ii.1, p. 287).
    The terms *pratiṣedha* and *nivṛtta* have the sense of negating or turning aside delusion (*moha*). They are aspects of the earlier vision that we have traced to Akṣapāda Gautama of knowledge as liberating.
11. *atha kathaṁ tattvajñānam utpadyate?* (*NBh*.IV.ii.38, p. 306).
    samādhiviśeṣābhyāsāt. (*NS*.IV.ii.38, p. 307).
    sa tu pratyāhṛtasyendriyebhyo manaso dhārakeṇa prayatnena dhāryamāṇasyātmanā saṁyogas tattvabubhutsāviśiṣṭah, sati hi tasminn indriyārtheṣu

buddhayo notpadyante. (Ganganath Jha, *The Nyaya-Sutras of Gautama with the Bhasya of Vatsyayana, and the Vartika of Uddyotakara,* rpt, Delhi: Motilal Banarsidass, 1984 [hereafter tr *NBh*] tr *NBh.*IV.ii.38, p. 1648.

12. *araṇyaguhāpulinādiṣu yogābhyāsopadeśah.* (*NS.*IV.ii.42, p. 308).

13. The causal link between the epistemological framework of Nyāya and the cycle is supplied by introducing the additional notion of *ahaṁkāra* (sense of I) as in the following: *doṣanimittānāṁ tattvajñānād ahankāra-nivṛttir iti uktam* ('true knowledge of the causes of character defects brings about the cessation of the sense of I'.) (*NBh.*IV.ii.38, p. 306, cf. *NBh.*IV.ii.1, p. 289).

14. Uddyotakara explores this in his introduction to *NS.*I.i.1. Vātsyāyana's description of the *pramātṛ* (the agent of the means of cognition) as *yasyepsājihāsāprayuktasya pravṛttiḥ sa pramātā* is significant in the context of *NS.*I.i.2 where *pravṛtti* (action) is listed as an element in the cycle which has to cease.

15. *kecit tu dṛṣṭāntam aparigṛhītaṁ hetunā viśeṣahetum antareṇa sādhya-sādhanāyopādadate—'yathā pradīpaprakāśaḥ pradīpāntaraprakāśam antareṇa gṛhyate tathā pramānāṇi pramāṇāntaramantareṇa gṛhyanta iti'.* (*NBh.*II.i.20, p. 88).

16. '*pratyakṣādīnām pratyakṣādibhir upalabdhāv anavastheti' cet, na. samvid-visayā nimittānām upalabdhyā vyavahāropapatteh. pratyakṣeṇārtham upalabhe anumānenārtham upalabhe upamānenārtham upalabhe āgame-nārtham upalabhe iti, pratyakṣam me jñānam, anumānikam me jñānam, aupamānikam me jñānam āgamikam me jñānam iti—saṁvittiviṣayaṁ saṁvittinimittaṁ copalabhamānasya dharmārthasukhāpavar gaprayojanas tatpratyanīkaparivarjanaprayojanaś ca vyavahāra utpadyate. so'yam tāvaty eva nirvartate. na ca vyavahārāntaram anavasthāsādhanīyam yena prayukto' navasthām upadadīteti.* (*NBh.*II.i.20, p. 89).

17. *pramāṇato' rthapratipattau pravṛttisāmarthyād arthavat pramāṇam. pramāṇamantareṇa nārthapratipattih. nārthapratipattim antareṇa pravṛttisāmarthyam* (*NBh.*I.i.1, p. 1).

18. The analogy between the *pramāṇa* and the lamplight is interpreted as exemplifying the philosophical position that the *pramāṇas* are self-justifying. This image is similar to Nāgārjuna's analogy of fire illuminating itself and the object. We connect both images and the philosophical positions they exemplify with the opponent whose position is described by Vātsyāyana in his remarks to *NS.*I.i.20. The opponent argues that if perception is justified by another perception, an infinite regress would follow (see note 15). However, Vātsyāyana takes the analogy as exemplifying his own position that *pramāṇas* are not self-justifying but in need of additional justification by another *pramāṇa* just as the light of the lamp is illumined by the eyes. Uddyotakara follows Vātsyāyana's interpretation.

19. Formal adequacy in this sense is a fairly modest claim about a philosophical system. We do not wish to claim material adequacy for Diṅnāga's definitions nor to defend any larger part of his phenomenalist project. Modern criticisms of latter-day phenomenalist programmes have shown how very difficult it is to carry them through in detail.

20. To say that Vātsyāyana offered a pragmatic justification is not to endorse that justification as sound. In the present case, a crucial premise is dubious: that success in action proves the truth of the underlying beliefs. Without going into the matter in detail, we note that throughout history people have acted on the basis of strange beliefs, often with considerable success in achieving their goals.
21. *yadṛcchāśabdeṣu nāmnā viśiṣṭo'rtha ucyate ḍittha iti.*
jātiśabdeṣu jātyā gauriti.
kriyāśabdeṣu kriyayā pācaka iti.
dravyaśabdeṣu dravyeṇa daṇḍi viṣāṇīti. (vr *PS*.I.i.3, p. 83).
22. *manobhrāntiviṣayatvād vyabhicāriṇaḥ.* (Masaaki Hattori, *Dignaga, On Perception, Being the Pratyakṣa-pariccheda of Dignaga's Pramanasamuccaya,* Cambridge: Harvard University Press, 1968, n.3.7, p. 122). Hattori concludes: 'This statement inclines us to believe that Diṅnāga attributed error to *manas.*'
23. *yadā tu bāhya evārthaḥ prameyas tadā*
viṣayākārataivāsya pramāṇam
tadā hi jñānam svasaṁvedyam api svarūpam anapekṣyārthabhāsataivāsya pramāṇam. yasmāt so'rthas
tena mīyate.
yathā yathā hy arthasyākāraḥ subhrāditvena jñāne pratibhāti (niviśate) tattadrūpam sa viṣayaḥ pratīyate. (Hattori, *Dignaga,* n.1.64, pp. 104–5).
24. Hidenori Kitagawa, 'A Study of a Short Philosophical Treatise Ascribed to Dignaga', *Sino-Indain Studies* 5: 3–4, rpt, Tokyo: Indo-koten Ronrigaku, 1965, p. 438.
25. See Radhika Herzberger, *Bhartṛhari and the Buddhists,* Dodrecht: D. Reidel and Co., 1986, chs 3 and 4.
26. *Piṇḍārtha* (Frauwallner, 'Dignaga', p. 141).
27. Kitagawa, *A Short Philosophical Treatise Ascribed to Dignaga,* p. 434.
28. Nāgārjuna, *Mulamadhyamakarika de Nagarjuna avec Prasannapada commentaire de Candrakirti,* St Petersburg: Bibliotheca Buddhica LV, 1931 (hereafter *MMK*), p. 492.
29. *samantāvaraṇam samvṛttih ajñānam hi samantāt sarvapadārthatattvāvacchādanāt samvṛttir ity ucyate. parassambhavanaṁ vā samvṛttir anyonyasamāśrayeṇa ity arthaḥ.* (The covering over from all sides is the veiling. Indeed it is ignorance which is called saṁvṛtti because it covers over from all sides the reality (tattva) of all categories of things. Alternatively, saṁvṛtti is mutual origination through mutual dependence—this is the meaning. Or, veiling is through mutual dependence—this is the meaning. Or, veiling is the worldly usage based on the relation between words and things (sanketa). Worldly usage has [the division between] names and what is named, knowledge and what is to be known as its characteristic feature.) vr *MMK,* pp. 492–3.
30. Nāgārjuna, *The Dialectical Method of Nāgārjuna (Vigrahavyāvartanī),* translated by Kamaleswar Bhattacharya, Delhi: Motilal Banarsidass, 1978, stanza L. See also n.29.

31. *NBh*.I.i.1.
32. *prāmāṇyam vyavahāreṇa*: *PV*.I.6, p. 6. Manorathanandin glosses the above as: *yasya sādhanajñānasya tādātmyād anubhūte'pi prāmāṇye saśankā vyavaharttāro'anabhyāsavaśād anutpannānurūpaniścayaḥ tatrārthakriyājñānena pramāṇaniścayaḥ.*
33. *PV*.II.1.
34. One might want to add that even in Diṅnāga's system perception is not a guarantee of truth, and wonder where the difference lies. The difference, we submit, lies in the way in which perceptual judgements are to be validated or invalidated. In Dharmakīrti's case the judgement: 'This is a sky flower' is invalidated on the basis of an argument which includes as premise a universal statement derived from the essential nature (*svabhāva*) of the sky flower. In Diṅnāga's case its invalidation would reside in the inability to produce an example fulfilling the requirements of the second and third rules of the *Trairūpya.*
35. *pararūpam svarūpeṇa yayā saṁvriyate dhiyā ekārthapratibhāsinyā bhāvān āśritya bhedinaḥ*. *PV*.III.68.
('Taking the support of distinct particulars, that understanding, which has the appearance of an unitary object, covers over the form of the other'.) Manorathanandin's gloss is conceptually similar to Candrakīrti's gloss on *samvritti* quoted in Note 29: *svapratibhāsena pareṣām svalakṣaṇānām rūpam sarvato vyāvṛttam saṁvriyate prāchhādyate, sā buddhiḥ samvṛttir ucyate* ('by its own appearance it covers over or completely hides the other, [namely] the absolutely distinct form of the *svalaksanas*'), p. 281.

REFERENCES

FRAUWALLNER, 'Diṅnāga, sein Werk und seine Entwicklung', *Weiner Zeitschrift für die Kunde Südasiens* Band III, 1959, pp. 83–164.

HATTORI, Masaaki, *Diṅnāga, On Perception, Being the Pratyakṣa-pariccheda of Diṅnāga's Pramāṇasamuccaya* (Harvard University Press, Cambridge, 1968).

HERZBERGER, Radhika, *Bhartṛhari and the Buddhists* (D. Reidel and Co., Dordrecht, 1986).

JHA, Ganganath, *The Nyāya-Sūtras of Gautama with the Bhāsya of Vātsyāyana, and the Vārtika of Uddyotakara* (rpt, Motilal Banarsidass, Delhi, 1984).

KITAGAWA, Hidenori, 'A Study of a Short Philosophical Treatise Ascribed to Diṅnāga', *Sino-Indian Studies*, vol. 5, nos 3–4 (rpt, Indo-koten Ronrigaku no Kenkyu, Tokyo, 1965).

NĀGĀRJUNA, *Mūlamadhyamakarikā de Nāgārjuna avec Prasannapadā commentaire de Candrakirti* (Bibliotheca Buddhica LV, St Petersburg, 1931).

Nāgārjuna, *The Dialectical Method of Nāgārjuna* (*Vigrahavyāvartanī*) translated by Kamaleswar Bhattacharya (Motilal Banarsidass, Delhi, 1978).

Nyāyabinduṭīka, ed. by Dalsukhbhai Malvania, (Tibetan Sanskrit Works Series, vol. II, Patna, 1955).

Nyāya-Sūtra of Gautama with the Nyayabhāsya of Vātsyāyana, ed. by Ganganath Jha (Poona Oriental Series, Poona, 1939).

Pramāṇavārttikam of Dharmakīrti, ed. by Dwarikadas Shastri (Bauddha Bharati Series-3, Varanasi, 1968).

TUCCI, Giuseppe, *Opera Minora*, parts I and II, Studi Orientali Publicati (A Cura Della Scuola Orientale, Roma, 1971).

VIDYABHUSANA, S.C. *A History of Indian Logic*, Calcutta, 1921.

Chapter 16

Śaṅkara on *Satyaṃ Jñānam Anantaṃ Brahma*

Julius Lipner

Words strain,
Crack and sometimes break, under the burden,
Under the tension, slip, slide, perish,
Decay with imprecision, will not stay in place,
Will not stay still

T.S. Eliot: ***Four Quartets, Burnt Norton***

As a master-poet, Eliot understood the power and elusiveness of words. Yet it was precisely in exploring and indeed seeking to control their elusiveness that Eliot made words puissant. The philosopher and theologian no less. In pondering their deep matters, if they cannot communicate a sense of reality and truth, they cease to be successful architects of the word. Their verbal constructions—strained, cracked, tensioned to breaking point—come crashing down, will not stay in place. This essay is a tribute to two thinkers—a classical Hindu theologian and a modern Indian philosopher—who have left us an enduring legacy of words and ideas. Their writings have given us the space to explore and understand our use of words and so to understand and explore ourselves. Bimal Matilal the philosopher would, I know, have approved of a sustained attempt to examine Śaṅkara's understanding of how words should be used of the supreme reality. And since Bimal Matilal was a friend of many years I am especially glad to be able to make this contribution.

The text in question appears under *Taittirīya Upaniṣad* (*TaiUp.*) II.1.1 'Om' says the Upaniṣad, 'The knower of Brahman attains the highest. So it has been declared: Brahman is reality (*satyam*), knowledge (*jñānam*), infinite (*anantam*). He who knows it as placed in the depth, as heaven in the highest, attains every desire and the knowing Brahman as well . . . '. It seems that what the Upaniṣad itself presents as an important declaration, '*satyaṃ jñānam anantaṃ brahma*: Brahman is reality, knowledge, infinite', has been regarded from earliest times by Vedāntic theologians as a *śodhaka-vākya*, a critical statement which when interpreted both reveals and determines one's theological standpoint. Śaṅkara gives the declaration elaborate treatment, and it may well be that the protracted analysis of this magisterial early Vedāntin helped create a hermeneutic precedent for subsequent establisher of Vedāntic schools.

Śaṅkara's interpretation of the declaration has not gone unremarked by modern scholars. In general it seems that one of two positions has been adopted on his exegesis. Either it is held that Śaṅkara treated the declaration as primarily if not exclusively a *definition* (*lakṣaṇa*) of Brahman,[1] or it is argued that for Śaṅkara it was chiefly a crucial instance of *oblique predication* (*lakṣaṇā*) about the nature of Brahman.[2] Indeed there is and has been much confusion in presenting the matter this way. In fact, I do not think that we do justice to the subtlety of Śaṅkara's interpretation by thinking of it in terms of this uncompromising disjunctive wherein one is required to choose between two mutually exclusive alternatives. It seems to me that Śaṅkara understood the declaration as—more or less equally—both a definition *and* an instance of oblique predication, and the purpose of this essay is to explore this possibility.

Our exegetical problem arises through Śaṅkara's persistent use of the expression *lakṣaṇārtha-*. This Sanskrit compound, which consists of two terms, may be broken up in either of two ways, viz. (a) *lakṣaṇa* + *artha-*, which can mean 'intended as (*artham*)/in the sense of, a definition (*lakṣaṇa*)', and (b) *lakṣaṇā* + *artha-*, the meaning here being 'intended as (*artham*)/in the sense of, oblique predication (*lakṣaṇā*)'. Since when either *lakṣaṇa* or *lakṣaṇā* forms a compound with *artha-*, the result morphologically is identical (viz. *lakṣaṇārtha-*), we are forced to attempt a resolution of any resulting semantic ambiguity by a close attention to the context. Now, as we shall see, Śaṅkara says repeatedly that the declaration *satyaṃ jñānam anantaṃ brahma* is *lakṣaṇārtha-*. What precisely does he mean by this? That the declaration is intended as a definition of Brahman, that it is meant to describe the nature of Brahman in some oblique or figurative way? As we have noted, modern scholars, in their interpretation of Śaṅkara, tend to see these senses as mutually exclusive alternatives. I shall argue that this is a mistake: for Śaṅkara these are not mutually exclusive alternatives. He intended *both* senses of *lakṣaṇārtha-* to apply, but in an exegetically calculated way which I shall clarify in the course of my analysis.[3] This analysis, of course, will have to condense Śaṅkara's exegesis along the lines that suit our particular inquiry.

The statement, 'The knower of Brahman attains the highest' says Śaṅkara, encapsulates the teaching of this whole section of the Upaniṣad. He continues as follows. Since to attain Brahman one must know Brahman, the Upaniṣad describes Brahman as 'reality, knowledge, infinite'. This description is *lakṣaṇārtham.* This is the first time this expression occurs, and it is not long into Śaṅkara's exegesis. What exactly does it mean here? Clearly at this stage of the exegesis, the predominant sense is 'intended as a definition'.[4] For to attain Brahman, the knower of Brahman—the *brahmavit*—needs to have an efficacious description of his goal, and the most efficacious description of Brahman is a definition (*lakṣaṇa*). For Śaṅkara a definition is to the point, for a definition is capable of referring to the proper form (*svarūpa*) of the *definiendum* or thing-to-be-defined by excluding it from everything else.[5]

Thus for Śaṅkara, a definition has a positive and a negative function. It is not an exhaustive description of the *definiendum*, but it describes literally what the *definiendum* essentially is (the positive function) so as to single it out absolutely, viz. intimate what it is not (the negative function).[6] We shall see how Śaṅkara implements this understanding of the definition as his exegesis proceeds.

However, nothing we have said so far prevents us from interpreting this instance of *lakṣaṇārtham also* in the second meaning specified, though at this stage in a recessive rather than dominant sense. To understand why Śaṅkara could well have intended this, we must inquire into two features of his underlying thought: (i) his understanding of Brahman as *advaya* or non-dual—as the sole, ultimate, ineffable reality, realization of which sublates all awareness of difference, and (ii) his position on *lakṣaṇā* or oblique predication.

(i) 'We maintain', says Śaṅkara, 'that this is the ascertained meaning of the whole Vedānta: that at all times we are but Brahman who is homogeneous in nature, non-dual, unchanging, unborn, undecaying, undying, immortal, tranquil, and of the essence of the Ātman'.[7] The ubiquitous awareness of difference in our lives and the good and bad practical consequences that follow—love and hate, peacefulness and strife, wanting, needing, competing, striving, tolerating, compromising—are only provisionally real; from the final standpoint they are of the stuff of *māyā* (Śaṅkara uses this concept occasionally) and *avidyā*, viz. deceptiveness and ignorance, which tend to lead us away from our true, final goal—the realization of what is already the case: our essential, spiritual identity with Brahman. By accredited means, viz. an appropriate birth (if not in this life, then in the next), tutelage and life-style, we are to realize this identity and achieve our spiritual destiny. We are to become *brahmavidaḥ*, knowers/realizers of Brahman, or of what is ultimately the same reality, of the *Ātman*, of our truest Self. When this happens, the divisiveness of the experience of difference is sublated and the serpents of *māyā* and *avidyā*, amid whose oppressive coils the unenlightened continue to live their lives, have lost their sting.[8]

What or who,[9] then, is Brahman, the 'highest'? Śaṅkara answers as follows:

It stands to reason that Brahman cannot be plainly expressed (*ucyate*) [even] by such [comprehensive] words as 'being' and 'non-being', for words are ordinarily used to reveal some object and when heard by their hearers conventionally make known their object by reference to genus, action, attribute or relation. We know this from everyday experience. Thus 'ox' and 'horse' denote by reference to genus, '(he) cooks' and '(he) reads' by reference to action, 'white' and 'black' by reference to attribute, and 'wealthy' and 'ox-owner' by reference to relation. But Brahman belongs to no genus, whence It cannot be plainly expressed by such words as 'being', nor does It possess attributes by which It may be plainly expressed by words denoting attributes, for It is attributeless (*nirguṇa*); nor is It expressible by action-words since It is actionless, for scripture says that Brahman is 'without parts, without action, serene' [*Śvetāśvatara Upaniṣad* VI.19]. And since Brahman is non-relational because It is one, and because It is non-dual and not the object [of anything] but is the Self [of everything], it is proper to say that It cannot be plainly expressed (*ucyate*) by any word, whence scriptural passages [such as *TaiUp.* II.9.1 say, when referring to Brahman:] 'Words turn back from It . . .'.[10]

So Brahman in itself is the Absolute as relationless and ineffable. Very well. There is a great deal more to say, of course, about Śaṅkara's theological viewpoint, especially concerning our own existence in a fragmented, differentiated world, and our relationship to Brahman. But we mustn't lose sight of the point here: that Brahman is the transcendent, utterly homogeneous One 'from which words turn back' rightly. Yet in our text the Upaniṣad—sacred scripture, which cannot be void—actually purports to *describe* Brahman, and through such *positive* terms as 'reality' and 'knowledge'. You perceive Śaṅkara's quandary: either Brahman is not the non-dual, inexpressible One or holy scripture is void in this formulation. When confronting the horns of a dilemma, one tries either to render one or both of the horns ineffectual, or to escape from between the horns. To appreciate Śaṅkara's strategy here, we must now consider his understanding of *lakṣaṇā* or oblique predication.

(ii) We do not translate *lakṣaṇā* as 'figurative predication' for a good reason. For Śaṅkara, *lakṣaṇā* does not *necessarily* refer to a trope or figure of speech like a metaphor, in which *secondary* senses of words apply.[11] In his commentary on the *Brahma Sūtras*, under 3.3.9, he comments on *lakṣaṇā* as follows:

> Where *lakṣaṇā* is concerned, there is either closeness or remoteness [of the primary sense of the relevant word/s in relation to the assumed meaning of the word/s]. In the case of superimposition (*adhyāsa*), because cognition of one object is superimposed on something else, the *lakṣaṇā* is remote. But in the case of specifiers (*viśeṣaṇa*), as when a word expressing a whole refers to a part, we have proximate *lakṣaṇā*. For words referring to wholes are commonly applied to the parts, as in the case of cloth or villages [in that we say, 'The cloth is burnt', when only a part of it is, and 'The village is on fire', when only some of its huts are].[12]

In the article cited (see note 11), I have tried to show that according to Śaṅkara what we call metaphorical predication is an instance of remote *lakṣaṇā*. Here the 'primary sense' of a term—the core sense that comes to the fore when a term is used *literally*—say, that of 'lion' in the statement, 'Devadatta was a lion on the battlefield', is occluded in favour of another, related sense applied to the referent (viz. Devadatta). Devadatta was not *literally* a lion on the battlefield; he displayed the characteristics of a lion. And it is only because everyone understands this perfectly well, viz. that the word 'lion' does not apply to Devadatta literally, that the metaphor works. In other words, it is only because the word's primary sense (which in the first instance or properly, applies to a flesh-and-blood lion) is understood not to apply to the referent (Devadatta), that a secondary related sense (or senses, viz. bearing on prowess, temperament, courage, fearsomeness, etc.) applies. Here there is a crucial hiatus between primary and secondary senses in respect of the relevant referents, and the *lakṣaṇā* or oblique predication is, according to Śaṅkara's terminology, 'remote'. In this instance, oblique predication is *figurative* predication (Devadatta is signified by a figure of speech—a metaphor).

But for Śaṅkara all oblique predication, that is, all *lakṣaṇā*, is not figurative predication. Sometimes we can succeed in referring obliquely without abandoning wholly a term's core or proper sense (*svārtha*).[13] Thus we say, 'The village [a whole entity] is on fire' when in fact a small part is burning—a few huts. Insofar as the burning huts are an integral part of the whole village, the core sense of the word 'village' is not entirely abandoned, and a 'whole-term' can function semantically in respect of a part of that whole. This is still 'oblique' predication in that a not entirely straightforward use of language occurs, as would be the case in 'Some huts of the village are on fire'. But there is 'closeness' rather than 'remoteness' in this instance of *lakṣaṇā* by virtue of a particular use of 'specifiers' (*viśeṣaṇas*): because the term specifying the whole (e.g. 'village') refers to the part (some huts), the core or proper meaning (*svārtha*) of the specifying term (viz. 'village') is not entirely abandoned in its application to the implied subject, viz. the part (some of the village huts). The *intrinsic* semantic overlap makes this an instance of proximate rather than remote *lakṣaṇā*. Let us return now to Śaṅkara's exegesis of the *Taittirīya* formula.

In the light of our discussion so far, our contention that Śaṅkara could well have intended both senses of *lakṣaṇārtha-* to apply even in its first occurrence perhaps becomes more plausible, granted, however, that the definitional sense is here dominant. The formulation, it is claimed, defines Brahman, that is, describes Brahman's proper form. But Brahman, we have seen, is for Śaṅkara transcendent and more or less ineffable. How then can such positive terms as *satyam* ('reality') and *jñānam* ('knowledge') describe Brahman except in some oblique way, i.e. by *lakṣaṇā*? In this first instance of *lakṣaṇārtha-* then, I suggest that Śaṅkara stresses the first, viz. definitional, sense, but signals the second or oblique sense. The subtlety of his continuing exegesis in this direction will become more apparent.

Śaṅkara first insists on two formal features of the expression as understood by him: (i) that by means of specifying or attributive words (*viśeṣaṇas*) it purports to specify an object (*viśesya*), and (ii) that it is an instance of the grammatical role of *sāmānādhikaraṇya.*

(i) Brahman as the thing-to-be known (*vijñeya*) is the object referred to (*viśesya*) by the three *viśeṣaṇas: satyam, jñānam* and *anantam.*

(ii) In other words, the statement is a semantic unit, enabling it to function as a definition. Its semantic unity is expressed grammatically by the specifier-terms having the same case-ending as the term to which they are related as specifiers, viz. the subject-term, *brahma.* This is what *sāmānādhikaranya* implies in its grammatical aspect: semantic unity manifested by the same case-ending. But there was also an ontological aspect to the traditional understanding of *sāmānādhikaraṇya* (which means literally, 'having/bearing on the same locus'). Traditionally, *sāmānādhikaraṇya* was defined as follows: 'The referring to one and the same object by [a statement of] two or more [specifying] words which have different grounds for their occurrence'.[14] In other words, the specifying terms must not be synonyms. Does this imply that the referent of a defining *sāmānādhikaraṇya* is necessarily differentiated essentially (because the attributive terms are not synonymous)? Śaṅkara, as we might expect, said no; Rāmānuja, a later major theological rival, said yes.[15] In any case, in his exegesis, Śaṅkara stresses the grammatical aspect of the definitional *sāmānādhikaraṇya* we are considering, as intimating a semantic unity.[16] The definition, he concludes, purports to describe Brahman by the three terms 'reality', 'knowledge', and 'infinite' just as the statement 'The lotus is blue, large and fragrant' purports to describe the lotus as simultaneously 'blue' (*nīlam*), 'large' (*mahat*) and 'fragrant' (*sugandhi*). So, by virtue of their specifying, viz. descriptive function, there is a similarity between the statements 'Brahman is reality, knowledge, infinite' and 'The lotus is blue, large and fragrant'. Both are semantic unities; both contain sets of attributives (*viśeṣaṇas*) which are not synonyms. Though it is to his advantage to advert to this important similarity at this juncture of the discussion, by the time he concludes the exegesis Śaṅkara will have been at pains to show that there is greater reason to *contrast* these two statements than there is to compare them. The process begins by bringing the *pūrvapakṣin* or imaginary critic into play.

The two statements (viz. about Brahman and the lotus) can hardly be compared in terms of their use of *viśeṣaṇas* or specifiers, says the critic, because 'it is only when a number of substances of the same kind are susceptible of being referred to by more than one *viśeṣaṇa*, that [predicating] the *viśeṣaṇa* has significance'.[17] Thus lotuses may be blue or red; in that case, according to prevailing circumstances, it makes sense to specify that the lotus is blue, for in other circumstances, the lotus referred to *could* have been red. A correctly predicated *viśeṣaṇa* is the right attributive from among a number of potentially competing attributives. But Brahman as the highest (*param*), the supreme reality, is unique; (to use extensional language involving quantification) there is not a class of Brahmans from which a particular Brahman may be distinguished.[18] Does it make sense, then, to speak of *viśeṣaṇas* in reference to Brahman in the context of the Upaniṣadic statement?

Yes, says Śaṅkara, because the *viśeṣaṇas* in that context function 'mainly *lakṣaṇārtha*-' and not as specifiers. *Viśeṣaṇas* in their ordinary usage may distinguish between members of the same class, but *viśeṣaṇas* in the case of a definition (the term *lakṣaṇa* is used), distinguish their subject *from everything else (sarvata eva)*, and, 'we have said that the statement [about Brahman] is *lakṣaṇārtham*'.

Śaṅkara has made an important move here. He is distinguishing between the use of *viśeṣaṇas* in ordinary usage (as in the example of the lotus), and *viśeṣaṇas* in the context of a definition. In the latter case, in accordance with his understanding of the negational import of the definition (see earlier), *viśeṣaṇas* have primarily an excluding rather than a specifying function. They distinguish the *definiendum* more by saying what it is not than by saying what it is. They still specify, but not in the straightforward sense of everyday perceptual usage. It is at this point that Śaṅkara has given a significantly negational twist to his exegesis of the Upaniṣadic statement about Brahman. Thus while it may still be true to say that *lakṣaṇārtham* here should be understood mainly in the sense of 'intended as a definition', its second meaning has in fact been implicitly reinforced. One detects almost a hint of irony in the use of *lakṣaṇārtham* in this instance ('And we have said that the statement is *lakṣaṇārtham*!'), rather like Mark Antony's continuing references in *Julius Caesar*, when he spoke after Caesar's murder, to Brutus and the others as being 'honourable men'. Śaṅkara proceeds to build on the imputed negational content of the Upaniṣadic statement about its unique, classless subject.

In fact, Śaṅkara continues, in respect of their semantic import, the *viśeṣaṇas satyam*, etc. are not related directly to each other but only indirectly, for they bear on their subject insofar as they are subordinate to an ulterior purpose, viz. the overall sense of the whole. In other words, the semantic unity of the statement is a function of each predicate's direct reference to the joint subject, so that the overall sense of the statement derives from the integrated meaning of three constituent sub-statements, viz. 'Brahman is reality', 'Brahman is knowledge', and 'Brahman is infinite'.[19]

Śaṅkara has now created the opportunity of driving home the negational import of the Upaniṣadic definition of Brahman. 'That whose nature is ascertained to be non-mutable is the real (*satyam*)' he affirms; 'that whose nature is known to be inconstant, is unreal (*anṛtam*). Hence change (*vikāra*) is unreal. . . . So "Brahman is reality" means that Brahman is unchanging'. And this means that it is natural for Brahman to be the original unchanging cause, the prime mover of everything, rather than a mutable effect. Is causal agency (*kāraṇasya kārakatva*) intrinsically unconscious; is Brahman an unconscious prime mover?[20] To obviate this doubt, continues Śaṅkara, we need the second proposition derived from the definition: *jñānaṃ brahma*, Brahman is knowledge. But 'knowledge' here must not be understood as pertaining to an empirical knower (*jnātṛ*) which is eminently mutable (*vikriyamāṇa*), else Brahman could not be characterized as either reality or infinite (since that is infinite which is utterly devoid of division[21]). 'Being a knower' is bound up with the divisions of what is known and the act and content of knowing. Knowledge in the definition is 'Knowing/Awareness *per se*', for the term *jñāna* denotes a state of being (*bhāvasādhana*).[22]

This reference to a 'state of being' is the basis for saying that in Brahman knowledge and reality are one; Brahman's knowing is identical with Brahman's being. For Brahman to be is to know, and to know is to be; there is no principle of division whatsoever in Brahman. This is what it means to say that 'Brahman is infinite' (*anantaṃ brahma*)—the third derivative proposition. And this is why the text adds 'infinite'.

Well, says our patient critic, with perhaps a gleam in his eye, in that case the *viśeṣaṇas satyam* and *jñānam*, at any rate, seem to have become synonyms. If Brahman's infinitude is such that for Brahman to be is to know and so on, then the terms *satyam* and *jñānam* lose their individual (empirically derived) senses. They cease to function as purveyors of meaning. In short, Śaṅkara's interpretation has not only violated the grammatical rule of *sāmānādhikaraṇya* which he himself has adduced, but it has also rendered scripture vacuous. The Upaniṣadic text has become senseless (*śūnyārtha*) like the following piece of nonsense: 'Having bathed in the waters of a mirage and donned a crown of flowers made from air, the barren woman's son marches on, carrying a bow fashioned from a hare's horn'!

'Not so', answers Śaṅkara,

> because the Upaniṣadic statement is *lakṣaṇārtha-*. And we have said that though they have a qualifying function, *satyam* etc. are primarily *lakṣaṇārtha-*. A defining statement is void only when the *definiendum* doesn't exist, but because the Upaniṣadic statement is *lakṣaṇārtha-*, we believe that it is not senseless. Moreover, since the terms *satyam* etc. retain their specifying function they do not abandon their proper

meanings (*svārtha*). If they were meaningless they could not act as controls (*niyantṛ*) of their subject; but because they are meaningful they function as [semantic] controls of their subject Brahman [by distinguishing it] from all other subjects with attributes opposed to it. Indeed, the word *brahma* is meaningful in its own right. So in the Upaniṣadic statement, the term *anantam* functions as a specifier by virtue of negating finitude, while the terms *satyam* and *jñānam* act as specifiers in so far as they retain their proper meanings.[23]

That Brahman exists we know from faith in the scriptures, which cannot talk about a reality which does not really exist, as if it were existing. And *brahman*, which derives from *bṛ*, *bṛhati* = to grow, increase, means 'the Great One'. Thus the Upaniṣad is engaged in defining something real, to wit, the highest reality. It does this by means of the three *viśeṣaṇas*, *satyam*, *jñānam* and *anantam*. All three bear upon or semantically control (*niyantṛtva*) their subject, but they do so in different ways. Here Śaṅkara distinguishes between the specifying function of the positive terms, *satyam* and *jñānam*, and the function of the negative term, *anantam* (infinite).

Satyam and *jñānam*, he contends, retain their proper meanings (*svārtha*). So they are not synonymous; but, as we have seen, they must be eviscerated of that empirical content which fails to do justice to the utter and simple (viz. divisionless) perfection of their referent, Brahman. In other words, though they are positive semantic controls, they do not connote any kind of existential or epistemic mutability. This emptying or kenotic function is signalled by the use of the term *anantam*. By negating any principle of division (*antavattva*) in Brahman, the term 'infinite' functions as a negative control and requires of us an epistemically purified understanding of the nature of Brahman. Brahman is real, but not real in the way empirical objects are; Brahman is knowing, but not in our way of knowing. Brahman, in fact, is infinite reality, infinite knowledge. Brahman is reality-knowledge *per se*. What exactly does this mean? The answer is that the epistemic content of this expression is methodologically elusive. The negative control *anantam* does not permit a straightforward empirical understanding of the terms *satyam* and *jñānam*—Brahman transcends that; nevertheless their empirical content does provide some positive purchase, projected by *anantam* into a purified, trans-empirical if systematically elusive insight. It is thus, through the correct hermeneutic process, that we are led to make an *informed* speech- and knowing-act concerning Brahman; in other words, to read the scripture aright.

So the exegesis of the Upaniṣadic definition has become predominantly but not entirely negational—negational, that is, of its empirical mutational content. For Śaṅkara insists (he must, else our understanding of Brahman would be vacuous and grave violation of hermeneutic rules would ensue) that the literal meanings (*svārtha*) of the positive *viśeṣaṇas, satyam* and *jñānam,* be retained. But in their overriding negational context how can it be doubted that they are to be understood except in an *oblique* sense, viz. in the sense of (proximate) *lakṣaṇā* distinguished earlier, wherein the terms are used literally yet not straightforwardly? Must we not say, that in the flurry of *lakṣaṇārthas* contained in the excerpt above, the second ('oblique') meaning of *lakṣaṇārtha-* comes to the fore, in tension with its definitional import? The impending climax of Śaṅkara' exegesis will vindicate our claim.

The critic is now directed to probe deeper into what is involved by referring to Brahman as 'knowledge' (*jñāna*). His argument, in brief, is as follows: Scripture teaches that one can know the innermost self (*ātman*) to be Brahman. In other words, the essence (*svarūpa*) of the knowing self is Brahman. But the knowing of the knowing self, as all acts of knowing, is a changeable, impermanent, dependent act. To know something entails dependence on the object of knowledge and other limiting conditions. The sense of the verbal root (*dhātvartha*), 'to know', then, is inalienably bound up with mutability, transience and dependence. The verb 'to know', in all its various cases and tenses, is invariably used in this way. How then can the scripture refer to Brahman as 'knowledge' without also implying impermanence and relationships of dependence in Brahman?

Śaṅkara answers that the inner Self's, viz. Brahman's knowing is not like our knowing, that insofar as Brahman's knowing or knowledge is non-different from Brahman's very essence (*svarūpa*) it *is* that very essence in the way illumination and warmth are of the very essence of the sun and fire respectively. No question of impermanence and dependence arises here. In fact all talk of agency, viz. in respect of acts of knowing etc., *vis-à-vis* Brahman or the inner Self, is *indirect* talk (*upacārāt*). 'Because knowledge is Brahman's very essence, in so far as being a knower is not distinguishable from Brahman's essence, and in so far as Brahman does not depend on causes and instruments of knowing, we must conclude that Brahman's knowledge is eternal (and changeless). Hence [in the statement 'Brahman is knowledge'] empirical connotations of the verbal root ['to know' etc.] do not apply to Brahman, for agency is not of Brahman's essence.'[24]

Śaṅkara's exegesis now reaches its denouement. Because, for reasons given, Brahman cannot be said straightforwardly to be an agent of knowledge (*jñānakartṛ*), 'Brahman cannot directly be expressed (*vācya*) by the word "knowledge" ... Brahman is intimated (*lakṣyate*), not plainly expressed (*na tūcyate*) by the word "knowledge", because Brahman is devoid of attributes [classifiable] as genus and the like which legitimate the straightforward/conventional application of words (*śabdapravṛttihetu*). Similarly for the term *satyam*. Because Brahman's essence is totally devoid of every distinction, Brahman is intimated (*lakṣyate*) by the expression "Brahman is reality" in so far as the word *satyam* bears generally on external (viz. empirical) reality. So Brahman is not directly expressible (*vācya*) by the word *satyam*'.[25]

So here we have it. Note the semantic contrast in this passage between the words *ucyate* (and *vācya*-) and *lakṣyate*. As the context makes very clear, *ucyate* is used of conventional referents, viz. referents that are empirically distinguishable and categorizable in terms of genus, species, action, property, etc. It is the word pointedly used by Śaṅkara to refer to conventional being in contrast to the reality that is Brahman. See, for example, the passage excerpted from Śaṅkara's commentary on *Gītā* 13.12 (note 10), where *ucyate* is clearly used in the context of this contrast. It would hardly make good sense to translate *lakṣyate* by 'is defined' in the passage above.[26] On the contrary, Śaṅkara has brought his exegesis to such a pass as to clearly deliver it finally of the second meaning of *lakṣaṇārtha*- distinguished earlier, viz. that Brahman is indirectly signified by the Upaniṣadic definitional formulation. In other words, Brahman can only be defined by oblique signification. But the oblique signification here is not metaphorical. Since the positive *viśeṣaṇas* do not abandon their proper meanings (*svārtha*),[27] they are to be understood literally yet in the purified sense described earlier, that is, in an overwhelmingly apophatic context.[28]

Must we not conclude, then, that our claim has been vindicated, and that from the beginning it was intended that *lakṣaṇ-ārtha-* should be understood in both senses, viz. the definitional and that of proximate *lakṣaṇā*, though in the subtly evolving way we have sought to analyse? There is much to gain, psychologically and theologically, from such gradual elucidation of meaning; exegetically it establishes plausibility and insight by its cumulative effect. It could also be that Śaṅkara was not keen to stress, at least at the outset, his oblique rendering of what appear to be direct scriptural passages; hermeneutic convention had it that indirect meanings could be resorted to only after straightforward interpretations were seen to be inappropriate (hence the emphasis on *proximate lakṣaṇā*, where literal meaning is retained).

Śaṅkara rounds off his exegesis most gratifyingly for our purposes. 'Therefore the words *satyam* etc., juxtaposed as they are and [semantically] controlling and being controlled with mutual effect, deny of Brahman what is conventionally expressed by the words *satyam* etc. and so become *lakṣaṇārtha-* [viz. function as a definition by means of proximate *lakṣaṇā*]. Thus is established such scriptural passages as [*TaiUp.* II.7.1], "Words turn back from It, for It is unattainable with the mind", . . . and that Brahman is inexpressible, and that Brahman cannot be the object of statements like the one about the blue lotus.'[29]

Śaṅkara the advaitin has plausibly interpreted, by a magisterial unfolding of the text, a *prima facie* positive scriptural statement about Brahman so as to allow it an advaitic or monistic tenor. Yet linguistic appearances have been respected (for the formulation is a positive statement about Brahman) and theological justice has been done (for we are afforded a positive, if elusive glimpse into the nature of Brahman), in that, where and as appropriate, the formulation retains a positive semantic purchase. Similar descriptions of the highest reality (e.g. *BAUp.* III.9.28: *vijñānam ānandaṃ brahma*, 'Brahman is consciousness, bliss'), one supposes, are to be understood in like manner. Brahman emerges, in Śaṅkara's exegesis, as utterly transcendent yet not as utterly unknowable. Our salvific path, through the interaction of scriptural understanding and contemplation, is assured. Our highest goal—

the *param*—is not beyond reach. But if one understands Śaṅkara aright, when all is said and done, one is left revering the ultimate mystery of the transcendent with *neti, neti*: 'It is not so, not so', for, 'Words, after speech, reach into the silence' (T.S. Eliot: ***Four Quartets, Burnt Norton***).

NOTES

1. See, e.g. M. Biardeau, 'La définition dans la pensée indienne', *Journal Asiatique* 245, 1957; C. Bartley, 'Interpreting "Satyaṃ Jñānam Anantaṃ Brahma"', in N.J. Allen, R.F. Gombrich, T. Raychaudhuri and G. Rizvi, eds, *Oxford University Papers on India, Vol. I, Part I*, Delhi: Oxford University Press, 1986, and apparently, J.M. van Boetzelaer, *Sureśvara's Taittirīyopaniṣadbhāṣyavārtikam*, Leiden: E.J. Brill, 1971, pp. 53–4.
2. See e.g. O. Lacombe, *L'Absolu selon le Vedānta*, Paris: Geuthner, 1937, pp. 79ff.
3. My position owes much to the teaching of my former mentor in Indology, Richard Desmet s.j. For a fine résumé of a lifetime's study of Śaṅkara, see his Pratap Seth Endowment Lecture, 1987, 'Forward steps in Śaṅkara Research' (*Darshana International* 26:3), in which the Upaniṣadic formulation is noted.
4. Or, 'intended as providing defining characteristics'. Even if one adopts this meaning, the ensuing discussion is not materially affected.
5. *sūtritasya brahmaṇo 'nirdhāritasvarūpaviśeṣasya sarvato vyāvṛttasvarūpaviśeṣasamarpaṇasamarthasya lakṣaṇasya abhidhānena svarūpanirdhāraṇya* . . . ; Sanskrit text taken from *Ten Principal Upanishads with Śaṅkarabhāṣya* (hereafter *TPU*), Delhi: Motilal Banarsidass, 1964, p. 282. Unless stated otherwise all translations in this essay are by the author.
6. In 'The Intensional Character of *Lakṣaṇa* and *Saṃkara* in Navya-Nyāya', *Indo-Iranian Journal* 8, 1964–5, pp. 85ff, B.K. Matilal says that in early Nyāya, 'the purpose of definition (*lakṣaṇa*) was to differentiate an entity from that which does not possess the nature or essence (*tattva*) of that entity. . . . In the new school ['usually considered to have begun . . . in the 14th century AD'] . . . the purpose of definition is to distinguish the definiendum (*laksya*) from all entities that are different from it (*itaravyāvartakatvam*)', p. 86. By stating that a definition of Brahman is meant to distinguish it from everything else (*sarvato vyāvṛtta-*; see note 5) Śaṅkara seems to be a very early forerunner of the later Nyāya view of definition.
7. *sarvadā samaikarasam advaitam avikriyam ajam ajaram amaram amṛtam abhayam ātmatattvaṃ brahmaiva sma ity eṣa sarvavedānta-*

niścito' artha ity evaṃ pratipadyāmahe. From Śaṅkara's commentary on the *Bṛhadāraṇyaka Upaniṣad* (hereafter *BAUp*), IV.4.6; see *TPU*, p. 920.

8. Śaṅkara has stated more than once, in his explanation of the text under consideration, that the highest Brahman and our inner Self (the *pratyagātman*) must be realized as identical.

9. Though we cannot go into this here, I would argue that in the context of Śaṅkara's thought it is more appropriate to describe Brahman as super-personal than as impersonal.

10. *upapatteś ca sadasadādiśabdair brahma nocyate iti sarvo hi śabdo'rthaprakāśanāya prayuktaḥ, śrūyamāṇaś ca śrotṛbhiḥ jātikriyāguṇasaṃbandhadvāreṇa saṃketagrahaṇaṃ savyapekṣo'rthaṃ pratyāyayati, nānyathā adṛṣṭatvāt. tad yathā gaur aśva iti vā jātitaḥ, pacati paṭhatīti vā kriyātaḥ, śuklaḥ kṛṣṇa iti vā guṇataḥ, dhanī gomān iti vā saṃbandhataḥ. na tu brahma jātimat ato na sadādiśabdavācyam. nāpi guṇavad yena guṇaśabdenocyeta nirguṇatvāt. nāpi kriyāśabdavācyaṃ niṣkriyatvāt 'niṣkalaṃ niṣkriyaṃ śāntam' iti śruteḥ. na ca saṃbandhi ekatvāt. advayatvād aviṣayatvād ātmatvāc ca na kenacic chabdenocyata iti yuktaṃ 'yato vāco nirvartante' ityādiśrutibhyaś ca.* From Śaṅkara's commentary on the *Bhagavadgītā* 13.12. See *Bhagavadgītā with Śāṅkarabhāṣya*, Delhi: Motilal Banarsidass, 1978, Reprint of 1929, ed., p. 203. Take note of the expression 'is (not) plainly expressed' (*ucyate*); it is of major import, as we shall see, in Śaṅkara's understanding of the way words refer to Brahman.

11. I have analysed and commented on Śaṅkara's understanding of metaphor in 'Śaṅkara on Metaphor with Reference to *Gītā* 13.12–18'; see R.W. Perrett, ed., *Indian Philosophy of Religion*, The Hague: Kluwer Academic Publishers, 1989.

12. *lakṣaṇāyām api tu saṃnikarṣaviprakarṣau bhavata eva. adhyāsapakṣe hy arthāntarabuddhir arthāntare nikṣipyata iti viprakṛṣṭā lakṣaṇā; viśeṣaṇapakṣe tv avayavivacanena śabdenāvayavaḥ samarpyata iti saṃnikṛṣṭā. samudāyeṣu hi pravṛttāḥ śabdā avayaveṣv api pravartamānā dṛṣṭāḥ paṭagrāmādiṣu.* From *Brahmasūtraśāṃkarabhāṣyam*, ed. by Ramachandra Shastri Dhupakar and Mahadeva Shastri Bakre, Bombay: Tukaram Javaji, 1904.

13. '*Lakṣaṇā* occurs [in one of two modes] depending on whether [the applied sense of a term] is connected to its proper sense by proximity or remoteness': *lakṣaṇā ca yathāsaṃbhavaṃ saṃnikṛṣṭena viprakṛṣṭena vā svārthasaṃbandhena pravartate*, Śaṅkara on *Brahma Sūtra* 4.1.6.

14. *bhinnapravṛttinimittānāṃ śabdānām ekasminn arthe vṛttiḥ sāmānādhikaraṇyam*, quoting Rāmānuja's Commentary on the *Brahma Sūtras*, 1.1.13.

15. See further, *The Face of Truth: A Study of Meaning and Metaphysics in the Vedantic Theology of Rāmānuja*, by J. Lipner, London: Macmillan, and New York: State University of New York Press, 1986, pp. 29ff.

16. Technically, *satyaṃ jñānam anantaṃ brahma* may also be rendered: 'Truly, Brahman is infinite knowledge' where *satyam* is understood as an

adverb ('truly'). By insisting that the statement consists of three attributives and a subject-term by way of *sāmānādhikaraṇya*, Śaṅkara rejects this interpretation.

17. *nanu viśesyaṃ viśeṣaṇāntaraṃ vyabhicarad viśesyate, yathā nīlaṃ raktaṃ cotpalam iti. yadā hy anekāni dravyāṇy ekajātīyāny anekaviśeṣaṇayogīni, tadā viśeṣaṇasyārthavattvam. TPU*, p. 282.

18. On the acceptability of using extensional, as opposed to intensional, language in translating Indian logic with special reference to Nyāya, see Matilal, 'The Intensional Character of Lakṣaṇa and Śaṅkara'; and Karl Potter, 'Astitva Jñeyatva Abhidheyatva', to *Wiener Zeitschrift für die Kunde Süd- und Ostasiens etc.* 12–13, 1968–9, pp. 275ff.

19. *satyādiśabdā na parasparaṃ saṃbadhyante, parārthatvāt viśesyārthā hi te. ata ekaiko viśeṣaṇaśabdāḥ parasparaṃ nirapekṣo brahmaśabdena saṃbadhyate—satyaṃ brahma jñānaṃ brahma anantaṃ brahmeti. TPU*, p. 283.

20. A scriptural example adduced by Śaṅkara, i.e. *Chāndogya Up.* VI.1.4, where the causal agency, viz. clay, is unconscious, might raise the question.

21. *yad dhi na kutaścit pravibhajyate tad anantam. TPU*, p. 283.

22. *jñānaṃ jñaptir avabodho—bhāvasādhano jñānaśabdo na tu jñānakartṛ. TPU*, p. 283. On *bhāvasādhana*, see E.G. Kahrs, 'Exploring the *Saddanīti*', *Journal of the Pali Text Society* 17, 1992, p. 13. Kahrs explains 'the category of *bhāva*' as 'being; state of action . . . state of being . . . the mere activity [denoted by the verbal root]'. Thus *jñānaṃ jñapti* in the quotation above refer to Brahman as a Thinking, not as an agent of thinking, and not as 'the intellectual faculty' (thus van Boetzelaer, *Sureśvara's Taittirīyopaniṣadbhāṣyavārtikam*, p. 56), nor does *jñānam* mean 'the *potency* of knowledge' (ibid., emphasis added). On the contrary, if one is to use Thomistic language, *jñānam* here means *actus purus* expressed as 'knowledge', viz. the full actuality of Thinking itself.

23. *na. lakṣaṇārthatvāt. viśeṣaṇatve'pi satyādīnāṃ lakṣaṇārthaprādhānyam ity avocāma. śūnye hi lakṣye'anarthakaṃ lakṣaṇavacanam. ataḥ lakṣaṇārthatvān manyāmahe na śūnyārthateti. viśeṣaṇārthatve' pi ca satyādīnāṃ svārthāparityāga eva. śūnyārthatve hi śatyādiśābdānāṃ viśesyaniyantṛtvānupapattiḥ. satyādyarthair arthavattve tu tadviparītadharmavadbhyo viśesyebhyo brahmaṇo viśeṣasya niyantṛtvam upapadyate. brahmaśabdo'pi svārthenārthavān eva. tatrānantaśabdo antavattvapratiṣedhadvāreṇa viśeṣaṇam. satyajñānaśabdau tu svārthasamarpaṇenaiva viśeṣaṇe bhavataḥ. TPU*, p. 284.

24. *vijñātṛsvarūpāvyatirekāt karaṇādinimittānapekṣatvāc ca brahmaṇo jñānasvarūpatve' pi nityatvaprasiddhiḥ. ato naiva dhātvarthas tadakriyārūpatvāt. TPU*, p. 285. See also note 22.

25. *ata eva ca na jñānakartṛ, tasmād eva ca na jñānaśabdavācyam api tadbrahma . . . jñanaśabdena tal lakṣyate, na tūcyate, śabdapravṛttihetujātyādidharmarahitatvāt. tathā satyaśabdenāpi. sarvaviśeṣapratyastamitasvarūpatvād brahmaṇo bāhyasattāsāmānyaviṣayena satyaśabdena lakṣyate satyaṃ brahmeti. na tu satyaśabdavācyam eva brahma. TPU*, p. 285.

26. Śaṅkara makes use of a similar linguistic contrast elsewhere. Desmet (op. cit.) adverts to *Upadeṣasāhasrī*, 18:29: *ābhāso yatra tatraiva śabdāḥ pratyagdṛśiṃ sthitāḥ lakṣayeyur na sākṣāt tam abhidadhyuḥ kathaṃcana*; 'Where there is a reflection of the inner witnessing [Self in the form of the ego], words that pertain to the former do so indirectly; they can never directly denote it'.
27. As Ānandagiri in his *Ṭīkā* on Śaṅkara's commentary on this text reiterates: *yadyapi satyādiśabdānāṃ brahmaṇā mukhyo'nvayas tathāpi . . . pārṣṇikānvayenetaretarasaṃnidhāv anyonyasya vṛttiniyāmakā bhavanti*: 'Although the words *satyam* etc. bear on Brahman literally, still by virtue of the mutual effect they have through their juxtaposition, they function as predicative controls/restrainers with respect to each other'. See the Ānandāśrama Series, vol. 12, p. 53.
28. Hence Biardeau cannot be right when she says ('La définition dans la pensée indienne', pp. 282ff) that the literal meaning retained by the *viśeṣaṇas* may well run counter to their being understood as *lakṣaṇā*. This is to overlook Śaṅkara's important distinction between proximate and remote *lakṣaṇā* (see earlier). Likewise Bartley's statement, 'It should be clear that attempts to show that Śaṅkara thought of the terms in the *Taittirīya* formulation as applying to Brahman in secondary senses—i.e. by *lakṣaṇā*—rest upon a fundamental misunderstanding', ('Interpreting "Satyaṃ Jñānam Anantaṃ Brahma"', p. 110) itself suffers from the fundamental oversight just noted.
29. *evam satyādiśabdā itaretarasannidhānād anyonyaniyamyaniyāmakāḥ santaḥ satyādiśabdavacyāt tannivartakā brahmaṇo lakṣaṇārthāś ca bhavantīti. ataḥ siddhaṃ 'yato vāco nivartante, aprāpya manasā saha' . . . iti cāvācyatvaṃ nīlotpalavad avākyārthatvaṃ ca brahmaṇaḥ. TPU*, p. 285.

Chapter 17

Some Indian Strands of Thought Relating to the Problem of Evil

Margaret Chatterjee

INTRODUCTION

The texture of what follows will be loosely woven, although the weave may tighten up when arguments are under consideration. But cultural phenomena straddle the distinctions logicians seek to place before us, and the attitudes evidenced in the *ethoi* of diverse peoples appear with the unfailing regularity of a hand-blocked design, recognizable no doubt, but not bearing the sharp definition that those possessed of *l'esprit de géometrie* would wish. For me, in other words, the cultural basket is the key. Not, however, that I see forms of life as bedrock, for they could only be such in societies that did not change at all, and there are probably none like this today. Indeed, in studying 'modern Indian society' the investigator is constantly made aware of the metamorphoses taking place *within* cultural baskets.

My reflections take off from the first of the Stephanos Nirmalendu Ghosh lectures delivered by Professor Matilal in 1980.[1] As a background to this essay on *duḥkha* it is useful to note what he said in his Inaugural Lecture in Oxford[2] about the common factors in religions. He lists the following: a sense that the unexamined life is not worth living, control of instincts, the cultivation of certain positive emotions, that external circumstances are not all there can be, people can be better than they are and experience more than they do, the reference to a higher plane of existence. He confesses that this is an impressionist view. Even so it is interesting on many counts, although one almost hesitates to take him up on what he himself counts as impressionistic. At first sight many of the considerations (and they belong to different categories of discourse) seem to refer to ethics. And then one notices that there is no mention of conduct towards others on the list, although the possibility of people being 'better' than they are would seem to involve their relations to others. His approach is further clarified in the *duḥkha* lecture in which a twofold definition of a religious act is given, namely one given or approved by a religious tradition, and one dominated by a concern for one's own good, e.g. *nirvāṇa*, salvation, mystical union or heaven. By contrast he defines a moral act as one showing concern for one's fellowmen and lack of self-interest.

The relation of ethics and religion opens up a whole syllabus of issues which cannot be gone into here. It may be pertinent to mention in passing that Professor Matilal was not inclined to make a sharp distinction between Hindu and Buddhist views in *certain* contexts (e.g. in considering art). The close connection between ethical and religious goals of life on the Buddhist view would seem to belie this if his characterization of the distinction between moral and ethical acts is taken as the 'standard' Hindu one (if there be one such). There is something Pickwickian about regarding the pursuit of *nirvāṇa* et al. as self-regarding, given the *anatta* doctrine and the general injunction to curb egoity. Resort to the cultural basket offers a measure of clarification. *Dharmaśāstras* and *mokshaśāstras*, we are taught, are to be distinguished. In this way an intertextual distinction is made the basis of distinguishing between the phenomenology of the ethical and the religious life. Or is it the other way round? In any case to demarcate the ethical from the religious (on the ground of what is *samājik* and non-*samājik*) constitutes a major complication if we are to try and isolate a so-called Hindu view of the problem of evil. I am not taking up two other strategies which are quite feasible, namely tracing the matter *historically*, or non-acceptance of Matilal's demarcation between the ethical and the religious on the ground that a study of Hindu culture reveals concern with the *ethico-religious* (which is my own view), a position which, it seems to me, can be satisfactorily worked out developmentally and with due noting of the comparatively late arrival of treatments of *mokṣa* in the philosophical texts. Let us turn to the lecture on *duḥkha*.

THE ARGUMENT CONCERNING *DUḤKHA*

In what follows I present Matilal's argument, proceeding closely according to his text:

(1) The 'pain-existence equation' (in Eliade's language) underlines the undesirability or non-finality of the worldly life for persons who strive for a transcendent truth beyond all this.
(2) The pain-thesis is not factual but evaluative (i.e. not a proposition but evaluative-exhortative).
(3) The pain-thesis is a prescription for those wanting to attain liberation.
(4) If we are *mumukṣu* we should attach a negative value to all varieties of happiness.
(5) *Duḥkha* loses its significance in the context of *nirvāṇa* et al.
(6) If the pain-thesis is non-factual 'it is not falsifiable by citation of any apparent counter-example'.
(7) The pain-thesis is a *satya*, i.e. the sort of expression that is 'used in the Indian context ambiguously for both factual truths and evaluative exhortations'.

Now I have several difficulties about this sequence of positions. Re (1), is there not a considerable difference between regarding worldly life as 'undesirable' and regarding it as non-final? The point is perhaps that the *mumukṣu* will need to regard life as *both* undesirable and non-final if he is to detach himself from his present condition. To be weighed down by present miseries or regard them as all that there is, is not to make oneself unfit for the pursuit of 'transcendent truth'. Next arises the question of the proper attitude the *mumukṣu* should have. This includes both his attitude to misery and his attitude to happiness. The classic (and classical) advice to look alike on weal and woe is consonant with the pain-thesis if we take the latter as evaluative. To attach a negative value to all varieties of happiness would enable the *mumukṣu* to avoid the temptation to disregard the pain-thesis. Such disregard would amount to ignoring a prescription, if indeed the pain-thesis amounts to a prescription. *Duḥkha* is said to lose its significance (I'm not sure what this means—lose its sting, lose its relevance—which?) in the context of *nirvāṇa*, etc. (presumably in the context of the *pursuit* of *nirvāṇa*, etc.) in that the *mumukṣu* knows there is more and that that more is not in the nature of *duḥkha.* I leave out the further puzzle as to how he can be said to *know* this while still in a state of *duḥkha.* Point (6) is a logical one, taking, however, it must be noted, a hypothetical form. But the insistence on non-factuality seems to be radically undermined by point (7), by the reference to *satya*, for if there are expressions which are ambiguously descriptive and prescriptive, and the pain-thesis is one such, the series of steps by which it is made out to be evaluative seems to be without point. The argument, in fact, seems to hover between maintaining that it is evaluative and maintaining that it is prescriptive. Yet not all evaluative sentences are prescriptive (unless we conflate persuasion and prescription) although *some* may be. A more plausible elucidation, if we wish to avoid saying that the pain-thesis is a factual statement, might be to maintain that it has the logical status of a hypothetical in that *if,* and only *if,* one is striving for transcendent truth, one should (in the manner of an imperative *bhāvaya*) be aware of the 'undesirability or non-finality of the worldly life'. *Duḥkha* will not 'lose its significance' for one who does not set his sights on *nirvāṇa*, etc. on such a view.

Now, quite evidently, not all persons strive for a transcendent truth beyond this worldly life. Such persons will not only be concerned with factual assertions about the perceptual world in the manner of the Carvakas, but also make their own evaluative judgements about what amounts to weal and woe, what causes them, and how the former can be attained and the latter overcome. The ordinary individual in all cultures finds that life contains both suffering and happiness, and that some sufferings can be remedied, or at least mitigated, and some not. The ordinary man also finds that the time factor works variously in the case of both suffering and happiness. Transience is a blessing when suffering comes to an end. Duration is a curse when sufferings are prolonged. The ordinary man, along with Goethe, bewails the transience of happiness. Above all, he understands the value of endurance, the Stoic and the *sthitaprajña* alike typifying just this. The *mumukṣu* could perhaps say, when reminded of the attitude of those who do not share his aspirations, that the religious stance (as shown in setting one's sights on transcendent truth) enables a man to see the empirical world with its joys and sorrows in proper perspective, i.e. as something to be valued negatively, i.e. discounted. This, however, would not be offered in consolation, for consolation is not under discussion. The question is rather whether there is anything *other* than what is our lot at the *vyāvahārika* level and which can be a proper goal for the *mumukṣu.*

And now a more serious difficulty raises its head. If *duḥkha* be taken as a counterpart of the concept of evil in other traditions does the pain-thesis contain a prescription (or advice) to disregard all that could be included under the rubric of evil in order that the *mumukṣu* avoid hindrance in the pursuit of his transcendent goal? Or, more plainly, does the concept of *duḥkha* actually accommodate the concept of evil at all or is it rather the nearest we can find to it, or, thirdly, is it a concept which functions *in lieu of* it? To consider this, albeit in somewhat minimal fashion, must occupy us next.

THE CONCEPT OF EVIL

Overtly monotheist religions (as against any that may include monotheist forms within a more general rubric), it seems hardly necessary to say, have found the presence of evil in the world a serious challenge to the posits of (a) omnipotence (b) goodness that usually accompany the monotheist standpoint. Does the fact of evil, or of suffering for that matter, pose a problem (whether religious, metaphysical or ethical) if the posits of the omnipotence and goodness of God are not maintained? At the level of myth, contrary powers are taken care of through what is virtually an exercise of imagination on the part of poets and sages. Gods and demons wrestle, and, if the prospect is not exactly edifying from the human perspective, at least there may be some satisfaction in finding human struggles matched by similar struggles on a cosmic scale. Vedic concerns were this-worldly to a degree, flights of mythopoeic imagination notwithstanding. The theme of *duḥkha* is scarcely mentioned, and propitiations/remedies are found through resort to ritual, through human participation in an equilibrium which needed careful tending by all engaged in it, whatever be their place in the cosmic hierarchy. In fact the maintenance of balance was seen as of such crowning importance that the beneficent is polarized to the destructive rather than to the evil. The theory of *guṇas* reinforced an attitude which set greatest store by harmony. The mythological perspective continues in some of the Upaniṣads, for example in the *Bṛhadāraṇyaka* and *Chāndogya Upaniṣads* good and evil can arise through the conflict of gods and demons. They can also arise through the fetters made by human beings' acts in this or previous lives (*MaitriUp.* IV.2). The law of *karma*, if not inexorable, which it doesn't seem to be, enters the scene in tandem operation with cosmic powers. The factuality of much that is undesirable, call it *duḥkha* or *pāpa* as you will, is taken for granted. Why it should occur at all (the question that troubles the theist) is of less import than how it should be tackled and brought to an end. Ritual performance, valour, ascetic rigour, ethical action, saintliness—all in turn appeared at different periods of history, and to different sets of people, as ways of countering the surd element in life. The difference in standpoint between the theist and the outlook outlined here cannot be overemphasized. It amounts virtually to the difference between why and how. To ask 'why' is to raise a question about justice. The Indian standpoint is centred on the 'how' of getting rid of the unwanted condition. Cosmic contrariness and the *karma* theory provided sufficient explanation of 'why'. Add the posit of divine *līlā* and you get an additional reason why questions cannot be pushed very far. The tendency to find the cause of sufferings/evil in one's own actions diverted attention from the possibility that much of human misery is caused by what *others* do. The searchlight of diagnosis is turned on human frailties, rather than on the injustice inflicted by a potentate-type god or on the unjust structures of society. Disasters and catastrophes were only what one might expect, the concept of a perfect world being a thoroughly alien idea. Divine beings underwent trials and tribulations, and epic heroes and heroines most certainly had a miserable time. It would all be worked out—although not necessarily in a felicific way—over cycles of time much vaster than anything economists could envisage, generations later, in their talk of 'in the long run'.

It is worth noticing, moreover, the role played, by default, by the deity according to the philosophical systems. On the Vaiśeṣika view, the role of God is confined to a reshuffling of constituents that already exist. Likewise the part he plays in meting out reward and punishment is never fully reconcilable with the law of *karma* which again, already operates. The Vedāntic combination of *līlā* and *karma* is hardly conducive to the raising of weighty questions about the whys and wherefores of human misery. Furthermore, since, on the Vedāntic view, human souls are not created, there is no creator who can in any way be made *responsible* for human destinies. It is hard to extract anything like a concept of Providence from the systems unless we turn, for example, to Śaiva Siddhānta. No doubt a concept of Providence faces other difficulties, especially concerning divine attributes, powers, limitations if any, and so on. A Hindu concept of Providence would have to be at the *saguṇa* level, located, let us venture to say, in the deity's *snigdha rūpa.* But the implication of his *snigdha rūpa* would not be omnipotence (although man would be more likely to prevail if he had god on his side) but rather his *accessibility.* In devotional literature in regional languages we find developed expositions of the doctrine of grace, but I am not very sure we have much that is analogous to Providence (this of course is neither a merit nor a demerit). For the latter, it seems to me, we need the concepts of creation, teleology, and the affirmation of continuous presence throughout history. Since teleology connotes tendency in a certain direction, and this seems to presuppose lack, Hindu thinkers found it incompatible with the nature of a divine being. *Līlā* did not connote waywardness so much as an overflowing inexhaustibility whose nature it was to be limitless. The multiple traditions of Hinduism do not, as far as I know, include looking upon God as a fellow-sufferer with man. But if some of the Vaiṣṇava analogies for the relation between God and man are followed through (the analogy of friendship, for example) this possibility is not excluded. At the everyday level the concept of Bhagavan, which is not treated in scholarly texts, seems to function very like Providence. But if one looks for the concept of a God who acts in *history* (as distinct from one who descends from time to time but in cosmic time rather than in historic time) one is hard put to find it. We are thrown back on the concept of *duḥkha* once more.

DUḤKHA RETHOUGHT

Whether or not *duḥkha* parallels the term evil in other religious traditions, there can be no doubt about its pivotal role in Hindu thought systems. If liberation is regarded as a religious goal rather than an ethical one, *duḥkha* is likewise regarded as something which has to be tackled through the religious life. If we follow Professor Matilal's analysis, this life is concerned with the quest for 'transcendent truth'. I wish to proceed in a rather different direction in what follows. We have two alternatives to *duḥkha*, namely *sukha* and *ānanda*. Let us examine the *duḥkha/sukha* polarity first. The ordinary man takes it for granted that *sukha* is preferable to *duḥkha* but knows through experience that pursuit of the one and avoidance of the other are often self-defeating. We are *overtaken* by experiences. They constitute our *Befindlichkeit*. Vātsyāyana does not draw our attention to the circumstance that all experiences involve an element of *duḥkha* as consolation. He could have done so. At least part of the message of Buddha in the mustard seed episode is to point out, in John Stuart Mill's phrase, nature's everyday performances, i.e. to reconcile us to the situation. Vātsyāyana does not seek to reconcile us to the situation. Let us move on.

A contrary reminder could be given. If '*duḥkhaṃ hina suhkhaṃ na bhavati*' is true one could also say '*sukhaṃ hina duḥkhaṃ na bhavati*'. But if there is no doubt about the first (temporality alone would provide sufficient verification of it since felicific situations come to an end) there is serious doubt as to the truth of the second. A great deal of suffering has no mitigating element of happiness in it whatsoever. To maintain that the patient racked with pain and the concentration camp victim undergoing torture experience anything other than unalloyed suffering would be a cruel mockery. There is something more than pain, namely agony, and most languages have a range of terms which doctors and their patients are all too familiar with. Agony, both physical and mental, scarcely fits into the classification given in the *Sāṃkhyakārikās*.

But the two Sanskrit tags invite further scrutiny. Do they not point to the human circumstance that (the cases of agony apart) 'the admixture of *sukha* and *duḥkha*' defies distinction? It would seem so. And here I think we need to pull out something from the cultural basket, namely this—admixtures are bad. This is writ large in Manu, for example. But if any should object that to admit admixture still does not justify saying *sarvam duḥkham* (*sarvam* being an exaggeration, as hedonistically inclined students sometimes like to point out) we can have recourse to a piece of proverbial wisdom in Bengali at least and possibly in other regional languages as well. When a situation is muddied, churned up, turgid, it is remarked that 'the water is stirred up' (a loose translation), i.e. the entire water is affected and not just the unclear surface. It is the muddiness which predominates. Here we have a telling example of the way philosophical insights can feed off proverbial wisdom. The opposite is, of course, often the case, when philosophers depart radically from what the ordinary man thinks to be self-evident. But the analogy must not be pressed too far. Sediment eventually settles and the pond appears relatively clear after some time. This does not happen in the case of human suffering. The existence of the patient (his *Befindlichkeit*) continues to be pervaded by pain.

A similar point can be made about the Buddhist analysis of transience. Moments of happiness are all too fleeting. *Duḥkha* also, no doubt, has its ebb and flow. When pain is a little less there is a modicum of relief, a neutral watershed rather than a moment of pleasure, but it is muddied by the knowledge that the pain will return.[3] Temporality comes to our rescue when *duḥkha* is somewhat less for a time. It can also be a scourge when miseries endure. So both Vātsyāyana and the Buddhists give no weight to mitigating moments of relief, and for the same reason. Such weightage would deflect one from realizing the vanity of existence and the need to cultivate detachment. What emerges from the foregoing is that in the case of the *sukha / duḥkha* polarity the two tags draw attention to not only polarity but admixture, and in terms of the cultural basket, admixtures are disvalues. Whether they count as *evil* is difficult to say for the reason that, it seems to me, philosophical tradition is content to make a general categorial distinction between positive and negative value, the further elucidation of the latter being left either to mythology or to the theory of *karma*.

It is clear at any rate that the *sukha / duḥkha* syndrome belongs to the *vyāvahārika* level and that there is no way out of it at that level. The factuality of this, I suggest, is not of no account, for unless a man is convinced that the unalloyed absence of *duḥkha* is not to be had in worldly (*weltlich*) existence (I phrase this so as to take in the Buddhist position as well) he will not be moved to look elsewhere. *Sarvam duḥkham* I therefore venture to say, is not so much 'ambiguous' as containing a two-staged sequence of insights, namely:

(a) *sukha / duḥkha* is both a polarity and something more. It indicates an admixture which is irremediable at the *vyāvahārika* level.

(b) a polarity of a different kind is possible, that between *duḥkha* and *ānanda / nirvāṇa.*

Each of the insights is open to expansion. Regarding the first, we need to agree that admixtures are bad, that what is unalloyed (suffering excepted) is preferable to what is mixed. Regarding the second, we need to understand that the nature of the second alternative to *duḥkha*, *ānanda / nirvāṇa*, is of a completely different order from *sukha.* The term *duḥkha*, therefore, has two different contraries. Indeed *ānanda* and *nirvāṇa* are also different from each other (not that one could verify this experientially) belonging as they do to different thought systems. However, they are both equally unspecifiable and indescribable through the use of ordinary language. Phenomenologically it is more than possible that the nature of *duḥkha* for the aspirant after higher things is also vastly different from what is experienced by the disappointed experiencer of the *sukha / duḥkha* syndrome. The intuition into there being something radically wrong with life as we normally live it is surely something further (other/more) than normal disgust at the pendulums of fortune and admixtures of experience that are our common lot. We have then, it could be said, something very like a *metabasis eis allo genos* in the shift from insight (a) to insight (b). But advance to (b) would need to begin with insight (a). Or, invoking Divided Line parlance, demarcation serves to point up the route of ascent. If there is any cogency at all in what has been sketched so far ambiguity resides in the umbrella term *duḥkha* and not in oscillation between factual and evaluative uses of language. The ambiguity of the word *duḥkha* consists in its dual usage in both polarities, i.e. *sukha / duḥkha*, and *duḥkha / ānanda / nirvāṇa*. It is small wonder that many German Indologists and some of their predecessors attributed a generalized *Weltschmerz* to all Indian philosophical thought. But it is time reflection turned to another leading strand in Indian thought, the theory of *karma*.

KARMA

The *karma* concept provides a neat explanatory hypothesis about all woeful phenomena, referring them to an individual's past acts, extending the connotation of past beyond the bounds of a single lifetime. Moreover it leaves open a range of possibilities for the future, providing determination without determinism. Radhakrishnan describes the law of *karma* as that 'by which virtue brings its triumph and ill-doing its retribution' and that it is 'the unfolding of the law of our being'.[4]

The problem is that this is just what it does not do. The wicked prosper and the good suffer, and even if things were to be evened out in the future, something which may well be stymied by further calamities, this provides no explanation, let alone justification, of unmerited suffering *now*. It provides no explanation of natural evils, including within this category the innumerable miseries caused by disease. Furthermore, we are left in the dark as to whether it is *moral* to accumulate good *karmas* or merely prudential. The chances are that it is *prudential* in that it is wise to cut short *saṃsāra*. But what about those sufferings which are due to others' actions or society's inaction? Could war, famine or the Holocaust, for example, be at all explained by reference to an individual's *karmas?* If not, we are led to the position that, on balance, individuals are *victims* rather than agents who suffer the effects of their own previous actions. There are other difficulties too. According to the *karma* theory it looks as if to exist at all is a punishment in the sense that, if all *karmas* had been worked through, *saṃsāra* would have come to an end. Does this mean that there is merit in not being born again or does it mean that in not being born again one has gone beyond the sphere of merit and demerit? Bringing in the posit of God's existence creates a further complication. The position presented in the *Śvetāśvatara Upaniṣad* (6, 16) that God is the ordainer of all deeds, seems irreconcilable with the *karma* theory. When the Vaiśeṣikas hold that God can interfere with the operation of *karma* in order to dispense justice does this reflect a sense that the law of *karma* is unjust? But would it make sense to speak of an impersonal law as just or unjust? The operation of such a law is not elucidated as far as its origin is concerned, although the possibility of going beyond it through leading a life free of attachments is set out as a possibility. As for the difficulty of combining the law of *karma* with the posit of God's existence, this may, in fact, underline the point that, in philosophical Hindu thought at least, the issue of whether God exists or not is not of paramount importance.

So if the law of *karma* is to bear the brunt of explanation of human woe the considerations given above would indicate that it is hardly able to do this. There were other elements in the cultural basket which dealt with surd factors in the cosmos, among them mythological beings and the *guṇas* being the chief, and it is here I suspect, that one would need to look for Hindu insights into the nature of evil. So far then, it looks as if *duḥkha* and evil are not coextensive terms. Nor can we say that it is evil which is the cause of *duḥkha* since *duḥkha* can be caused by chance happenings in the natural world which cannot be described as either good or evil. As far as philosophical traditions were concerned one might hazard the opinion that while a great deal of attention was bestowed on the analysis of *woe* and how it was to be tackled there were concepts and trends of thought in a long history of philosophical thought to work against the recognition of radical evil (*Böse*). Among these were the concept of *māyā* (limited to Advaita Vedānta no doubt), the value set on detachment, the yogic life-style centred on a target beyond good and evil, seeing the main causes of *duḥkha* as internal (*lobh, krodh* and the like), and not to be missed, the distinction between the *vyāvahārika* and the *pāramārthika.* This wide scatter of concepts and tendencies is matched by the wide range of rubrics under which the treatment of evil can be found in the relevant literature, ranging over suffering, bondage, sin, ignorance, demons, etc. That the Zoroastrian 'solution' of the problem of evil did not form part of the corpus of ways of seeing 'contrary' phenomena is a strange quirk of cultural history. The only strict dualism to be found among the systems is in the Sāṃkhya and the contending principles therein bear no resemblance whatsoever to Ahriman and Ahura Mazda. Deriving from the Arabic/Persian, the word 'shaitan' is in common usage in North India, downgraded into the meaning 'naughty' and used mainly to upbraid small children. At village level the word can still be used to refer to ne'er-do-wells, but there are other more colourful terms that fit them more neatly. A study of expletives and terms of abuse is not irrelevant in this connection as these give an idea of popular understanding of various shades of iniquity. The difficulty of unravelling the fine filaments of popular imagination and linguistic usage may seem a far cry from the conceptual venturing of philosophers but it is a task which I believe is worth undertaking. For example to catch the resonances of the common question one Bengali may ask another, 'How is your *kajkormo*?' reveals a whole thought-world and life-world regarding the scope of human action and what might accrue from it. But I must pass on.

MODERN HINDU THINKERS

An attempt will be made to show that the innovative thinking of 'non-professionals' in the modern era took reflection on the problem of evil forward in unexpected ways—unexpected given the corpus of concepts thrown up by 'the tradition'. It is the thinkers of the so-called Renaissance that were bold enough to *specify* evils and regard them as targets for combat whether it be colonialism, poverty, superstition or the host of practices against which reformers took up cudgels in the nineteenth century. Lest it be thought that this was a purely ethical crusade we have only to look at Bankim Chandra Chatterji's utilization of the myths of the people, in particular the people of Bengal, to recognize its religious quality. It hardly needs stressing that an activist interpretation of the *Gītā* enabled Tilak, Sri Aurobindo and Gandhi to treat action other than as a means of adding to bondage. This cleared the way for looking on bonds as factors to be fought rather than escaped or avoided. The problem of how separate karmic lines could permit of action to alleviate others' woes was met by a new stress on service (*seva*). The possibility of fighting evils jointly was a major development out of this insight. One could even say that it provided a measure of intellectual underpinning for the nationalist movement. Gandhi's distinction between evil and the evil-doer challenged a well-known teaching in the *Bṛhadāraṇyaka Upaniṣad* (4, 4, 5.): 'According as one acts, according as one behaves, so does he become. The doer of good becomes good. The doer of evil becomes evil.' His key concept in political philosophy—that of *satyāgraha*—centres around the notion that suffering voluntarily undergone is not evil but contains persuasive power.

If it be objected that the undergoing of voluntary suffering also characterized the life-style of the traditional ascetic it would be pertinent to recall that there is a big difference between asceticism undergone for the sake of individual self-perfection and the *tapasya* which is aimed at converting the heart of the 'adversary', changing unjust structures and shifting the balance of forces in situations which are in a state of gridlock. Gandhi used several synonyms for the word 'evil', e.g. what is *adharma*, *nāpāk* or unholy or satanic (the last of these could have been derived from evangelical vocabulary or from William Blake). Gandhi was confident that men could rise to heights of great heroism, but he was rather less able to see that there were also immense depths to which they could fall. This provides the ground of his disagreement with Martin Buber in respect of the tragic events on the Continent in this century.[5] The scale of calamities and catastrophes that can take place in the human world is no doubt described in the epic literature. But if creation and destruction are to be expected in recurring sequences does this not blunt the edge of disaster, encouraging a sweeping up of *sufferings* under the general rubric of *suffering*? Yet the reformist thinkers of the nineteenth and twentieth centuries, *in spite* of 'the tradition', had an eye for individual sufferings, and strengthened an activist interpretation of *karma* which had always been implicit in its open-ended character but tended to be overlaid by the sense of the dead weight of past *karmas*. But did any of them have any conception of radical evil (*Böse*) as against woe (*Übel*)? Tagore perhaps did, in his paintings and his last poems.

CONCLUSION

We return then to the concepts of good and evil. It is primarily to Plato that we owe the insight that the ultimate in metaphysics and religion is also the ultimate in ethics. When thought through, the implications are startling. The True and the Good (no less than the Beautiful) cannot but be one and the same. If so, then the False (cf. Gandhi's 'untruth') and the Evil are likewise the same. The insights of the Hindu sages, if one does not do them an injustice by generalizing, are rather different. Logical considerations prompt the standpoint that what is beyond good and evil cannot properly bear the epithet good. We are often told, however, that even so, the ultimate is *sattvika* rather than otherwise. But strictly speaking, that of which we speak in venturing to say anything at all of the ultimate, is not *sattvika* either in that this term bears the implication of contrast with two others.

Now since time began, more than one cultural basket, has borne a shadowy intimation into the possibility that light might *contain* darkness. The cosmos provides the root metaphor out of which this intimation springs. The twenty-four-hour cycle contains both day and night, and 'day' often means the whole twenty-four hours rather than just the hours of daylight. Is this not why, in order to speak of evil, poets have had to invoke a principle *banished* from the firmament, fallen from the sphere of light into *outer* darkness, something far more terrible than the darkness which, as a daily occurrence, gnaws into the light only to disappear and reappear? Falsehood/the Lie, consists in a rupture of truth, just as evil consists in a rupture of good. To admit the reality of radical evil is to probe an abyss which yawns beyond the natural polarity of light and darkness. It requires recognition of an outer darkness into which human beings can sink and from which evil erupts through their deeds. I now pose the question that must come next. Were Indian thinkers at all familiar with the distinction between *Übel* and *Böse?* Or was a world-view that tended either to stoicism or to celebration or which passed this by?

I cannot answer the question I have posed but will approach it obliquely. Let me invoke my original metaphor of a loose texture, reminding myself that pulling together loose threads may have the effect of rendering a fabric thin or even tattered. In the many-structured mansions of Hindu culture philosophical thinking is not necessarily the key to the rest, important though it be as a *phenomenon* partially revelatory of the culture concerned. In any case Hindu philosophical thought is fascinatingly diverse. While in some cultures philosophical thought appears to break from the realm of myth 'successfully' (a process which, in my view, often impoverishes it), the root metaphors embedded in language still resonate with the life-experiences of generations long forgotten. In recalling the names of ancestors on ritual occasions the Hindu overtly recognizes a presence which on other occasions still speaks through idiom, gesture and attitude. Shifting analogies, we explore a palimpsest (as Nehru once remarked in writing of India's history), different layers of which reveal themselves to the explorer, but none of which are totally erased from archetypal consciousness.

The dualism of the Persians sprang out of the starkness of their life in the desert. It had no appeal for the people of Āryavarta for whom light filtered through the trees of the forest in endless multiplicity of dappled forms and for whom diversity and unity were eventually to surface as major thought-forms. The factuality of human suffering appeared all the more paradoxical in conditions of relative plenitude (conditions which rarely obtained in desert life away from the great river-systems). Now a pluralist world-view readily accommodates contrary powers which in turn are also seen to be plural. The unitary terms formulated generations later through abstract thought represent a strenuous pulling together of elements already deeply rooted in everyday experience. It so happened that in undertaking this exercise with reference to the surd elements in life the resulting concept for the Hindus was not 'evil' but 'suffering'. This product of the churning of minds is eloquent and poignant witness to the rooting of Hindu thought in *Erlebnisse,* something not sufficiently granted by those who detect therein mainly speculative excursion. Is it not a fact that human life is beset with sufferings?

What I would like to draw attention to next is the *leap* of thinking accomplished by the reformist men of the Indian Renaissance, given the tradition they inherited, and the circumstance that even the impact of Persian and Muslim thought had not brought about any such leap. From one point of view their particularization of sufferings, in the working out of strategies for combating them, recaptures the pluralist insights of pre-philosophical times. But there was more. The analysis of specific sufferings enabled them to diagnose the causes of ills and discontents, to pinpoint the loci of wrongs, to lay bare the structure of institutions which needed changing. In so doing they virtually paralleled the labours of reformers in other parts of the world, and here I refer to the unpacking of 'evil' into 'evils' to be identified and combated.

Now the treatment of suffering and sufferings, evil and evils, is part of a wider consideration, that of 'spirituality', another term for which we seek in vain a precise Hindu equivalent.[6] But this much seems clear—in treating of evil/suffering we are at grips with human life itself, including its heights and depths. Most cultural baskets accommodate a cosmic dimension herein, even those which seem to have abandoned it. For no *Bildung* can be without a mytho-poeic element. The surd elements in life present themselves today both in terms of intransigence and opportunity, despair and hope, polarities which persist to the end, defying synthesis. Looking back at the whole caravan of concepts which make their way through Hindu intellectual history in respect of these surd elements we discover a perception of the human dilemma which is second to none in terms of subtlety. It comes to terms with the dilemma in a distinctive way which calls for understanding against the background of an entire spiritual landscape[7] and which defies assimilation in thought patterns of an alien form.

NOTES

1. Vide B.K. Matilal, *Logical and Ethical Issues of Religious Belief*, University of Calcutta, 1982.
2. *The Logical Illumination of Indian Mysticism*, Oxford University Press, 1977.
3. Cf. the Buddhist thesis '*duḥkha-duḥkha, anityatā-duḥkha, saṁskāra-duḥkha*'.
4. *The Principal Upaniṣads*, London, 1953, pp. 113ff.
5. Vide my *Gandhi and his Jewish Friends*, Macmillan, 1992, pp. 115ff.
6. Vide my *The Concept of Spirituality*, Allied Publishers Private Limited, 1987.
7. Vide my use of this phrase in the above and also in my *The Religious Spectrum*, Delhi: Allied Publishers, 1984.

Chapter 18

Outline of an Advaita Vedāntic Aesthetics

Eliot Deutsch

I want to draw a sketch, as it were, of what an Advaita (non-dualistic) Vedāntic aesthetics would consist of. It is certainly the case that Śaṅkara (*c.* 830) and his followers were not interested in any central way in developing a philosophical aesthetics (either in the Western sense of that term or, say, as understood by the astute Indian thinker Abhinavagupta, who did share with Śaṅkara a number of basic metaphysical ideas); therefore I clearly will not be explicating or excavating an existing body of ideas that would constitute as such an Advaitic aesthetics. I propose rather to 'reconstruct' (if that term may still be allowed in this era of deconstruction) that aesthetics; which is to say, I will try to draw out implications from certain established Advaitic metaphysical principles which may form the core of an Advaitic aesthetics.

Many writers have often noted the kinship of art and spirituality in India, so it is most appropriate, it would seem, to carry out such a reconstruction. Also, although Śaṅkara and other Advaitins did not pay much attention to the arts of India, they certainly did not exhibit, on the other hand, any hostility to them and might very well have taken it for granted that the kinship (albeit certainly not identity) between aesthetic experience and spiritual insight or *jñāna* was indeed so central as not to have to remark upon it. In any event, I would like to examine what a theory of creativity might look like from the standpoint of Vedānta and something about the nature of aesthetic experience and the ontological status of the work of art that relates to it.

It should also be noted, I think, at the very beginning that an Advaitic aesthetics must of necessity be 'exemplary', which is to say it must be essentially *prescriptive* rather than *descriptive* in character. Its (reconstructed) theory of creativity is not intended to tell us how artists of various kinds (painters, poets, musicians . . .) in different cultures (Asian, Western and other) living at different times (medieval, modern) have actually gone about their business, rather it is intended to tell us how an Advaitic-minded artist would work, insofar as he adhered to the basic (metaphysical) principles of Vedānta, and how his or her work ought to be understood and experienced in that context. I shall not, therefore, be concerned with how Advaita would prescribe Indian art as such, but rather with what kind of generalized notions about art and its creativity it would affirm as following from, or at least being most compatible with, its metaphysics.

I

In addressing the question, why and how is there a world of appearance which we ordinarily take to be real? non-technically (that is, without reference to specific ideas about causality; *satkāryavāda, vivartavāda*) Śaṅkara appeals to the doctrine of *līlā*—the sportive play of the creator god. 'Those who theorize about creation (*sṛṣṭi*)', Śaṅkara writes, 'think that creation is the expansion of Īśvara'.[1] And further: 'The activity of the Lord . . . may be supposed to be mere sport (*līlā*) proceeding from his own nature without reference to any purpose.'[2]

As I have noted before: 'The concept of *līlā*, of play or sport, seeks to convey that Īśvara creates (sustains and destroys) worlds out of the sheer joy of doing so. Answering to no compelling necessity, his creative act is simply a release of energy for its own sake. Creation is not informed by any selfish motive. It is spontaneous, without any purpose. No moral consequences attach to the creator in his activity, for *līlā* is precisely different in kind from all action which yields results that are binding upon, and which determine, the actor. It is simply the divine's nature to create just as it is man's nature to breathe in and out.'[3]

But insofar as Īśvara may be said to exist, then just so far is man called upon to imitate that divine creativity in his or her own life and being. 'There are no gods apart from the Self' is an oft-heard Advaitic theme. We are called upon, in our own creativity, to be as free, yet disciplined, insightful beings. Let us see what this involves.

J.A.B. van Buitenen in a note to his study of Rāmānuja's *Vedārthasaṃgraha* has rightly pointed out that 'the important conception of God's sport is best understood by its opposite *karman.* It contains a free action (an action not resulting from a preceding action in an endless retrogressive succession) performed to no purpose at all: no purpose that of necessity would result in new *phalas* [fruits] for the agent to suffer or enjoy'.[4] *Līlā* is thus closely allied to *karmayoga* as set forth, for example, in the *Bhagavadgītā.* The artist, and this is in keeping with much traditional Indian aesthetic theory, must be a kind of *karmayogīn*—one who acts without attachment to the fruits of his or her action, but nevertheless in a manner of loving concern or devotion (*bhakti*) which is informed by knowledge (*jñāna*) of reality. If creativity is to be a free act then it must arise from the special spiritual 'disinterestedness' associated with insight into reality; in short with the realization of *mokṣa.* Genuine spiritual creativity, for the gods and for us, is possible for Advaita only as it is attendant upon reality—for it is only with this attendance that the world of our ordinary experience is shown to be *māyā.*

Various play theories of art have been developed in the West (from Schiller to Gadamer) which likewise stress the 'disinterested' nature of the free act which is delighted in for its own sake and which, although rule constituted and regulated, defines the criterion of its own performance; but what distinguishes an Advaitic theory from these is its grounding in the doctrine of *māyā.* For Advaita, play—as disinterested action, as it is an expression of *karmayoga,* is possible and meaningful only to the extent that a radical discontinuity obtains between Reality (Brahman) and everything in our otherwise experienced world. Play requires an ontological space, as it were, from which consciousness may be at once a master of appearance, the *sākṣin,* the *māyin,* and detached from it; indeed, Advaita would insist, one can be a master—not in the sense of dominance but of skill—only when one has attained that detachment which rests on the realization of self as Brahman.

The artist must, then, if he is to act in play, be aware of both sides, as it were, of the great ontological divide: he or she must work *with* the materials of her art *from* the non-obsessive stance that comes from the realization of the self.

Ananda K. Coomaraswamy has stated that:

> It is true that if the artist has not conformed *himself* to the pattern of the thing to be made he has not really known it and cannot work originally. But if he has thus conformed himself he will be in fact expressing *himself* in bringing it forth. Not indeed expressing his 'personality', himself as 'this man' So-and-So, but himself *sub specie aeternitatis*, and apart from individual idiosyncracy. The idea of the thing to be made is brought to life in him, and it will be from the supraindividual life of the artist himself that the vitality of the finished work will be derived.[5]

Classical Indian aesthetic theory makes much of the necessity of the artist to so transcend himself that he becomes an 'instrument', as it were, of the spiritual power of being. This is not meant as some kind of simple passivity, but rather, in keeping with *karma-yoga* concentrating one's own creative energy (*śakti*) so as to be at-one with the very vitality of being.

Advaita would carry this one step forward and would regard the master-work as a kind of *śruti*—an authorless revelation of being. Revelation theories of art are not conspicuously present today—and in the past some of them were no doubt distorted by a sentimental wayward romanticism that exalted the artist beyond all human being as a 'priest of beauty' while still having him retain a powerful ego—nevertheless, Advaita would argue, the spiritual possibilities of art can be realized only when the artist is able to attain a speaking *from*, rather than a talking *to*, reality.

This in turn requires a profound listening. Creativity calls for the artist to know in his or her own being the silent rhythms of being.

Creativity is a kind of *adhyāsa*, a 'superimposition' of the finite on the infinite, the spoken upon the ineffable, a confounding of self and non-self, but unlike ordinary perception and conceptualization where the perceiver-conceiver is ignorant of the activity and therefore becomes a victim of it, with Advaitic-informed creativity there is a knowing awareness and hence once again freedom. Form (*rūpa*) is brought into existence, not fortuitously but with a knowing skill or accomplishment (*yukti*).

Skill in action, then, requires a spontaneity that is genuinely efficacious. The artist must be a master of the conditions of his act and not their victim. She must, to be begin with, be a master of herself. The spontaneity involved in creativity is, therefore, to be sharply distinguished from impulsiveness. Impulsiveness is groundless, which is to say that it works from the strongest force within one at the moment. Spontaneity, on the other hand, as involved in skilful creativity, is grounded in the deepest structures of one's being and is, accordingly, a non-egoistic expression of one's spiritual potentiality.

Creative actions that are skilful, which are performed with spontaneity, result then in the achievement of that form which appears to be inevitable and in which, it is said, we are able to rejoice. This leads us to some brief discussion about the nature of aesthetic experience and the ontological status of the work of art.

II

M. Hiriyanna, the noted scholar of Indian philosophy, devotes a couple of pages in his book *Art Experience* to what he calls the 'aesthetic theory of the Vedānta'. He says there that the word *ānanda* ('bliss') 'contains the clue' to this theory, but that, in true Advaitic fashion, 'perfect beauty which is identical with the ultimate reality is revealed only to the knower. We perceive only its outward symbols and we may describe them as beautiful in a secondary sense, since we experience *ānanda* at their sight'.[6] Hiriyanna goes on to say then that

> we may well compare the person appreciating art to a *jīvanmukta*. He does indeed get a foretaste of *mokṣa* then; but it is not *mokṣa* in fact because it is transient, not being based upon perfect knowledge.[7]

J.L. Masson and M.V. Patwardhan in their work *Śāntarasa and Abhinavagupta's Philosophy of Aesthetics* refer to the distinction drawn between *rasāsvāda* and *brahmāsvāda* in the *Śrībhagavadbhaktirasāyanam* (I.12), where it is said that

> When this (consciousness limited by worldly objects) becomes manifest in the mind, it turns into *rasa*, although owing to its being mixed with

insentient objects it is somewhat less (than the joy of pure consciousness).[8]

Classical Indian aesthetics (from Bharata in his *Nāṭya Śāstra* to Abhinavagupta and beyond) developed the central concept *rasa*, the 'taste' or 'flavour' associated with the depersonalized, transformed emotion appropriate to different forms of expression in art. *Rasa* is at once the essence of art and the experience of it. *Rasa* is said to be *alaukika*, 'extraordinary' just insofar as the everyday emotions, the *bhāvas*, of love, fear, anger and so on are transformed and made the object of disinterested pleasure.

So a strong kinship is noted between aesthetic experience, understood in terms of *rasa*, and spiritual consciousness, as defined by Vedānta—with the latter granted a superior status. Among the points then of presumed similarity between *rasāsvāda* and *brahmāsvāda* which Masson and Patwardhan note are:

- In both cases, the distance between the subject and the object is removed.
- Time and space disappear for the duration of the experience.
- In both cases special preparation is necessary.

And among the differences noted are:

- The drama [art] is not expected . . . to change one's life radically. To have a profound aesthetic experience is simply satisfying and does not imply that one will be in any sense profoundly altered.[9]

Now it is certainly the case that Advaita would not allow the aesthetic to be accorded equal value with the full realization of Self, the achievement of *mokṣa*: the latter has an altogether unique position. Nevertheless it would, I think, want to bring aesthetic experience (and here I will confine this to the experience of art, not of beauty in nature) as closely as possible to Brahman-realization. It would, I believe, want to take the points of alleged similarity which I have selected from the Masson–Patwardhan discussion together with the supposed point of difference which they set forth and argue as follows:

Whereas Masson and Patwardhan see in both aesthetic and spiritual experience a 'removal of the distance between subject and object', Advaita would nevertheless insist on there being a qualitative difference between the different states of consciousness with respect to this factor. Advaita, I think, would see aesthetic experience as a heightened concentration upon an artwork in such a way that the experiencer is made unaware of himself *qua* ego while at the same time being intensely aware of the object of his or her perception. In aesthetic experience it is not so much that the distance between subject and object is removed as it is radically transformed, thereby enabling the subject to be most acutely attentive to the qualities and meanings proffered by the object. In other words, the experience of art—although sometimes culminating in what Abhinava referred to as *śāntarasa*, the 'peaceful' *rasa* which closely approximates the Advaitic overcoming of subject/object distinctions—must be seen in its own terms as a kind of *pramāṇa*, if you will—a way of acquiring knowledge of a special kind.

In recent Western aesthetics a great deal of work has been done concerning 'truth' and 'meaning' in art, with something of a consensus emerging both among hermeneutically minded thinkers (Gadamer, Ricoeur) and analytically oriented writers (Goodman, Danto) that both truth and meaning in art is *sui generis*, not reducible to, or parasitic upon, other epistemic accounts, approaches or experiences. The fundamental office of a metaphor, for example, is to create a new meaning of its own kind and not to extend an otherwise fixed literal meaning. Concentrating on an artwork in its own terms means to see formal qualities and relations, including conventional representative elements if present, as aesthetic *content*, as that which is intrinsic to the work. Advaita would reject a 'beauty is in the eye of the beholder' approach, just as it rejected Buddhist subjective idealism. Phenomenally, the subject–object relation holds, according to Śaṅkara, throughout experience, albeit in very different ways.

The aesthetic, as a kind of *prāmaṇa*, then discloses aspects of objects (artworks) which in turn enables us to see new dimensions of the world. It organizes, in Nelson Goodman's terms, a 'world-making', a constituted vision which presents entirely new possibilities of experience.

One thing that is certainly characteristic of Advaita is its insistence upon understanding different levels and kinds of experience and knowledge in their own integrity, for being precisely the kinds of experience and knowledge that they are. The aesthetic is not simply an intimation of pure consciousness, although it is certainly that, but it offers itself as well as an avenue of understanding and enrichment ever as valuable (and for many of us indeed even more so) than simple perception and reasoning. The aesthetic demands an *integration* of otherwise disparate, and sometimes opposed, orders of feeling and intellect. It yields, through this very integration and consequent concentration, insight into that *nāma-rūpa* configuration which we call 'world'.

This leads us to that alleged second point of similarity between the aesthetic and the spiritual which we have selected from the Masson–Patwardhan account. Time and space, they say, 'disappear for the duration of the experience'. A closer Advaitic analysis would, I think, show something like this. In aesthetic experience it is certainly the case that our ordinary involvement with serial, public time and geometrically measurable space, however these are to be understood philosophically, is indeed suspended; but rather than time and space disappearing, we have with the aesthetic the realization of very different spatial and temporal relationships.

I have argued elsewhere that the primary meaning of time is that which is intrinsic to an activity, with artwork time being its consummate achievement.[10] Works of art do not so much exist in time as they create the temporality that goes to define their being. Music, dance, even the so-called spatial arts of painting and architecture, give rise to temporal relations that exist nowhere else except in the specific concrete works. And the experience of them is very much the apprehension of those relations that constitute the work. Concentrating attention on what is present in a work, the experiencer of it takes on as his own consciousness the intrinsic rhythms and complex patterns established by the work.

Advaita, in its general theory of perception (*pratyakṣa*) argues that perception is a taking on of the 'form' of the object. Perception in general is a kind of gestalt assimilation, the mind 'becoming' that which it experiences. Perception, for Advaita, is an active engagement and not a passive reception. This is all the more so in that concentrated perceptual intensity which is the aesthetic: the time, the space, intrinsic to the vital object, the artwork, becomes the very content of consciousness. A qualitatively different kind of temporality and spatiality, rather than their disappearance as such, becomes part and parcel of aesthetic experience.

Masson and Patwardhan have said that for both aesthetic and spiritual experience 'special preparation is necessary'. Yes, but of very different kinds. According to classical Indian aesthetics, the respondent to art must be at once sensitive and knowledgeable; he must be a *sahṛdaya*, 'one of similar heart'. Abhinava defines the *sahṛdaya* as

> those people who are capable of identifying with the subject matter, as the mirror of their hearts has been polished through constant repetition and study of poetry, and who sympathetically respond in their own hearts.[11]

Advaita would hardly take exception to this, for it is entirely in keeping with its own claim of equality, as it were, between individual and universal consciousness. It would, however, stress one factor which might otherwise be overlooked and that is that the equality that obtains in spiritual experience, unlike with the aesthetic, is one of identity; the realization that individual consciousness in its own depth is the same as universal consciousness in its ground (*tat tvam asi*). The *sahṛdaya*, on the other hand, must be equal to the work in the sense, not of identity with it, but of being open to what it presents. Its content becomes a part of his consciousness and transforms it, but, once again, the artwork is not his consciousness as such. This leads us to a final consideration with regard to aesthetic experience.

As I have indicated before, among the *differences* between the aesthetic and the religious (or purely spiritual) that Masson and Patwardhan noted was the supposed fact that 'to have a profound aesthetic experience . . . does not imply that one will be in any sense profoundly altered'. Now I think Advaita would strongly disagree with this claim. Not only are all experiences transfigurative in a *karmic* sense—one being affected always by the substance and style of one's actions—but especially those where there is that concentrated participation of subject with object which we have seen to be essential to the aesthetic. It is not the case, as Masson and Patwardhan allow, that 'to have a profound aesthetic experience is simply satisfying . . . ', for indeed the more profound it is the more it changes one's past values and present understanding, and one's capacity for future awareness. Aesthetic experience is not just additive, one self-sufficient thing at a time, but like spiritual practice generally it becomes part and parcel of what Confucius was pleased to call 'self-cultivation' or what Advaita might understand in terms of *manana*, the inward appropriation of that which is heard (*śravaṇa*).

To appropriate something, to be influenced by it, means to bring about a reintegration of consciousness, a taking on, in artwork experience, of something of that ordered integration which is the work. Becoming equal to it, the *sahṛdaya* meets the vitality, the infused power of the work and is enhanced thereby.

This does not mean, however, that art, aesthetic experience, is to be a kind of therapy, a healing of a divided or distraught consciousness. Just as yoga (*jñāna* or other) is genuinely possible only for one who has already achieved a high level of selfless awareness, so the transformative power of the aesthetic is realizable only by one who is open, available and receptive to it in a sensitive knowing way.

III

An artwork is *māyā* par excellence, for it is brought forth to be a semblance, an 'illusion'. But unlike a Platonic reading of this which would regard the artwork as an inferior thing, an imitation of an imitation, the Advaitin would see it as providing the basis for a very special ontological status to be accorded the work. It is precisely its own awareness, as it were, of being illusory that lifts an artwork from a binding *avidyā* to a kind of liberating *jñāna*—a self-shining *abhāsa*, an illusion so aware of being so that it becomes a radiant presence. Let me explain.

For Advaita, a work of art, like everything other than Brahman, pure consciousness itself, is sublatable, but as a product of *līlā* it is rather like the world understood in terms of *pariṇāmavāda* in relation to *vivartavāda*, a real transformation until such time as seen otherwise—and a real transformation, once again, of a special kind, for an artwork plays what we might call a participatory role. A masterful artwork not only points to Reality, as the creative symbol that it is, it participates in that Reality as partaking of its very essence.

The silence, the plenitude which is Brahman, not only is the background context, if you will, of all expression, it is a fundamental intrinsic feature of art, of an art which listens as well as speaks. As an articulation of form, an artwork by its very nature presents both itself as the creative expression that it is and the unsaid, the inarticulate, inexpressible Being in which it participates. Hans-Georg Gadamer, very much in keeping with what an Advaitin would hold, has said that 'our sensitive-spiritual existence is an aesthetic resonance chamber that resonates with the voices that are constantly reaching us, preceding all explicit aesthetic judgement'.[12] By which he means that 'the nature of [what he calls] the hermeneutical experience is not that something is outside and desires admission. Rather, we are possessed by something and precisely by means of it we are opened up for the new, the different, the true'.[13]

A masterful work of art brings always as part and parcel of its being *saccidānanda*, the blissful consciousness of being itself. For waking consciousness, ordinary things are just there 'at hand' for us; they are the Other, as objects to us. For the deeper, or at least higher integrated consciousness associated with aesthetic experience, the 'object', the artwork is seen, experienced, recognized to be grounded in, and imbued with, spiritual being. It becomes for us, like Īśvara, like *saguṇa* Brahman, the supreme, but self-surpassing and fulfilling illusion.

In sum, I have tried to present what I think to be some of the main features of a reconstructed Advaita aesthetics. Its theory of creativity would rely heavily upon the concept of *līlā*, sportive play, seen in the context of *karmayoga*, skilful action performed for its own sake as a kind of 'offering' and thus as a kind of 'ritual' practice. Being a freely constituted act, creativity becomes as well an insightful act grounded in reality, and hence becomes a kind of *śruti*-making, an authorless revelation.

The experience of art, for Advaita, while having an integrity of its own, is closely allied to spiritual experience. The aesthetic involves a concentrated attention upon its object in such a way that the distinction between them is radically transformed. The subject–object distinction is not abolished, for aesthetic experience is always, as the phenomenologist would say, an experience *of* its object. It discloses the inherent meaning and truth presented by the object and opens up for us new possibilities of understanding and experiencing the world. It shows the qualitative character of temporal and spatial relations and brings about profound changes in the consciousness of the sensitive-knowing experiencer.

The ontological status of art, for Advaita, would—like for every other phenomenal thing—be *māyic*; but, most importantly, of a very special kind. For art in its creativity, experience and very being is a highly self-conscious illusion or semblance which because of that very fact grounds it in Reality in a participatory way. It gives the artwork then something like the status of Īśvara, that playful spirit, who remains for Advaita the supreme illusion.

NOTES

1. Śaṅkara, Commentary on *Māṇḍūkya Kārikā*, I.7, in *Select Passages from Śaṃkara's Commentary on Māṇḍūkya Upaniṣad and Kārikā*, trans. by T.M.P. Mahadevan, Madras: Ganesh & Co., 1961.
2. Śaṅkara, *Brahmasūtrabhāsya*, II.1, 33; trans. by George Thibaut (Sacred Books of the East Series, vol. XXXIV), Delhi: Motilal Banarsidass, 1973.
3. *Advaita Vedānta: A Philosophical Reconstruction*, Honolulu: East-West Center Press, 1969, ch. 4.
4. Rāmānuja, *Vedārthasaṃgraha*, ed. and trans. by J.A.B. van Buitenen, Poona (Deccan College Monograph Series: 16), 1956, p. 192.
5. Ananda K. Coomaraswamy, *Christian and Oriental Philosophy of Art*, New York: Dover Publications, 1956, p. 36.
6. M. Hiriyanna, *Art Experience*, Mysore: Kavyalaya Publishers, 1954.
7. Ibid., p. 10.
8. J.L. Masson and M.V. Patwardhan, *Śāntarasa & Abhinavagupta's Philosophy of Aesthetics*, Poona: Bhandarkar Oriental Research Institute, 1969.
9. Ibid., pp. 161–2.
10. *Personhood, Creativity and Freedom*, Honolulu: University of Hawaii Press, 1982.
11. Ibid., p. 78, n.4.
12. Hans-Georg Gadamer, *Philosophical Hermeneutics*, trans. and ed. by David E. Linge, Berkeley: University of California Press, 1976, p. 8.
13. Ibid., p. 9.

Chapter 19

Religiophilosophical Meditations on the Ṛgvedic Dictum: *Ekam Sad Viprā Bahudhā Vadanti*[1]

R.A. Mall

I INTRODUCTION

That human beings think, judge, feel and act differently is a general observation amply verified. But the difference is not total and fully exclusive, for in that case we cannot even articulate such a total difference. That human beings meet to differ and differ to meet seems to point at some deep-rooted anthropological similarity as the very basis for the possibility of understanding and communication among philosophies, cultures and religions. In this respect there is a point in talking about some form of a transcendental hermeneutic anthropology.

The idea of a common humanity may be considered a fact of empirical experience. It can also be given the status of a postulative norm which stands first for a methodological beginning and second for a regulative idea to be realized in the discourses of philosophies, religions and cultures. The universality we just hinted at must not be taken in a strong sense. It really makes its appearance in and through the overlapping centres which are there due to different reasons.[2] The *ekam sad* . . . of the Ṛgvedic discovery transcends the local, cultural, philosophical and religious differences and yet helps us to comprehend them.

The mutual understanding among different cultures, philosophies and religions is made possible only when we avoid the two fictions of total identity and complete difference. Whereas total identity makes understanding really redundant, total difference makes it impossible. It is true that some taste, aptitude and a bent of mind are prior to the taste for a particular religion and philosophy. So the overlapping centre for mutual communication must be located in a sentiment which stands for a mentality over and above the particular conventions.

The hermeneutics we are badly in need of today and which alone can do justice to the *de facto* existing worldwide hermeneutic situation takes the two elements of understanding others and being understood by them seriously. Thus the desire to understand and the desire to be understood go hand in hand. And the Vedic dictum in the sense of a living transcendental attitude makes such a mutual communication possible. The term 'transcendental attitude' stands here not only for a methodological and formally logical possibility but also, and in our present context even more so, for a higher level of reflection and intuition accompanying our own particular philosophical conventions, cultures and religions.

The mentality to think in an either or logic, i.e. in a logic in terms of the principle of contradiction, is worst suited to solve the problems facing interculturality and interreligiosity today, for such a logic fails to understand the complexity of the subject at hand. The two-valued logic has led more to demarcation than to co-operation. The ideal of negation of contradiction may be useful as a formalistic goal to be pursued, but in the field of humanities (*Geisteswissenschaften*), where local and cultural patterns are of central importance, contradictions (at least in their less severe forms) are what we have to learn to mitigate and to live with. Monistic patterns appear in different masks in religion, philosophy, culture and politics and they are always more or less intolerant, for they necessarily tend to negate what is not like what they are. The Vedic dictum is an antidote for any such monistic way of thinking. It pleads for unity in terms of *ekam sad* . . . but does not stand for uniformity. It is against a universal reading of reality and allows for a plurality of interpretations.

II THE ORIGINAL CONTEXT

The Ṛgvedic dictum that that which is, is One Supreme Reality; only the wise men name it differently, is, no doubt, cryptic, aphoristic and also enigmatic. It, nevertheless, represents a reservoir for further religious and philosophical interpretations and speculations. And this is true not only within the context of Indian tradition but also in the context of the world today.

The Vedic saying as such seems to be an answer to a question which the Vedic seers must have put to themselves as to the real nature of the one truth which is intuited and realized in some spiritual, mystic experience. The Vedic seers feel the primacy of questions over answers, in spite of the fact of their mystic experience. It must have been this tension between critical thinking and reflection (*manana*) and spiritual, mystic experience (*nididhyāsana*) that led them to formulate this dictum which stands for some form of monistic pluralism or pluralistic monism. I do not wish here to introduce some types of neologism. All that the above phrase stands for is the inner unshakeable intuitive experience and conviction that there is one Supreme Reality which appears under different names. It is the conviction that there are different ways to one and the same goal. It may also stand for a postulate, for a regulative idea with which we should start whenever we come in contact with other philosophies, cultures and religions. That which unites us is nearer the truth than that which separates us. The different names are verbal, and sometimes even trivial if they thereby claim to be exclusively in possession of the One Truth.

The context in which this Vedic sentence originally occurs begins with a documentation of the result of the self-introspection. The Vedic writer describes the beauty and grandeur of the sun and the moon. He says that he is ignorant and aspires for knowledge. He asks about the First Born (*ko dadarś prathamaṃ jāyamāna*).[3] Besides the desire to know the unborn One, the Vedic seer also displays the presence of some doubt and scepticism in his attitude which is so characteristic of philosophic thought whether in the East or in the West.[4]

There is much controversy whether this Vedic saying is just anthropomorphic. The one Supreme Truth which was there from eternity is beyond names, for words and names came into being after human beings. That one (*tadekam*)[5] is beyond all analogies known to human mind. The Vedic seers may thus truly be interpreted to have formulated a metaphysical and religious intuition with a wide possibility of application in different fields of human activities transcending the narrow historico-temporal contexts in which the one truth under different names is expressed and exemplified.

III THE IMPORT OF THE VEDIC DICTUM TODAY

It is undoubtedly true that nothing happens without a context which may be of various natures. Since the different contexts are not just watertight compartments comparable to windowless monads, the philosophical and religious interpretations take place both intra- and interculturally and intra- and interreligiously. The Vedic dictum, put to such a test fares quite well.

Not only Western scholars but even some Indian scholars also tend to interpret the above saying in an anthropomorphic and polytheistic sense. Since they are convinced of a progressive, linear development from polytheism to monotheism and monism they fail to see the well-grounded possibility of interpreting the Vedic dictum as the deep-rooted conviction of the Vedic sages that different gods are just different manifestations of the one supreme God. Max Müller took much pains and construed a term for a position which, according to his reading of the Vedic dictum, is neither just a blunt polytheism nor a full-fledged monotheism. Henotheism is the term Max Müller invented, and he meant a transitional stage from polytheism to monotheism. Seen from the Christian theological framework, such a reading might be eligible, but it hardly does justice to the Indian religious and philosophical framework which, though not radically different, does believe in the real possibility of one God under different names. The richness and fullness of one God do in no way suffer from being called by different names. It is the greed of narrow human thinking which claims the One exclusively. The Indian religious tradition, in contrast to the Christian, does underline the fact that the one God remains the same even after being worshipped under different names. Any tradition which disapproves of such a possibility tends to be monistic leading to a very narrow hermeneutics which defines understanding in terms of self-understanding. The *ekam sad* . . . of the Vedic dictum is, of course, in need of language, but it also defies language in the sense of not being exhausted by its names. 'Neither polytheism nor henotheism nor even monotheism', writes Sharma, 'can be taken as the key-note of the early Vedic philosophy.'[6]

A very pertinent question regarding the philosophical and theological import of this Vedic dictum relates to the problem whether the One is more than the sum total of the many names given to it or whether the many names make and exhaust the One. If we accept the latter reading, it may lead to some form of uniformitarian thinking in philosophy and religion if the many names individually and exclusively claim the One. The first reading is more liberal and leaves room for transcendence with the further quality of being comprehensive, tolerant and adaptable. The Hindu-Trinity of the three gods in One Supreme God does not exhaust the One. The One is, no doubt, available to all the three, but it is fully available to none of them exclusively. Exchanging names for one and the same thing includes some underlying metonymic relation. Any theory of philosophical and theological hermeneutics today needs to be more pervasive, tolerant and open-minded, reading the philosophical and religious utterances of the past in the spirit of the Vedic dictum. The above-quoted Vedic dictum depicts a religiophilosophical framework of the Hindu mind, for which the religious and philosophical truth presents itself in manifold ways. The Vedic dictum thus allows us to read the Christian saying of there being many mansions in God's house as also there being many paths leading to God's house. Such a reading paves the way for a better mutual understanding among religions, philosophies and cultures. This implies the concept of an 'analogous hermeneutics' which allows the One to be called by different names. An analogous hermeneutics abstains from a hermeneutics for which the relation of identity is of paramount importance. It also rejects a hermeneutics which takes the differences very radically. Monistic patterns are, therefore, less tolerant, for they plead for a hermeneutics which either aims at converting or neglecting the other. The recorded human history is full of examples substantiating the thesis mentioned above. An analogous hermeneutics takes for its departure overlapping centres among philosophies, cultures and religions and tries to enlarge and deepen them for an ever better understanding.

IV THE IDEA OF '*RELIGIO PERENNIS*'

For an idea of *religio perennis,* all the truths in other fields of human life are derived from only one principle which is called God. This points to the ultimate theological anchorage which is amply testified by the theological realism of Thomas Aquinas and also by the claim of Catholic metaphysics to be the true form of *philosophia perennis.* Such one-sided and exclusive claims are not absent in Indian tradition. The term which shows some overlapping with the idea of a *religio perennis* in the Indian religious framework is '*sanātana dharma*', the one eternal truth or religion (*dharmaḥ sanātanaḥ*).

It is of course true that in some of its historical forms Hinduism itself claims to be this eternal religion. This claim is not legitimate, though. Such a claim became dominant during the nineteenth and twentieth centuries when Hinduism came in contact with Christianity. Tilak, Gandhi, Tagore, Ram Mohan Roy, Dayananda subscribed more or less to such a reading of *sanātana dharma.*

Sanātana or *Śāśvatadharma* is not itself a *dharma* in the sense of being a positive religion with its own cult and other religious practices. *Sanātana dharma* is like the transcendental ego which, too, is not something existing over and above the empirical and yet given to us at a higher level of consciousness. One who ascribes to such a reading of *sanātana dharma* is committed to no one particular religion as the only true religion. Its neutrality is its strength which must not be confused with disinterestedness.

To be true to the spirit of the Vedic dictum it is not justified to identify the '*tadekam*' with the positive religion of Hinduism. This would really mean committing the mistake of *pars pro toto.* The richness and tolerance of Hinduism lie in its conviction that the different religions, including Hinduism itself, are the different expressions of the one eternal *dharma.* All that Hinduism may claim as its own original contribution to the philosophy of comparative religion is its insight to see the *religio perennis* as the eternal religious wisdom lying at the back of all positive religions.[7] This additional insight itself is no positive religion one may really belong to. Nobody can call it exclusively his own. This insight is like a shadow which accompanies or should accompany all the positive religions reminding them all the time of the Vedic dictum of *ekam sad*. . . . It is also the very core of a religiously tolerant hermeneutics which alone is in the position to further the cause of a dialogue among religions. *Sanātana dharma* can, thus, be interpreted in the real spirit of the Vedic dictum as a regulative, basic religiosity which is the very cornerstone of interreligiosity. It may also function as a guiding principle on the methodological level.[8]

V THE VEDIC DICTUM AND THE IDEA OF *PHILOSOPHIA PERENNIS*

We have briefly hinted at the religious overlapping between the *sanātana dharma* and *religio perennis.* A similar overlapping and conceptual concordance may be worked out between the *ekam sad* . . . and the notion of a *philosophia perennis.* 'Every one', Jaspers writes, 'possesses philosophy only in his historical form, and this is of course, so far as it is true, an expression of *philosophia perennis,* which no one possesses as such.'[9] The term '*philosophia perennis*' was first introduced by A. Steuchus in 1540 for those fundamental truths which are taken to exist in all races and cultures beyond the limitations of time and space.[10] In spite of this liberal attitude, philosophers have not stopped claiming to locate *philosophia perennis* in some particular race, culture and philosophical convention.[11]

As there is no one positive religion which can and should be identified with the all-encompassing one universal, eternal religion, so also there is no one philosophical tradition, whether in the West or in the East, which can and should exclusively claim to possess the one *philosophia perennis* as such. The claim to possess *philosophia perennis* has been explicitly or implicitly made in nearly all traditions. Among the modern Western philosophers there are three names with the initials H (Hegel, Husserl and Heidegger) who in opposition to Schopenhauer, Deussen, Scheler and Jaspers identify the very idea of philosophy with the European philosophy. Heidegger even goes so far as to deny the application of the predicate philosophy to any other tradition. He writes: 'The often heard expression the "Occidental-European philosophy" is in truth a tautology'.[12] The reason for this, Heidegger's rather *a priori*, non-empirical and linguistic claim, is the conviction that philosophy is only of Greek origin. The adjective 'European' in the phrase 'European philosophy' gets a privileged treatment by Heidegger.[13] But such a one-sidedness is also found on the Indian side claiming that philosophy is essentially spiritual and Indian.[14]

A. Huxley uses the term perennial philosophy more in its cultural and historical aspects than Steuchus and aims at an integral thesis which takes the one *philosophia perennis* to express itself in various cultural and historical garbs. Huxley lays emphasis particularly on the converging elements.[15]

The Vedic dictum of one truth under different names rightly seems to plead for a liberal attitude that the one *philosophia perennis* is not the exclusive possession of any one. It further makes clear that translations are made and that rightly so. The thesis of the impossibility of translation is thus groundless, in spite of the fact that there are numerous difficulties in translating from one language and culture into another. The meeting of different philosophies, cultures and religions today is badly in need of such an attitude. We require such an attitude not only interculturally but also intraculturally. Such an intercultural and interphilosophical attitude methodologically takes into consideration the new global hermeneutic situation and supports a renewal of philosophical dialogue. The philosophical conviction, characteristic of such an attitude, pleads for the thesis of philosophy having different places of origin. Philosophy *qua* philosophy is not totally culture-bound though it needs the soil of a culture to take root and grow. The one *philosophia perennis*, which is rightly said to be universal, cannot have any prejudice for or against a particular philosophical tradition. The Vedic dictum might be taken to be a testimony for this. The *ekam sad* . . . has no mother tongue.

Another characteristic feature of the Vedic dictum can be seen in its philosophical modesty, for it does not overrate the importance of a logocentric orientation. Such modesty may lead to an epistemological attitude to abstain from claiming total understanding of all that is there to be understood. It also prevents us from hypostatizing one eternal idea of philosophy and philosophical truth.

There is a tendency in certain philosophical conventions to define the philosophical truth in terms of a particular tradition and that particular tradition in terms of philosophical truth. This is a strategy which is not only circular but also vacuous and merely nominalistic. *Philosophia perennis* as the very overlapping centre is the *sine qua non* for philosophical understanding beyond the narrow borders of one's own tradition and within the broader context of world philosophy. 'It is the *philosophia perennis*', Jaspers writes, 'which provides the common ground where most distant persons are related to each other, the Chinese with the Westerners, the thinkers before 2500 years with those of the present'.[16]

No philosophical convention can empty the endless reservoir of the *philosophia perennis.* It has ever been the dream of many metaphysicians and theologians to possess the one eternal truth in its entirety and in all concreteness. This greedy illusion has led to an endless misunderstanding among different philosophical '*Weltanschauungen*'. In the spirit of the phenomenological method, one may rightly view the *philosophia perennis* and the *ekam sad* . . . of the Vedic dictum as the ideal noematic correlate of the noetic acts of human beings belonging to different races, cultures, religions, etc. Such a phenomenological deconstruction of *philosophia perennis* does not fully reject the idea of the universality of the *philosophia perennis*; it rather puts it in an overall context of the universality of the noetico-noematic correlationship.

The *syādvāda* logic of the Jaina philosophy supplies us with a very effective tool of liberal deconstruction of every point of view which places itself in an absolute status and claims to be exclusively in possession of the one truth. Nothing is given to us in the absence of perspectives. The logic of *syādvāda*, of course, has to remain satisfied with the different perspectives taken without any further claim to an absolute reconciliation of different perspectives in one grand perspective. The Jainas might have taken their own

Weltanschauung as an all-encompassing view but that is, strictly speaking, no philosophy; it is rather a matter of personal decision and liking. Such is also a lesson we should draw from the Vedic dictum of *ekam sad*. . .

VI TOWARDS A METONYMIC THEORY OF AN ANALOGOUS HERMENEUTICS OF THE VEDIC DICTUM

No two philosophies, cultures and religions are totally commensurable or fully incommensurable. This is true interculturally as well as intraculturally. If the one supreme truth is to be found exclusively in one philosophical or religious tradition, the term diversity of cultures and religions loses every meaning. Such an understanding of *philosophia perennis* and *religio perennis* also hampers mutual understanding among cultures, philosophies and religions. The hermeneutics which is at work here is that of identity which, to put it strongly, asserts that to understand Śaṅkara is to be a Śāṅkarite, to understand Plato is to be a Platonic and so on. A hermeneutics of identity has the further disadvantage that it relegates other philosophies, cultures and religions to some primitive forms on the way of the progressive march to the telos of one philosophy, culture and religion. Hegel could be cited here as a fitting example.

In opposition to the hermeneutics of identity, there is one of total difference, which posits radical differences among cultures, religions and philosophies, to the extent that even the general terms philosophy, culture and religion are not applicable to other cultures, religions and philosophies. One wonders whether such a total difference can at all be articulated. The hermeneutics of total difference nips the very possibility of mutual understanding in the bud. On closer analysis, we find both the types of hermeneutics (one of total identity and the other of total difference) as two sides of the same coin, for they both establish their way of understanding as the most paradigmatic.

The Vedic dictum of one truth under different names tries to steer clear between these two extreme and fictional positions and allows us to accept and respect both similarities and differences making room for the common expression of the one Supreme Truth in different philosophies, religions and cultures. The one ultimate reality is undoubtedly in need of linguistic expression, but, at the same time, it resists every attempt at total expressibility.[17]

The science of metonymy deals with the fact of using different names for things which are associated. It uses one word for another expecting the different names to suggest one and the same thing. It thus points to the indubitable consciousness of difference between the name and the named, between concept and reality. The thing named is always in need of being named although it is not totally taken up with its names. The metonymic figure of speech, in its ultimately metaphysical and religious aspect, stresses the fact that the one Supreme Reality, though in need of expression, is not totally available to language and expression. This must have been the deep-rooted conviction of the Vedic seers when they put it in their famous words '*ekam sad viprā bahudhā vadanti*'.

In the spirit of Vedic dictum, we are better suited to overcome the naïve, mundane dogmatism which easily forgets that every point of view expresses a limited perspective.[18] No authentic philosopher or genuine believer can generate a dialogue among philosophies, cultures and religions unless he or she is personally convinced of the truth that the One Truth is nobody's possession alone. The discovery of one truth which the wise call by different names is one of the greatest contributions of Indian thought to the world.

The Vedic dictum can even be extended to political, social and ethical fields. All that we need is the cultivation of an inner attitude which allows to read and let read, to interpret and let interpret, to believe and let believe and so on. To discover and possess this culture of interculturality and interreligiosity on the edge between critical reflection (*manana*) and meditation (*nididhyāsana*) is what we are badly in need of today, for much hope rests on philosophy that it be not only a way of thought but also a way of life.[19]

The Vedic dictum formulates a religious and philosophical conviction (you may as well call it insight, intellectual intuition or some sort of mystic experience) which is beyond mere historicism and relativism. The Vedic dictum does accept the truth of relativism, but it is a relativism which makes the different names relative to the One Supreme Truth and not the names among themselves. It thus provides us with an overarching frame of reference.

NOTES

1. *Ṛgveda*, I, 164, 46. 'To What is One, Sages Give Many a Title', R.T.H. Griffith, trans., *The Hymns of Rigveda*, 2 vols, Benares: E.J. Lazarus and Co., 1920–6, 3rd edition.
2. R.A. Mall, *Philosophie im Vergleich der Kulturen* (Interkulturelle Philosophie-eine neue Orientiorung), Darmstadt: Wissenschaftliche Buchgesellschaft, 1995.
3. *Ṛgveda*, I, 164, 4.
4. Ibid., X, 6, 7.
5. Ibid., X, 129, 2.
6. C. Sharma, *A Critical Survey of Indian Philosophy*, Delhi: Motilal Banarsidass, 1976, p. 15.
7. Cf. P.A. Schlipp, ed., *The Philosophy of Radhakrishnan*, New York: Tudor Publishing Corporation, 1952, p. 80.
8. For a more extensive information and discussion see: W. Halbfass, *India and Europe*, New York: State University of New York Press, 1990; *Sanatana Dharma. An Advanced Text-Book of Hindu Religion and Ethics*, Madras, 1940 and B.K. Tirtha, *Sanatana Dharma*, Bombay, 1964.
9. 'Jeder Mensch besitzt Philosophie nur in seiner geschichtlichen Gestalt, diese ist doch, sofern sie wahr ist, ein Ausdruck der philosophia perennis, die als solche niemand besitzt'. K. Jaspers, *Weltgeschichte der Philosophie*, München: R. Piper and Co., 1982, pp. 20–1.
10. Cf. A. Steuchus, *De perenni philosophia*, with a new Introduction by Ch. B. Schmitt, London: Chatto and Windus, 1974.
11. Cf. R.A. Mall and H. Hülsmann, *Die drei Geburtsorte der Philosophie. China, Indien, Europa*, Bonn: Bouvier Verlag, 1989; and R.A. Mall, *Die orthaft ortlose philosophia perennis und die interkulturelle Philosophie, in: Das Begehren des Fremden*, ed. by L.J.B. Duala-M'bedy, Essen: Die Blaue Eule, 1992.
12. 'Die oft gehörte Redeweise von der "abendländisch-europäischen Philosophie" ist in Wahrheit eine Tautologie', M. Heidegger, *Was ist das—die Philosophie?*, Pfullingen: Neske, 1963, p. 13.
13. R.A. Mall, *Meditationen zum Adjektiv 'europäisch' aus interkultureller Sicht, in: Der technologische Imperativ*, München/Wien: Profil Verlag, 1992.
14. Cf. J.N. Mohanty, *Reason and Tradition in Indian Thought*, Oxford: Clarendon, 1992, p. 290.
15. A. Huxley, *The Perennial Philosophy*, London: Borgo Press, 1990.

16. 'Es ist die philosophia perennis, welche die Gemeinsamkeit schafft, in der die Fernsten miteinander verbunden sind, die Chinesen mit den Abendländern, die Denker vor 2500 Jahren mit der Gegenwart'. K. Jaspers, *Weltgeschichte der Philosophie*, München: R. Piper and Co. Verlag, 1982, p. 56 (my translation).
17. Cf. R.A. Mall, 'Metonymic Reflections on Śaṅkara's Concept of Brahman and Plato's Seventh Epistle', *Journal of Indian Council of Philosophical Research* 10:3, 1993.
18. Cf. F.C. Copleston, *Philosophies and Cultures*, Oxford: Oxford University Press, 1980, pp. 2–5.
19. Cf. R.A. Mall, *Philosophie als Denk- und Lebensweg*, in *Probleme philosophischer Mystik*, Sankt Augustin: Academia Verlag, 1991.

Chapter 20

The Buddhist Theory Concerning the Truth and Falsity of Cognition

Masaaki Hattori

In the Jaiminidarśana chapter of the *Sarvadarśanasaṃgraha*, a section is devoted to the elucidation of the doctrine that truth is intrinsic to a cognition (*svataḥprāmāṇya*), which forms one of the important subjects in the Mīmāṃsā system. The Naiyāyika, who is disinclined to accept the Mīmāṃsaka view that the cognition derived from the Vedic Scriptures is intrinsically true, points to the fact that there prevails a wide divergence or view concerning the truth and falsity of cognition, with the following verses:

> *pramāṇatvāpramāṇatve svataḥ sāṃkhyāḥ samāśritāḥ |*
> *naiyāyikās te parataḥ saugatāś caramaṃ svataḥ ||*
> *prathamaṃ parataḥ prāhuḥ prāmāṇyaṃ vedavādinaḥ |*
> *pramāṇatvaṃ svataḥ prāhuḥ parataś cāpramāṇatām ||*[1]

The Sāṃkhyas assert that both truth and falsity of cognition are intrinsic. The Naiyāyikas hold that both extrinsic, and the Bauddhas state that the latter [i.e. falsity] is intrinsic while the former, i.e. truth, is extrinsic. Teachers of the Vedas [i.e. the Mīmāṃsakas] maintain that truth is intrinsic while falsity is extrinsic.

A very similar statement is found in the *Mānameyodaya*, a Mīmāṃsaka manual composed in the seventeenth century.[2]

That the truth and falsity of cognition are determined *intrinsically* (*svataḥ*) means that a cognition is endowed by nature with truth or falsity, that is to say, that a cognition is determined to be true or false by itself. On the other hand, that the truth and falsity of a cognition is determined *extrinsically* (*parataḥ*) means that there are, apart from the cognition itself, some other factors, such as merit (*guṇa*) or defect (*doṣa*) belonging to the causes of cognition, which function to make a cognition either true or false. According as truth and falsity are considered either intrinsic or extrinsic, there can be the following four different views:

(1) both truth and falsity are intrinsic,
(2) both truth and falsity are extrinsic,
(3) falsity is intrinsic while truth is extrinsic, and
(4) truth is intrinsic while falsity is extrinsic.

In the above-cited verses these four views are ascribed respectively to (1) the Sāṃkhya, (2) the Naiyāyika, (3) the Bauddha, and (4) the Mīmāṃsaka. The Naiyāyika and the Mīmāṃsaka are known to have expounded (2) and (4) respectively, but it is doubtful whether (1) and (3) were really advocated by the Sāṃkhya and the Bauddha, since these views are nowhere attested in the works of the said schools. The Sāṃkhya might have maintained (1), for it is in harmony with the Sāṃkhya doctrine that an effect, whether it is physical or psychical, is immanent in its cause, but the Bauddha is known to have formed a view different from (3), as will be examined below.

Kumārila Bhaṭṭa (*c.* 600–60) was perhaps the first to discuss the problem concerning the truth and falsity of cognition. In the second chapter of his *Ślokavārttika* (*ŚV*), he treats this problem in detail with a view to proving that the cognition derived from the Vedic injunction (*codanā*) is never untrue. It was his intention to establish the unconditioned authority of the Veda. Some of the verses in the *ŚV* are found quoted in the *Tattvasaṃgraha* of Śāntarakṣita[3] (*c.* 725–88) and the commentary on it by Kamalaśīla (*c.* 740–95), and also in the *Nyāyamañjarī* by Jayanta Bhaṭṭa[4] (tenth century). No reference is made in these treatises to any other earlier work dealing with this subject.

Kumārila's discussion on the truth and falsity of cognition is set forth in his elucidation of Śabarasvamin's commentary on the *Jaiminisūtra*, 1.1.2 (*codanā-sūtra*). Śabarasvamin makes reference therein to the view of an opponent who does not accept the Mīmāṃsaka doctrine that the *dharma* which is the means to attain the ultimate bliss (*niḥśreyasa*) is indicated by the Vedic injunction. The opponent argues that the Vedic injunction might sometimes tell what does not conform to the reality (*atathābhūta*), just like such an ordinary statement as 'there are fruits on the bank of the river'. To this objection Śabara gives the following answer:

With the words it tells (*bravīti*) and what is not true (*vitatha*), something contradictory is expressed. It tells means it makes one know (*avabodhayati*), or it is an efficient cause for a knower. When a man knows [something] on the presence of a certain thing as an efficient cause, this latter makes him know it (*avabodhayati*). If, on the presence of an injunction (*codanā*), it is known that through the *agnihotra* there results heaven, how can it be said that it is not so. If it is not so, how can it be known. To say that he knows a thing which is unreal (*asantam artham*) is contradictory. Furthermore, it is not that from the statement he who desires heaven should perform sacrifice something dubious is known in the form: Does there result heaven or not? And, when this is definitely known, it would not be false. Because, a false knowledge is such one that, having appeared, perishes [with the notion] 'it is not so'. However, this [knowledge in question] is not reversed at another time, in another person, in another state, nor at another place. Therefore it is not untrue.[5]

Kumārila gives full support to Śabara's idea expressed in this passage through the examination of the nature of cognition, which begins with the following verse:

> *sarvavijñānaviṣayam idaṃ tāvat parīksyatām /*
> *pramāṇatvāpramāṇatve svataḥ kiṃ parato 'tha vā //*[6]

Concerning all cognitions, the following should, first of all, be examined: is truth or falsity [of a cognition] intrinsic or extrinsic?

The four alternative views mentioned above are listed by Sucaritamiśra under this verse in his *ŚV-Kāśikā*.[7] Kumārila rejects the first and the second views,[8] and then introduces the arguments put forward by the upholders of the third view, viz. the view that falsity is intrinsic and truth is extrinsic. The fundamental argument made by them is as follows:

> *aprāmāṇyam avastutvān na syāt kāraṇadoṣataḥ /*
> *vastutvāt tu guṇais teṣām prāmāṇyam upajanyate //*[9]

Since the falsity (*a-prāmāṇya*), [being the absence (*abhāva*) of truth (*prāmāṇya*),] is non-entity, [it cannot be produced. Therefore,] it would not result from the defects (*doṣa*) belonging to the causes [of cognition]. On the other hand, the truth, being a real entity, is produced by the merits (*guṇa*) belonging to these [causes].

The import of this verse is formulated by Sucaritamiśra into two kinds of syllogism:

A:	(Proposition)	The falsity [of a cognition] is without a cause.
	(Reason)	Because it is unreal.
	(Example)	Whatsoever is unreal is without a cause, as for instance, a hare's horn.
B:	(Proposition)	The truth [of a cognition] has its cause.
	(Reason)	Because it is real.
	(Example)	[Whatsoever is real has its cause, as] for example, a pot.[10]

Regarding the syllogism B, Sucaritamiśra makes the remark that, since there is, for the Bauddha, no real entity which is not effected [from a cause] (*akārya*), the Reason in this syllogism is not inconclusive (*anaikāntika*).[11] It is thus understood that the Bauddha was recognized as the upholder of the third view. In the *Sarvadarśanasaṃgraha* and also in the *Mānameyodaya*, this view is atributed to the Bauddha.

Since the Bauddha maintains that all phenomenal existences are conditioned (*saṃskṛta*) [by causes], the attribution of the third view to the Bauddha might be appropriate. However, no Bauddha work prior to Kumārila is known to us in which the problem concerning the truth and falsity of cognition is treated. Moreover, in the logico-epistemological school of the Bauddha a theory which contradicts the third view is known to have been formulated.

Neither Diṅnāga nor Dharmakīrti discussed whether a cognition is true intrinsically or extrinsically. Śāntarakṣita was the first among the Bauddhas who took up the problem concerning the truth and falsity of cognition. In his *Tattvasaṃgraha*, he devoted a chapter to this problem[12] and through a severe criticism against Kumārila's theory he clearly established the Bauddha position. A number of verses from Kumārila's *ŚV* and *Bṛhaṭṭīkā*[13] are cited by him and also by his disciple Kamalaśīla in his commentary on the *Tattvasaṃgraha*.

Śāntarakṣita first introduces the Mīmāṃsaka theory as follows:

meyabodhādike śaktis teṣām svābhāviki sthitā /
na hi svato 'satī śaktiḥ kartum anyena śakyate //[14]

The capacity to bring about the apprehension of a cognizable object, etc. belongs to them (=cognitions) by their very nature, since a capacity which does not exist intrinsically cannot be produced by anything else.

He then examines the meaning of '*svābhāvika*' (by their very nature) and states that he would accept the view expressed in this verse insofar as it means that the said capacity is produced by the very causes of cognition itself and not by some other causes which may arise later. However, he does not agree with the Mīmāṃsaka who maintains that the truth of cognition is ascertained intrinsically. The said capacity, he states, inheres somewhere in a cognition, but it cannot be apprehended by itself.[15]

It is admitted by the Bauddha that a cognition is cognized by itself. It, however, cognizes itself simply as an apprehension (*anubhavamātra*), but not as the apprehension corresponding to the reality. Even a false cognition is cognized by itself, but without regard to its falsity. For instance, a person with eye-disease, who cognizes a hair-tuft which in reality does not exist, is aware of his cognition, but not of its falsity. Thus the truth and falsity of a cognition is not ascertained intrinsically, i.e. by the cognition itself. Śāntarakṣita draws a conclusion with the following verse:

> *tasmād arthakriyājñānam anyad vā samapeksyate /*
> *niścayāyaiva na tv asyā (=śaktyās) ādhānāya viṣādivat //*[16]

Therefore, for the purpose of ascertainment (*niścaya*), the cognition apprehending the production of an effect or some other factor is required, as in the case of poison, etc., but not for supplying it (=capacity) [to the cognition].

The 'production of an effect' (*arthakriyā*) is, according to Dharmakīrti and his followers, the essential attribute of a real thing. The real fire has the power of producing an effect, such as cooking, burning, etc., while a cluster of *aśoka* flowers mistakenly cognized from afar as fire has no such power. Whether the cognition of fire is true or false is ascertained by the subsequent cognition by which the object cognized is known to produce the expected effect. A poisonous food or an intoxicating drink may not be distinguished from normal food or drink in appearance, but it can be discerned by perceiving its effect, viz. senselessness, perspiration or inarticulate speech. Similarly, a true cognition, which at first cannot be distinguished from a false one, is later recognized as true because of its production of an effect. The term '*anyat*' in the verse is interpreted by Kamalaśīla as meaning the 'cognition of the unstainedness of causes' (*hetuśuddhijñāna*). It is thus known that the Bauddha maintained the doctrine that the truth and falsity of a cognition is ascertained extrinsically, i.e. by the subsequent cognition.

This, however, is not the final conclusion reached by the Bauddha. Since Kumārila has pointed out, in his discussion on the truth and falsity of a cognition, a difficulty to be found in the doctrine that the truth is extrinsic to a cognition, the Bauddha has to be ready to meet it. Śāntarakṣita first introduces Kumārila's own idea which is clearly expressed in the following verse:

> *ātmalābhe ca bhāvānāṃ kāraṇāpekṣitā bhavet /*
> *labdhātmanāṃ svakāryeṣu pravṛttiḥ svayam eva tu //*[17]

Things are in need of causes when coming into existence, but once they are in existence the activity towards their own effects are done by themselves.

For instance, a pot is in need of a lump of clay, potter's stick, wheel and other things when it is produced, but, once being produced, it does not need those causes for the activity of containing water. In the same way, a cognition requires such causes as eye, object, light, etc. for its production, but once it is produced it does not depend on any cause for the activity of apprehending an object. It is the very nature of a cognition to apprehend an object, that is to say, a cognition is endowed with the capacity of apprehending an object. This means that the truth of a cognition is caused by nothing other than the cognition itself. Unless there is any element which affects its nature, a cognition remains true. If a cognition depends, for its truth, on any factor other than itself, e.g. the excellence (*guṇa*) of its causes, then its truth is to be ascertained by a subsequent cognition which apprehends the healthiness of the sense-organ, etc. Since a cognition is true intrinsically, it does not require another cognition to ascertain its truth.

The difficulty found by Kumārila in the doctrine of extrinsic ascertainment of the truth of a cognition is that it would inevitably be led to an infinite regress (*anavasthā*). If, he argues, the truth of a cognition were to be ascertained by a subsequent cognition, then, this second cognition would also require, for the ascertainment of its truth, the third cognition, which again would require the fourth. Thus there would follow an infinite regress, and the truth of the initial cognition would never be ascertained. If, on the other hand, it were held that a subsequent cognition is intrinsically true, then the first cognition should also be admitted to be true intrinsically, because there is no essential difference between the two cognitions.[18]

Against Kumārila's doctrine that the truth is intrinsic to a cognition, Śāntarakṣita states that a cognition which is admitted even by the Mīmāṃsaka to be momentary,[19] does not, after being produced, continue to exist so that it could assume the activity of apprehending an object. Since a cognition becomes non-entity immediately after coming into existence, it can hardly be maintained that a cognition, after being produced, functions towards its own effect all by itself (*svata eva*).[20] There is another defect to be found in Kumārila's doctrine: inasmuch as a cognition is intrinsically true, there would be no room for doubt (*saṃdeha*) and erroneous cognition (*viparyāsa*).[21]

After thus criticizing Kumārila's doctrine, Śāntarakṣita rejects the objection raised by Kumārila against the doctrine of extrinsic ascertainment of the truth of cognition. According to him, that the truth of a cognition is ascertained by the subsequent cognition does not necessarily entail the fault of infinite regress (*anavasthiti*). The second cognition, which ascertains the truth of the initial cognition, is ascertained to be true intrinsically and does not require the third cognition for the ascertainment of its truth. Why then is the third cognition not required? This question is answered by Śāntarakṣita with the discussion based on Dharmakīrti's concept of truth. Dharmakīrti characterizes the truth of a cognition as non-contradiction (*avisaṃvāda*) with the reality (*vastu*), and explains that non-contradiction means the constancy with regard to the production of an effect (*arthakriyāsthiti*).[22] For example the cognition of fire is known to be true insofar as the object cognized as fire produces without fail the effect such as burning (*dāha*). Śāntarakṣita accepts this truth concept, and maintains that the cognition reflecting the production of an effect (*arthakriyābhāsa-jñāna*)[23] is the distinctive mark of the truth. This cognition, he states, is clearly apprehended, and is determined [to be true] by the reflective consciousness (*āmarśanacetas*) which arises solely from that cognition.[24] Since the production of the effect, such as quenching thirst or cooking rice, is clearly known by a person who acts on his initial cognition of water or fire, and since he has a feeling of certainty about the truth of that knowledge, there is no need for another cognition which ascertains its truth. Thus the truth of the cognition reflecting the production of an effect is intrinsic.

There are other kinds of cognition which are also intrinsically true. For instance, an expert jeweller has a clear and distinct cognition of gems and coins. His cognition does not require another cognition for the ascertainment of its truth, because through repeated experience (*abhyāsa*) he is completely free from errors. Similarly the cognition of an object by a person who has repeatedly cognized the same object is also intrinsically true. Kamalaśīla enumerates five kinds of cognition which he recognizes as intrinsically true: (1) self-awareness of pleasure, pain, and other mental states, (2) yogin's cognition, (3) cognition reflecting the production of an effect, (4) inference, and (5) repeated perception.[25]

The Bauddha theory concerning the problem as to whether the truth of a cognition is intrinsic or extrinsic is clearly summarized by Kamalaśīla as follows: —Four alternative theories are mentioned by the Mīmāṃsaka, and, excepting the one maintained by himself, the other three are repudiated. But the Bauddha is not affected at all, because he does not accept any one of these four alternatives. The four theories are formulated with the restriction (*niyama*) that the truth or falsity of a cognition is *fixedly* intrinsic or *fixedly* extrinsic. The Bauddha on the other hand sets forth the theory of unrestriction (*aniyama*): he maintains that the truth and falsity are sometimes intrinsic and sometimes extrinsic. It, therefore, is not appropriate to mention only four theories. There is, as the fifth, the theory of unrestriction, which is maintained by the Bauddha.[26]

It seems that due attention was not paid to this Bauddha theory of unrestriction. In later times the theory that the truth is extrinsic and the falsity is intrinsic to a cognition is ascribed to the Bauddha, as mentioned at the beginning of this essay. This ascription is found already in the *ŚV-Vyākhyā* by Umveka, whose dates are earlier than those of Kamalaśīla. Both

Pārthasārathimiśra (eleventh–twelfth centuries) and Sucaritamiśra (twelfth century), who lived later than Kamalaśīla, also attributed that theory to the Bauddha, and do not refer to the theory of unrestriction expounded by Śāntarakṣita and Kamalaśīla. The author of the *Sarvadarśanasaṃgraha* must have simply followed this Mīmāṃsaka tradition.

NOTES

Abbreviations

ŚV — *Ślokavārttika* of Kumārila Bhaṭṭa (with *Kāśikā* of Sucaritamiśra), ed., Sāmbaśiva Sāstrī, Trivandrum: CBH Publications, 1990 (Reprint of Trivandrum Skt. Ser., XC, XCIX).

TS, TSP — *Tattvasaṃgraha* of Śāntarakṣita, *TS-Pañjikā* of Kamalaśīla. (1) Ed., Dwarikadas Shastri, Varanasi, 1968 (Bauddha Bharati Ser., 2). (2) Ed., E. Krishnamacharya, Baroda: Oriental Inst., 1988 (GOS, 31). Verse number and page number are given in accordance with (1).

1. *Sarvadarśanasaṃgraha of Sāyaṇa-Mādhava,* ed. by V.S. Abhyankar (Govt. Oriental Ser., A-4), Poona: Bhandarkar Oriental Research Inst., 1951, Second edition, p. 279.
2. *Mānameyodaya of Nārāyaṇa,* ed. by C. Kunhan Raja and S.S. Suryanarayana Sastri (The Adyar Lib. Ser., 105), Adyar, 1975, Second edition, p. 179.
3. Cf. E. Frauwallner, 'Kumārila's Bṛhaṭṭīkā', *WZKSO* VI, 1962, p. 84 (*Kleine Schriften,* p. 329).
4. *Nyāyamañjari,* ed. by Pt. Sūrya Nārāyaṇa Śukla (Kashi Skt. Ser., Nyāya Section, 15), Benares, 1936, pp. 147ff.
5. E. Frauwallner, *Materialien zur ältesten Erkenntnislehre der Karmamīmāṃsā,* Wien: Hermann Böhlaus Nachf., Kommissionsverlag der Österreichischen Akademie der Wissenschaften, 1968, pp. 16.18–18.2.
6. *ŚV,* II.33.
7. *ŚV-Kāśikā,* p. 79.23–5.
8. *ŚV,* II.34–7.
9. Ibid., II.39.
10. *ŚV-Kāśikā,* p. 85.22–3: *eṣa cātra prayogo bhavati—aprāmāṇyam na kāraṇavat. avastutvāt. yad yad avastu tat tan na kāraṇavat, yathā śaśaviṣāṇam.* p. 86.5: *prayogas ca bhavati—prāmāṇyam kāraṇavat. vastutvāt. ghaṭavat.*

11. Ibid., p. 86.5–6.
12. Svataḥprāmāṇyaparīkṣā (*TS*, 2810–3122).
13. Frauwallner, *Kumārila's Bṛhaṭṭīkā*, pp. 84ff.
14. *TS*, 2812 (d: *pāryate* instead of *śakyate* in Bauddha Bharati Ser. ed.). The latter half is taken from *ŚV*, II.47cd.
15. Cf. ibid., 2832: *etāvat tu vadanty atra sudhiyaḥ saugatā ime / jñāne kvacit sthitāpy eṣā (=śaktiḥ) na bodhuṃ śakyate svataḥ //*
16. Ibid., 2835.
17. *ŚV*, II.48 = *TS*, 2847 (b: *iṣyate* instead of *bhavet*; c-d: *labdhātmānaḥ . . . vartante* instead of *labdhātmanām . . . pravṛttiḥ*). Different interpretations given by Uṃbekabhaṭṭa and Pārthasārathimiśra on *ŚV*, II.47–8 are carefully examined in J. Taber, 'What did Kumārila Bhaṭṭa Mean by *svataḥ prāmāṇya*?' *JAOS* 112, 1992, pp. 204–21.
18. Cf. *TS*, 2852–3; *ŚV*, II.49–51.
19. Cf. *TS*, 2921–3. Śāntarakṣita cites here *ŚV*, IV (Pratyakṣa), 54–5, which are meant for explaining the meaning of the term '*buddhi-janman*' in *Mīmāṃsāsūtra* I.1.4: *satsamprayoge puruṣasyeindriyāṇāṃ buddhi-janma tat pratyakṣam. . . .* According to Kumārila, '*buddhi-janman*' implies that a cognition functions simultaneously with the birth of a cognition. Since, he states, a cognition does not continue to exist even for a single moment (*na . . . tat kṣaṇam apy āste*), it cannot be assumed that a cognition functions to apprehend an object some time after being produced. Thus he admits that a cognition is momentary. In his commentary on *TS*, 2924, Kamalaśīla quotes the following passage from the *Śābarabhāṣya* (Bib. Ind. ed., p. 9.17; Frauwallner, *Materialien*, p. 28.19–20): *kṣaṇikā hi sā (=buddhiḥ), na buddhy antarakālam avasthāsyate.*
20. Cf. *TS*, 2926: *ataś ca śakyate vaktuṃ svata eva na vartate / paścāt pramā svakaryeṣu nairūpyād gaganābjavat //*
21. Cf. ibid., 2928.
22. *Pramāṇavārttika*, II (Pramāṇasiddhi), la-c. Cf. S. Katsura, 'Dharmakīrti's Theory of Truth', *JIPh* 12, 1984, pp. 215–35; V.A. van Bijlert, *Epistemology and Spiritual Authority*, Wien: Arbeitskeris für Tibetische und Buddhistiche Studien, Universität Wien, 1989, pp. 120ff.
23. Cf. *TS*, 2958. Instead of the term '*arthakriyābhyāsajñānād*' in both editions, read '*ābhāsa*' (Tib., don byed par snaṅ ba'i śes las). Śāntarakṣita here takes into consideration the Vijñānavāda doctrine that an object in the external world is in fact the appearance of an object in a cognition. Hence, instead of attributing *arthakriyā* to a thing in the external world, he attributes it to an object appearing in the cognition.
24. *TS*, 2959–60: *arthakriyāvabhāsaṃ ca jñānaṃ saṃvedyate sphuṭam / niściyate ca tanmātrabhāvyāmarśanacetasā // atas tasya svataḥ samyak prāmāṇyasya viniścayāt / nottarārthakriyāprāptipratyayaḥ samapekṣyate //* The term '*āmarśanacetas*' in 2959d is interpreted by Kamalaśīla as '*parāmarśajñāna*'.
25. *TSP* ad *TS*, 2944 (p. 938.11–13): *taiḥ kiṃcit svataḥ pramāṇam iṣṭam, yathā svasaṃvedanapratyakṣam, yogijñānam, arthakriyājñānam, anu-*

mānam, abhyāsavac ca pratyakṣam, tad dhi svata eva niścīyate, abhyāsabalenāpahastitabhrāntikāraṇatvāt. In his *Pramāṇaviniścayaṭīkā*, Dharmottara (*c.* 740–800) states that *anumāna* and *arthakriyājñāna* are intrinsically true, cf. E. Steinkellner and H. Krasser, *Dharmottaras Exkurs zur Definition gültiger Erkenntnis im Pramāṇaviniścaya*, Wien: Verlag der Österr. Akad. d. Wiss., 1989, pp. 42–3, 83. For the dates of Dharmottara, see H. Krasser, 'On the Relationship between Dharmottara, Śāntarakṣita and Kamalaśīla', *Tibetan Studies: Proceedings of the 5th Seminar of the International Association for Tibetan Studies NARITA 1989*, Naritasan Shinshoji, 1992, vol. 1, pp. 151–8.

26. Cf. *TSP*, p. 981.12–15: *yat tu pakṣacatuṣṭayam upanyasya pakṣatraye doṣābhidhānaṃ kṛtam, tatrāpi na kācit bauddhasya kṣatiḥ, na hi bauddhair eṣām caturnām ekatamo 'pi pakṣo ' bhīṣṭaḥ, aniyamapakṣasyeṣṭatvāt. tathā hi—ubhayam apy etat kiṃcit svata kiṃcit parata iti pūrvam upavarṇitam. ata eva pakṣacatuṣṭayopanyāso 'py ayuktaḥ, pañcamasyāpy aniyamapakṣasya saṃbhavāt.*

List of Publications by B.K. Matilal

A. BOOKS

The Navya-nyāya Doctrine of Negation, the Semantics and Ontology of Negative Statements in Navya-nyaya Philosophy, Harvard Oriental Series, vol. 46 (Harvard University Press, Cambridge, Mass., 1968).

Epistemology, Logic and Grammar in Indian Philosophical Analysis, Janua Linguarum series no. 111 (Mouton & Co., The Hague, 1971).

Nyaya-Vaisesika (A Historical Survey), vol. VI of A History of Indian Literature, general editor: Jan Gonda (Otto Harrassowitz, Wiesbaden, 1977).

The Logical Illumination of Indian Mysticism (Oxford University Press, Oxford, 1978; tr. into Yugoslavian, Kulture Istoka, 1990, Zagreb, pp. 37–41).

The Central Philosophy of Jainism (Anekanta-vada) (L.D. Institute of Indology, L.D. Series 74, Ahmedabad, 1981).

Logical and Ethical Issues of Religious Beliefs (Calcutta University, Calcutta, 1982).

Śaśadhara's Nyāyasiddhānta-dīpa (A critical edition with Introduction and Notes) (L.D. Institute of Indology, L.D. Series, Ahmedabad, 1976).

Logic, Language and Reality (Motilal Banarsidass, New Delhi, 1985), 2nd edition, 1990. 2nd edition and paperback edition (under new subtitle: *Indian Philosophy and Contemporary Issues*).

Perception: An Essay on Classical Indian Theory of Knowledge (Clarendon Press, Oxford University, Oxford, 1986).

Nīti, Yukti O Dharma (in Bengali) (Ananda Publishers, Calcutta, 1988).

Confrontation of Cultures, Sakharam Deuskar Lecture at Centre for the Study of Social Sciences, Calcutta, 1988 (K.L. Bagchi Press, Calcutta, 1988).

The Word and the World: India's Contribution to the Study of Language (Oxford University Press, Delhi, 1990).

Under Preparation

The Indian Ethos: Studies in Ethics and Moral Emotions in the Great Epics and Narrative Literature of India, 5 of projected 12 chapters are in ms. (first draft).

The Development of Logic in India, commissioned by Institut International de Philosophie (Kluwer Publishers, Paris).

Collected Papers, ed. by H. Tiwari and J. Ganeri (OUP, Delhi).

B. ARTICLES

I. Indian Logic

'The Doctrine of Nyāyābhāsa (Pseudo-Reason)', *The Calcutta Review*, Calcutta, 1959.

'The Doctrine of Karaṇa in Grammar and Logic', *Proceedings and Transactions of All-Indian Oriental Conference*, Bhuvaneswar. Also reprinted in *Ganganath Jha Research Institute Journal*, Allahabad, India, 1961.

'A Discourse on Self-Contradictory Terms', *The Calcutta Review*, Calcutta University (January) 1961.

'The Intensional Character of Lakṣaṇa and Saṃkara in Navya-nyāya', *Indo-Iranian Journal*, The Hague, 8, 1964.

'Gaṅgeśa on the Concept of Universal Property (Kevalānvayin)', *Proceedings of the Third International Congress for Logic, Methodology, and Philosophy of Science*, ed., B. van Rootselaar. Also published in *Philosophy East and West* (Honolulu, Hawaii, 1968).

'Diṅnāga's Remark on the Concept of Anumeya', *Ganganath Jha Research Institute Journal*, Umesh Misra Commemorative Volume, Allahabad, 1969–70.

'Reference and Existence in Nyāya and Buddhism', *Journal of Indian Philosophy*, 1, 1970.

'A Note on the Nyāya Fallacy: Sadhyasama and Petitio Principii', *Journal of Indian Philosophy*, 2, 1974.

'On the Navya-nyāya Logic of Property and Location', *Proceedings of the 1975 International Symposium of Multiple-valued Logic* (Indiana University, Bloomington, USA, 1975).

'Jagadiśa's Classification of Grammatical Categories', *Indological Studies in Honour of Dr V. Raghavan*, eds, K.V. Sharma and R.N. Dandekar, Delhi, 1978.

Three essays on a) Vacāspati Miśra, b) Bhāsarvajña, and c) Udayana, *in History of Indian Logic and Metaphysics*, ed., K.H. Potter (Princeton University Press, USA, 1978).

'Aristotle and the Question of his Influence on Indian Logic', *Proceedings of a Round Table Conference* (UNESCO celebration of 2500th death anniversary of Aristotle), Paris, 1979.

'Double Negation in Navya-nyāya', *Sanskrit and Indian Studies: Essays in Honour of Daniel H.H. Ingalls*, eds, M. Nagatomi et al., 1980.

'A Note on "The Difference of Difference", *A Corpus of Indian Studies: Essays in Honour of Gaurinath Sastri*, eds, G. Bhattacharya et al., Calcutta, 1980.

'Pramana as Evidence', *Philosophical Essays in Honour of Dr Anantalal Thakur, The Asiatic Society Journal*, Calcutta, 1987.

'Debate and Dialectics in Ancient India', *Philosophical Essays in Honour of Dr Anantalal Thakur, The Asiatic Society Journal*, Calcutta, 1987.

'On the Theory of Numbers and Paryāpti Relation', *The Asiatic Society Journal* (Pandit Madhusudan Nyayacharya Commemoration Issue, Calcutta, 1987).

'Dharmakirti and the Universally Negative Inference', *Conference Papers at the 2nd International Dharmakirti Conference in Vienna*, June 1989, ed., E. Steinkellner, 1990.

II. Philosophy of Language in India

'Indian Theorists on the Nature of the Sentence (vākya)', *Foundations of Language*, D. Reidel, 2, 1966.

'The Notion of Substance and Quality in Ancient Indian Grammar', *Oriental Studies II* (University of Tartu, Estonia, USSR, 1973).

'The Ineffable', in *Language and Indian Philosophy*, ed., H.B. Coward, Canadian Society for the Study of Religion, 1978.

'On the Notion of the Locative in Sanskrit', *Indian Journal of Linguistics*, Calcutta, 10, 1983.

'Some Comments of Patañjali under 1.2.64', *Proceedings of the International Seminar on Studies in the Aṣṭādhyāyī of Pāṇini*, eds, S.D. Joshi and S.D. Laddu, Pune, 1983, pp. 119–20.

'Grammaticality and Meaningfulness', *Amṛtadhārā*, Prof. R.N. Dandekar Felicitation Volume, ed., S.D. Joshi, Delhi, 1983.

'The "Context Principle" in Indian Philosophy of Language', Lecture at Tel-Aviv University, published by Sackler Institute of Advanced Studies, Tel-Aviv, 1985.

'On Bhartrhari's Linguistic Insight', in *Sanskrit and Related Studies*, eds, B.K. Matilal and P. Bilimoria, Delhi, in press since 1988.

'The Context Principle and Some Indian Controversies Over Meaning', with P.K. Sen, *Mind*, January 1988.

'Bhavānanda on "What is a Kāraka?" ', S.D. Joshi Felicitation Volume, Poona, eds, M. Deshpande et al., in press since 1988.

'Awareness and Meaning in Navya-nyāya', in *Analytic Philosophy in Comparative Perspective*, eds, B.K. Matilal and J.L. Shaw (Reidel Pub. Co., Dordrecht, 1985). Also included in *The Proceedings of the International Conference on Meaning*, at Jadavpur University, Calcutta, 1984.

(1) 'Indian Philosophy of Language', (2) 'Sphoṭa', (3) 'Word vs. Sentence, (4) 'Jayanta', in *Sprachphilosophie: Ein Hanbuch Zeitgenossischer Forschung*, eds, K. Lorenz, et al. (Walter de Gruyter, Berlin), forthcoming.

'Sign Conceptions of India', in *Handbook of Semiotics*, ed., R. Posner, with J. Panda (Walter de Gruyter, Berlin), forthcoming.

'Śābdabodha and the Problem of Knowledge-representation in Sanskrit', *Journal of Indian Philosophy*, 15, 1988.

'Ineffability: Some Problems of Indian Logic and Language', in *An Anthology on Mysticism and Language III*, ed., S. Katz, Notre Dame Press, in press. A later version in the Conference Papers, 1988, edited by Bruce Alton, University of Toronto, in press.

III. Buddhism

'A Critique of the Mādhyamika Position', in *The Problems of Two Truths in Buddhism and Vedānta*, ed., M. Sprung (D. Reidel, Dordrecht, 1973).

'A Critique of Buddhist Idealism', in *Buddhist Studies in Honour of Dr I.B. Horner*, eds, L. Cousins, A. Kunst and K.R. Norman (D. Reidel, Dordrecht/Boston, 1975).

'Ontological Problems in Nyāya, Buddhism and Jainism: A Comparative Analysis', *Journal of Indian Philosophy*, 5, 1977. Reprinted in *Philosophies of Existence*, ed., P. Morewedges, Fordham University Press, 1982.

'The Enigma of Buddhism: Duhkha and Nirvāṇa', *Journal of Dharma*, 2, 1977.

'Transmigration and the Causal Chain in Nyaya and Buddhism', in *Development in Buddhist Thought: Canadian Contributions to*

Buddhist Studies, ed., R.C. Amore, Canadian Society for the Study of Religion, 1979.

'A Note on Avidyā in Buddhism', *Buddhist Studies in Honour of W. Rahula*, eds, S. Balasooriya et al., Gordon Fraser, 1980.

'Ālayavijñāna, Transmigration and Absolutism', *M. Kuppuswami Sastri Birth Centenary Volume*, ed., S. Janaki, Madras, 1981.

'Problems of Self-Awareness (Sva-saṃvedana)', *Studies in the History of Indian Thought*, ed., M. Hattori (Kyoto University, Kyoto, 1985).

'Buddhist Logic and Epistemology', *Introductory Essay in Buddhist Logic and Epistemology*, eds, B.K. Matilal and R.D. Evans (D. Reidel, Dordrecht, 1986).

'Uddyotakara on Diṅnāga's Apoha', *Journal of Buddhism, Inaugural Issue*, 1989.

'Madhyamika', *The Annals of the Bhandarkar Oriental Research Institute* (R.G. Bhandarkar 150th Birth Anniversary Volume), Poona, 1987.

IV. Jainism

'A Note on the Jaina Conception of Substance', *Sambodhi* (A.N. Upadhye Commemoration Volume), 5, 1976.

'Saptabhaṅgī: The Jaina Doctrine of Sevenfold Predictions', in *Self, Knowledge and Freedom: Essays in Honour of Kalidas Bhattacharya*, eds, J.N. Mohanty et al. (World Press, Calcutta, 1978).

'Memory (as a pramāṇa)', in *Studies in Indian Philosophy: Commemoration Volume for Pandit Sukhlalji Sanghavi*, eds, D. Malvania et al. (LD Series 84, Ahmedabad, 1981).

'Jainism', *The World Encyclopaedia of Peace* (Pergamon Press, Oxford, 1987).

V. Epistemology and General Problems of Indian Philosophy and Religion

'Indian Theories of Knowledge and Truth', *Philosophy East and West*, Honolulu, October, 1968.

'Dialectical Materialism Viewed through an Indian Eye', *The Calcutta Review*, Calcutta, 1959.

'Is Inherence Relation Perceptible?' (in Sanskrit), Golden Jubilee issue of Sanskrit Sahitya Parishat Patrika, 50, 1969.

'On Marxist Dialectics: Comments on Hao Wang's Articles', *Philosophy East and West*, Honolulu, 24, 1974.

'Causality in the Nyāya-Vaiśeṣika School', *Philosophy East and West*, Honolulu, 25, 1975.

'Mysticism and Reality: Ineffability', *Journal of Indian Philosophy*, Dordrecht, 3, 1975.

'Karma: A Metaphysical Hypothesis of Moral Causation', in *Contemporary Indian Philosophies of History*, eds, T.M.P. Mahadevan et al. (World Press, Calcutta, 1977).

'Problems of Inter-faith Studies' and 'Towards Defining "Religion" in the Indian Context', in *Meetings of Religions: New Orientations and Perspectives*, ed., T. Aykara (Dharmaram Institute, Bangalore, 1978).

'Religion and the Problem of Integration of Mankind', *Proceedings of the International Conference of Scientists and Religious Leaders on the Future of Mankind*, ed., Migami, Kyoto/Hiroshima, 1979.

'India: Contemporary Developments in Philosophy Since 1945', in *Handbook of World Philosophy*, ed., J.R. Burr (Greenwood Press, Connecticut, USA, 1980).

'Error and Truth: Some Classical Indian Theories', *Philosophy East and West*, Honolulu, 39, 1981.

'Guṇaratna Sūri as a Commentator on Śaśadhara', in *D.R. Bhanderkar Birth Centenary Volume*, ed., S. Bandyopadhyay (University of Calcutta, Calcutta, 1982).

'An Exercise in Scepticism', in *Pramā-Pramāṇa and Knowledge-Justification: Empirical Knowledge and Evidence*, ed., P. Mukhopadhyay (Jadavpur University, Calcutta, 1982).

'Indian Philosophy: Is there a Problem Today?', in *Indian Philosophy: Past and Future*, eds, S.S. Rama Rao Pappu et al. (Motilal Banarsidass, Delhi, 1982).

'Religion und Moral', in *Sehnsucht nach dem Ursprung zu Mircea Eliades*, ed., H.P. Deurr (Syndikat, Frankfurt, 1983).

'On Omnipotence', *Our Heritage* (Sanskrit College, Calcutta, 1983).

'Moral Dilemmas and Religious Dogmas', *Bulletin of the R.K. Mission Institute of Culture*, Calcutta, 35, 1983.

'Knowing that One Knows', *Journal of the Indian Council of Philosophical Research*, Delhi, 1984.

'Analytical Philosophy in Comparative Perspective: An Introduction', in *Analytical Philosophy in Comparative Perspective*, eds, B.K. Matilal et al. (D. Reidel, Dordrecht, 1985).

'Scepticism and Mysticism', *Journal of the American Oriental Society*, D.H.H. Ingalls Felicitation Volume, 1987. A Japanese translation (by Professor Nagasaki) in the Otani University Journal, 1985. Also published (an earlier version) by the Sackler Institute of Advanced Studies in Humanities, Tel-Aviv University, 1985.

'On the Thesis of the Universality of Suffering', in *Suffering in World Religions*, ed., K. Tiwari (Motilal Banarsidass, Delhi, 1987).

Six entries in *Encyclopaedia of Religion*, eds, M. Eliade et al. (Macmillan, New York, 1987) (Jñāna, Lokāyata, Mīmāṃsā, Gauḍapada, Nimbārka Vijñānabhikṣu).

'Between Peace and Deterrence', *Oxford Project for Peace Studies*, ed., E. Kay (Rex Collins, Oxford, 1987).

'Bhāratīya Darśan-carcā O Mahilara' (in Bengali), *Sananda*, 10 September 1987, pp. 69–74.

'The Nyāya Critique of Buddhist Doctrine of Non-Soul', in *Freedom, Transcendence and Reality* (Kalidas Bhattacharyya Commemoration Volume), ed., P.K. Sengupta (ICPR, Delhi, 1988). Also published in *Journal of Indian Philosophy*, 1989, March, pp. 61–79.

'Hindudharma bahujanasevita baṭavṛkṣer chāyā' (in Bengali), *DESH*, 16 April 1988, 1989, pp. 31–8.

'Maulavād Ki O Kena' (in Bengali), *DESH*, 30 June 1990, pp. 15–18.

'Knowledge, Truth and Pramātva', *J.N. Mohanty Felicitation Volume*, ed., Dayakrishna, forthcoming.

Edited and contributed 14 new entries to the Oriental Religion Section of Longman's (single volume) *Encyclopaedia of Religion*, Columbia University Press, 1989.

'A Realist View of Perception', *P.F. Strawson Felicitation Volume*, eds, R.R. Verma et al., OUP, forthcoming.

Seventy-five entries for *A Dictionary of Philosophical Quotations*, ed., A.J. Ayer, Blackwell (in press).

VI. Indian Literature, History, and Others

'A Love Story from Eighth-Century India', with J. Masson, *Journal of Jadavpur University Comparative Literature*, Calcutta, 1967.

Three entries:

a) 'Bhaṭṭa Nārāyaṇa'

b) 'Kalhaṇa', and

c) 'Mahimabhaṭṭa'

in *Dictionary of Oriental Literature*, eds, J. Prusek and D. Zbavital (Basic Books, New York, 1974).

'Rama's Moral Decisions', *Adyar Library Bulletin*, K. Kunjunni Raja Felicitation Volume, 1982.

'Radhakrishnan, Sarvepalli', and 'Dasgupta, Surendranath', two contributions in *Twentieth Century Thinkers*, ed., E. Burkland (St James Press, London, 1988).

'Rāma and Kṛṣṇa in the Great Epics' (in Bengali) *DESH*, 6 chapters, serialized June–December 1987.

'Relevance of the Study of Indian Philosophy Today', Oxford Majlis Commemoration Volume, Oxford, 1987.

'On Dogmas of Indology', in *Studies in Ancient Indian History*, D.L. Sircar Commemoration Volume, eds, K. Dasgupta et al., Calcutta, 1988.

'Images of India: Perceptions and Problems', in N.V. Banerji Commemoration Volume, ed., M. Chatterjee, Delhi, 1989.

'Ethical Relativism and Confrontation of Cultures', in *An Anthology, Relativism*, ed., M. Krausz, Notre Dame University Press, 1989.

'Moral Dilemmas: Insights from Indian Epics', in *An Anthology on Moral Dilemmas and the Mahābhārata*, ed., B.K. Matilal, Shimla, 1989.

'Caste, Karma and the Gītā', in *Philosophy of Indian Religion*, ed., Roy Perrett (Kluwer, Dordrecht, 1989), pp. 195–201.

'Dharma and Rationality', in *Rationality in Question*, eds, S. Biderman and Ben-Ami Scharfstein (Leiden, Brill, 1989), pp. 191–216.

'Gāndhārī' (in Bengali), *DESH*, 1 September 1990, pp. 68–74.

'Kṛṣṇa: In Defence of a Devious Divinity', in *An Anthology: Essays on the Mahābhārata*, ed., A. Sharma, Brill & Co., forthcoming.

'Ideas and Values in Radhakrishnan's Thought', in *The Radhakrishnan Centennial Volume*, eds, D.P. Chattopadhyay et al., Delhi, 1989.

'Bankimchandra on Hinduism and Nationalism', in a volume to celebrate 150th Birthday of Bankimchandra, eds, S. Bhattacharya et al. (JNU, Oxford University Press, Delhi), forthcoming.

'The Problem of Inter-faith Studies', *Sophia*, Special 100th Issue, ed. by P. Bilimoria (Australia, 1995), vol. 34, no. 1, pp. 167–72.

C. EDITORIAL WORK

Journal of Indian Philosophy (since 1970, vols I–XVI), vol. XVII, 1989, nos 1, 2 and continuing.

Analytical Philosophy in Comparative Perspective (an anthology edited with J.L. Shaw) (D. Reidel, Synthäse Library Series, Dordrecht, 1985).

Buddhist Logic and Epistemology (an anthology edited with R. Evans) (D. Reidel, Studies of Classical Indian Series, Dordrecht, 1986).

Sanskrit and Related Studies (an anthology edited with P. Bilimoria) (Indian Book Centre, Sri Satguru Pub., 1990).

General Editor of the Studies of Classical Indian Series (D. Reidel, Dordrecht, Holland).

Vol. 1, *Philosophy and Argument in Late Vedanta* (Phyllis Granoff), Foreword by B. Matilal.

Knowing from Words (an anthology edited with A. Chakrabarti) (Synthäse Library Kluwer, 1994).

Vol. 2, *Sanskrit and Indian Studies, Essays in Honour of D.H.H. Ingalls*, eds, M. Nagatoni, B. Matilal, J. Masson and E. Dimock, 1980.

Vol. 3, *The Oceanic Feeling*, J. Moussaieff Masson, 1980.

Vol. 4, *The Structure of the World in Udayana's Realism*, M. Tachikawa, 1981, Foreword by B. Matilal.

Vol. 5, *The Twelve Door Treatise of Nāgārjuna*, tr. by H. Cheng, 1982.

Vol. 6, *Tradition and Argument in Classical Indian Linguistics*, J. Bronkhorst, 1986.

Vol. 7, *Buddhist Logic and Epistemology*, eds, B.K. Matilal and R. Evans, 1986.

Vol. 8, *Bhartṛhari and the Buddhists*, R. Herzberger, 1986, Foreword by B. Matilal.

Vol. 9, *Dinnaga on the Interpretation of Linguistic Signs*, R. Hayes, 1988.

Vol. 10, *Śabdapramāṇa: Word and Knowledge*, P. Bilimoria, 1988.

Moral Dilemmas in the Mahābhārata (an anthology of the papers at the colloquium), Preface (IIAS, Shimla, 1989).

Vol. 11, *The Ontology of the Middle Way: Candrakīrti*, P. Fenner, 1990.

D. REVIEWS

Review of R. Das's *Introduction of Shankara* (Firma K.L. Mukhopadhyay, Calcutta, 1968), *Journal of the American Oriental Society*, New Haven, January–March 1971, pp. 156–7.

Review of E. Deutsch's *Advaita Vedanta* (East-West Center Press, University of Hawaii, Honolulu, 1969), *Philosophy East and West*, vol. 71, no. 3, July 1971, pp. 332–5.

Review of D. Sharma's *The Differentiation Theory of Meaning in Indian Logic* (Mouton & Co., The Hague, 1969), *Foundations of Language*, vol. 8, 1972, pp. 578–83.

Review of C. Goekoop's *The Logic of Invariable Concomitance in the Tattvacintamani* (D. Reidel Pub. Co., Dordrecht/Holland, 1967), *Journal of the American Oriental Society*, New Haven, vol. 9, no. 1, 1972, pp. 169–73.

Review of *The Concept of Duty in South Asia*, eds, W.D. O'Flaherty and J. Duncan M. Derrett, *Journal of the Royal Asiatic Society*, 1979, pp. 78–9.

Reviews of (1) *Frits Staal: Word Order in Sanskrit and Universal Grammar (Reidel)*, (2) *J.V. Bhattacharya; Nyāyamañjari, The Compendium of Indian Philosophical Logic* (B. Motilal), *Journal of the Royal Asiatic Society*, 1980.

Reviews of (1) *P.S. Jaini: The Jaina Path of Purification* (Univ. of California Press), (2) *K. Bhattacharya: The Dialectical Method of Nagarjuna* (B. Motilal), *Bulletin of the School of Oriental and African Studies*, 1980.

Review of R. Panikkar's *The Vedic Experience*, New Fire, 1980.

Review of *The Sāṃkhya Sūtras of Pañcāsikha and the Samkhyatattvaloka* (of Swami Hariaharananda Aranya), *International Journal of the Philosophy of Religion*, vol. 12, 1980, p. 125.

Review of A. Piatigorsky's *The Buddhist Philosophy of Thought, Times Literary Supplement*, 4 May 1984.

Review of J.L. Brockington's *Righteousness Rāma: The Evolution of an Epic* (in Bengali) (*Ananda Bazaar Patrika*, Calcutta, 23 December 1985).

Review of B.W.J. van der Kuip's *Contributions to the Development of Tibetan Buddhist Epistemology*, Wiesbaden, 1983, BSOAS, 1985, part 1, pp. 161–3.

Review of Anand Amaladass S.J.'s *Philosophical Implications of Dhvani*, Vienna, 1984, BSOAS.

Review of David Kalupahana's (1) *Nāgārjuna: The Philosophy of the Middle-Way*, (2) *A Path of Righteousness: Dhammapada*, BSOAS, 1987, pp. 151–3.

Review of Eli Franco's *Perception, Knowledge and Disbelief* (Jayarasi Tattvapaplavasimha), Hamburg, 1988, in *Journal of the American Oriental Society*, forthcoming.

E. SANSKRIT COMPOSITIONS

Raṭharajjuḥ (one-act play in Sanskrit), translation of R.N. Tagore's *Kaler Jatra* (Sanskrit Pustak Bhander, Calcutta, 1961).

Janmadivaṣīyam, Tradition of Tagore's poem 'Panchisi Baisakha' (Sanskrta Sahitya Parisat Patrika, 1962).

Translation in Sanskrit verses of some short poems from R.N. Tagore's *Kanika* (Sanskrta Sahitya Parisat Patrika, 1962), October, 1990.